METHODS IN MOLECULAR BIOLOGY

For further volumes:
http://www.springer.com/series/7651

Plant Proteostasis

Methods and Protocols

Edited by

L. Maria Lois

Development Program, CRAG (CSIC-IRTA-UAB-UB) Edifici CRAG-Campus UAB, Bellaterra (Cerdanyola del Vallés), Barcelona, Spain

Rune Matthiesen

Computational and Experimental Biology Group, National Health Institute Dr. Ricardo Jorge, IP, Lisbon, Portugal

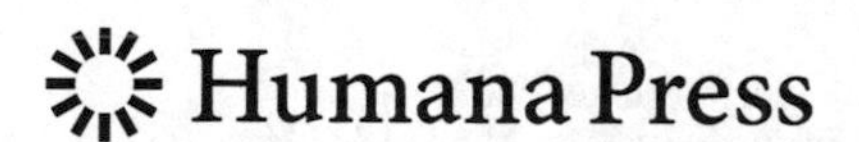

Editors
L. Maria Lois
Development Program
CRAG (CSIC-IRTA-UAB-UB) Edifici CRAG-
Campus UAB
Bellaterra (Cerdanyola del Vallés)
Barcelona, Spain

Rune Matthiesen
Computational and Experimental Biology Group
National Health Institute Dr. Ricardo Jorge
IP, Lisbon, Portugal

ISSN 1064-3745 ISSN 1940-6029 (electronic)
Methods in Molecular Biology
ISBN 978-1-4939-8131-1 ISBN 978-1-4939-3759-2 (eBook)
DOI 10.1007/978-1-4939-3759-2

Printed on acid-free paper

This Humana Press imprint is published by Springer Nature
The registered company is Springer Science+Business Media LLC New York

Preface

Plant responses to environmental stimuli and developmental transitions are regulated by complex regulatory networks that deliver the specific physiological outcome to assure plant survival. These networks include transcriptional regulation but also sophisticated posttranslational modifications that aim to regulate protein activity. In contrast to transcriptional regulation, which involves de novo protein synthesis, posttranslational modifications modulate protein activity in short time periods facilitating rapid cell responses. The molecular consequences of posttranslational modifications on the protein target are highly variable and include changes in protein structure, subcellular localization, activity, partner interactions, stability, or solubility.

Proteins can be modified by a wide array of compounds that vary in their nature, size, and conjugation mechanism. As such, reactive oxygen species induce protein oxidation independently on enzymatic catalysis, while other posttranslational modifications involving the addition of small organic groups (i.e., phosphate or methyl groups) are regulated by enzymes dedicated to the addition or removal of the specific modifier. Finally, one of the most complex posttranslational modification groups is represented by the ubiquitin (Ub) and ubiquitin-like (Ubl) modifiers, which are small proteins that are conjugated to protein targets through a cascade of three enzymatic steps and deconjugated by specific peptidases.

The branching complexity of post-translational modifications, together with their labile nature and the need of custom-tailored molecular tools, make their analysis really challenging. In plants, the absence of well established commercial tools, the more complex plant cell manipulation required for biochemical studies, and the gene amplification displayed by many members of these regulatory components, result in a higher difficulty degree of biochemical and genetic studies. The analysis of protein homeostasis is even more complex in non-plants models since specific protocols and tools are poorly or not developed.

In this book, we have collected detailed protocols describing state-of-the-art approaches that will facilitate the understanding of protein homeostasis in plant stress responses and development. Some findings made in this area of plant research could become valuable molecular tools in selection processes for improving agronomic performance, but also for contributing to address next challenges in agriculture such as precision horticulture.

Part I contains protocols focusing on the study of ubiquitin-dependent posttranslational modifications. While Chapter 1 describes a protocol for studying a novel ubiquitin conjugation mechanism independent of lysine residues, the other chapters focus on different aspects of the classical ubiquitin-dependent protein degradation system. Chapter 2 provides methods for analyzing the in vivo dynamics of cullins, key components of RING E3 ligases catalyzing ubiquitin conjugation to substrates. Chapter 3 describes the study of F-box proteins, another component of RING E3 ligases, as plant hormone receptor, which has become a key step in triggering hormone signaling. As many posttranslational modifications, ubiquitination is a reversible modification and Chapter 4 focuses on approaches for the study of enzymes involved in ubiquitin removal from its substrate. Chapters 5 and 6 address the generation of substrates for analyzing the in vivo ubiquitin/proteasome system and the N-rule pathway for protein degradation, respectively. Finally, Chapter 7 extends the study of the N-rule pathway through methods for identifying E3 ligases.

Part II is dedicated to protocols focused on the study of Ubl posttranslational modifications, including in vitro SUMO chain formation (Chapter 8), the kinetic analysis of SUMO conjugation machinery (Chapter 9), and the in vitro analysis of SUMO proteases involved in SUMO maturation and SUMO removal from substrates (Chapter 10). In addition, Chapter 11 addresses the analysis of cellular distribution of SUMO conjugation machinery members as a strategy to get insights into their in vivo role. Another Ubl modification involved in many aspects of plant stress responses and development is autophagy, and biochemical and cell biology protocols for its study are described in Chaps. 12 and 13.

The study of protein homeostasis requires a broad variety of protocols that go beyond the analysis of enzymatic activities responsible for posttranslational modifications, and some of these protocols are comprised in Part III. A very useful and rapid approach to study protein stability consists in the expression of recombinant protein in plant protoplasts as is described in Chapter 14. Another emerging field in plant protein homeostasis is the study of protein aggregate formation in response to environmental stress, and their purification, described in Chapter 15, is the first step into their analysis. Chapter 16 provides a protocol for the study of another phenomenon occurring in response to stress consisting in protein oxidation under reactive oxygen species generation and the determination of proteasome activity. When the aim is to identify global changes in protein homeostasis during physiological responses, comparative proteomics based on iTRAQ are to be used (Chapter 17). Chapter 18 describes methods for the study of protein binding to phosphatidylinositol as a modulation mechanism of protein homeostasis. Also, organelle purification is recommended in order to reduce the complexity of the sample when performing proteomic studies in cell compartments, and Chapter 19 describes methods for the study of chloroplast proteome. Finally, Chapter 20 focuses on a general but also essential technique when trying to determine fluctuations in protein levels between samples, which is western blotting normalization.

Finally, Past IV encloses protocols for the in silico analysis of different aspects of proteostasis. Chapter 21 describes a protocol for identifying the genes encoding specific protein families and investigating their syntenic relationship. Chapter 22 focuses on methods for performing phylogenetic analysis, as a means of inferring functional conservation in different plant species. The last chapter (Chapter 23) describes the use of bioinformatics tools for data mining, focusing on the SUMO gene network.

We are thankful to the authors who have contributed to make this book possible. Also, we thank John Walker, the series editor, for his advice and the colleagues at Humana Press for producing this book. This book is based upon work from COST Action (PROTEOSTASIS BM1307), supported by COST (European Cooperation in Science and Technology).

Barcelona, Spain ***L. Maria Lois***
Lisbon, Portugal ***Rune Matthiesen***

Contents

Contributors

VICTOR A. ALBERT • *Department of Biological Sciences, SUNY-University at Buffalo, Buffalo, NY, USA*
MONTSE AMENÓS • *Confocal Microscopy Facility, CRAG (CSIC-IRTA-UAB-UB), Edifici CRAG-Campus UAB, Bellaterra (Cerdanyola del Vallés), Barcelona, Spain*
ASCENSIÓN ANDRÉS-GARRIDO • *Instituto de Bioquímica Vegetal y Fotosíntesis, Consejo Superior de Investigaciones Científicas (CSIC), Universidad de Sevilla, Sevilla, Spain*
HERLÂNDER AZEVEDO • *CIBIO, InBIO—Research Network in Biodiversity and Evolutionary Biology, Universidade do Porto, Campus Agrário de Vairão, Vairão, Portugal*
ANDREAS BACHMAIR • *Department of Biochemistry and Cell Biology, Max F. Perutz Laboratories, University of Vienna, Vienna, Austria*
CHRISTOPHE BAILLY • *Institut de Biologie Paris-Seine, Unité Mixte de Recherche 7622, Paris, France; Biologie du Développement, Institut de Biologie Paris-Seine, Unité Mixte de Recherche 7622, Centre National de la Recherche Scientifique, Paris, France*
DIANE C. BASSHAM • *Department of Genetics, Development and Cell Biology and Plant Sciences Institute, Iowa State University, Ames, IA, USA*
THOMAS BRYLOK • *Botanik, Department Biologie I, Ludwig-Maximilians-Universität, Planegg-Martinsried, Germany*
LAURENT CAMBORDE • *Laboratoire de Virologie Moléculaire, Institut Jacques Monod, CNRS, Univ. Paris-Diderot, Sorbonne Paris Cité, Paris, France; Laboratoire de Recherche en Sciences Végétales, CNRS, Univ Toulouse Paul Sabatier, Toulouse, France*
LORENZO CARRETERO-PAULET • *Department of Plant Systems Biology, VIB, Ghent University, Ghent, Belgium*
LAURA CASTAÑO-MIQUEL • *Development Program, CRAG (CSIC-IRTA-UAB-UB), Edifici CRAG-Campus UAB, Bellaterra (Cerdanyola del Vallés), Barcelona, Spain*
PEDRO HUMBERTO CASTRO • *Biosystems and Integrative Sciences Institute (BioISI), Plant Functional Biology Center, University of Minho, Campus de Gualtar, Braga, Portugal*
QIAN CHEN • *State Key Laboratory of Plant Genomics, National Center for Plant Gene Research, Institute of Genetics and Developmental Biology, Chinese Academy of Sciences, Beijing, P. R. China; Graduate University of Chinese Academy of Sciences, Beijing, P. R. China*
NÚRIA S. COLL • *Centre for Research in Agricultural Genomics (CSIC-IRTA-UAB-UB), Bellaterra, Spain*
JOSÉ L. CRESPO • *Instituto de Bioquímica Vegetal y Fotosíntesis, Consejo Superior de Investigaciones Científicas (CSIC), Universidad de Sevilla, Sevilla, Spain*
NICO DISSMEYER • *Leibniz Institute of Plant Biochemistry (IPB), Halle (Saale), Germany; ScienceCampus Halle – Plant-Based Bioeconomy, Halle (Saale), Germany*
HAYAT EL-MAAROUF-BOUTEAU • *Institut de Biologie Paris-Seine, Unité Mixte de Recherche 7622, Sorbonne Universités, Paris, France; Biologie du Développement, Institut de Biologie Paris-Seine, Unité Mixte de Recherche 7622, Centre National de la Recherche Scientifique, Paris, France*

Lennart Eschen-Lippold • *Leibniz Institute of Plant Biochemistry (IPB), Halle (Saale), Germany*
Frederik Faden • *Leibniz Institute of Plant Biochemistry (IPB), Halle (Saale), Germany; ScienceCampus Halle – Plant-Based Bioeconomy, Halle (Saale), Germany*
T. Farmaki • *CE.R.T.H.-IN.A.B., Thessaloniki, Greece*
Anna Franciosini • *RIKEN Plant Science Center, Yokohama, Kanagawa, Japan*
Antje Hellmuth • *Department of Molecular Signal Processing, Leibniz Institute of Plant Biochemistry (IPB), Halle (Saale), Germany*
Erika Isono • *Department of Plant Systems Biology, Technische Universität München, Freising, Germany*
Isabelle Jupin • *Laboratoire de Virologie Moléculaire, Institut Jacques Monod, CNRS, UMR 7592, Univ. Paris-Diderot, Sorbonne Paris Cité, Paris, France*
Kamila Kalinowska • *Department of Plant Systems Biology, Technische Universität München, Freising, Germany*
Maria Klecker • *Leibniz Institute of Plant Biochemistry (IPB), Halle (Saale), Germany; ScienceCampus Halle – Plant-Based Bioeconomy, Halle (Saale), Germany*
Ping Lan • *Institute of Plant and Microbial Biology, Academia Sinica, Taipei, Taiwan; Institute of Soil Science, Chinese Academy of Sciences, Nanjing, China*
Saul Lema A • *Centre for Research in Agricultural Genomics (CSIC-IRTA-UAB-UB), Bellaterra, Spain*
L. Maria Lois • *Development Program, CRAG (CSIC-IRTA-UAB-UB), Edifici CRAG-Campus UAB, Bellaterra (Cerdanyola del Vallés), Barcelona, Spain*
Alexandre Papadopoulos Magalhães • *Biosystems and Integrative Sciences Institute (BioISI), Plant Functional Biology Center, University of Minho, Campus de Gualtar, Braga, Portugal; CIBIO, InBIO–Research Network in Biodiversity and Evolutionary Biology, Universidade do Porto, Campus Agrário de Vairão, Vairão, Portugal*
Abraham Mas • *Development Program, CRAG (CSIC-IRTA-UAB-UB), Edifici CRAG-Campus UAB, Bellaterra (Cerdanyola del Vallés), Barcelona, Spain*
Patrice Meimoun • *Institut de Biologie Paris-Seine, Sorbonne Universités, Unité Mixte de Recherche 7622, Paris, France; Biologie du Développement, Institut de Biologie Paris-Seine, Unité Mixte de Recherche 7622, Paris, France*
Augustin C. Mot • *Leibniz Institute of Plant Biochemistry (IPB), Halle (Saale), Germany; ScienceCampus Halle – Plant-Based Bioeconomy, Halle (Saale), Germany; Faculty of Chemistry and Chemical Engineering, Babes-Bolyai University, Cluj-Napoca, Romania*
Marie-Kristin Nagel • *Department of Plant Systems Biology, Technische Universität München, Freising, Germany*
Christin Naumann • *Leibniz Institute of Plant Biochemistry (IPB), Halle (Saale), Germany; ScienceCampus Halle – Plant-Based Bioeconomy, Halle (Saale), Germany*
Catharina Nickel • *Botanik, Department Biologie I, Ludwig-Maximilians-Universität, Planegg-Martinsried, Germany*
Beatriz Orosa • *School of Biological and Biomedical Sciences, University of Durham, Durham, UK*
María Esther Pérez-Pérez • *Instituto de Bioquímica Vegetal y Fotosíntesis, Consejo Superior de Investigaciones Científicas (CSIC), Universidad de Sevilla, Sevilla, Spain*
Marc Planas-Marquès • *Centre for Research in Agricultural Genomics (CSIC-IRTA-UAB-UB), Bellaterra, Spain*

SÉVERINE PLANCHAIS • *Laboratoire de Virologie Moléculaire, Institut Jacques Monod, CNRS, Univ. Paris-Diderot, Sorbonne Paris Cité, Paris, France; Adaptation des Plantes aux Contraintes Environnementales, IEES Paris, Univ. Pierre-et-Marie-Curie, Paris, France*
YUNTING PU • *Department of Genetics, Development and Cell Biology, Iowa State University, Ames, IA, USA*
ARI SADANANDOM • *School of Biological and Biomedical Sciences, University of Durham, Durham, UK*
MIGUEL ÂNGELO SANTOS • *Biosystems and Integrative Sciences Institute (BioISI), Plant Functional Biology Center, University of Minho, Campus de Gualtar, Braga, Portugal; CIBIO, InBIO-Research Network in Biodiversity and Evolutionary Biology, Universidade do Porto, Campus Agrário de Vairão, Vairão, Portugal*
WOLFGANG SCHMIDT • *Institute of Plant and Microbial Biology, Academia Sinica, Taipei, Taiwan; Institute of Plant and Microbial Biology, Academia Sinica, Taipei, Taiwan*
SERENA SCHWENKERT • *Botanik, Department Biologie I, Ludwig-Maximilians-Universität, Planegg-Martinsried, Germany*
GIOVANNA SERINO • *Dipartimento di Biologia e Biotecnologie "C. Darwin", Sapienza Università di Roma, Rome, Italy*
AGNIESZKA SIRKO • *Institute of Biochemistry and Biophysics, Polish Academy of Sciences, Warsaw, Poland*
ANJIL SRIVASTAVA • *School of Biological and Biomedical Sciences, University of Durham, Durham, UK*
RUI MANUEL TAVARES • *Biosystems and Integrative Sciences Institute (BioISI), Plant Functional Biology Center, University of Minho, Campus de Gualtar, Braga, Portugal*
KONSTANTIN TOMANOV • *Department of Biochemistry and Cell Biology, Max F. Perutz Laboratories, University of Vienna, Vienna, Austria*
ISABEL CRISTINA VÉLEZ-BERMÚDEZ • *Institute of Plant and Microbial Biology, Academia Sinica, Taipei, Taiwan*
LUZ IRINA A. CALDERÓN VILLALOBOS • *Department of Molecular Signal Processing, Leibniz Institute of Plant Biochemistry (IPB), Halle (Saale), Germany*
TUAN-NAN WEN • *Institute of Plant and Microbial Biology, Academia Sinica, Taipei, Taiwan*
QIONG XIA • *Institut de Biologie Paris-Seine, Unité Mixte de Recherche 7622, Sorbonne Universités, Paris, France; Biologie du Développement, Institut de Biologie Paris-Seine, Unité Mixte de Recherche 7622, Centre National de la Recherche Scientifique, Paris, France*
QI XIE • *State Key Laboratory of Plant Genomics, National Center for Plant Gene Research, Institute of Genetics and Developmental Biology, Chinese Academy of Sciences, Beijing, P. R. China; Graduate University of Chinese Academy of Sciences, Beijing, P. R. China*
XIAOYUAN YANG • *State Key Laboratory of Plant Genomics, National Center for Plant Gene Research, Institute of Genetics and Developmental Biology, Chinese Academy of Sciences, Beijing, P. R. China*
GARY YATES • *School of Biological and Biomedical Sciences, University of Durham, Durham, UK*
IONIDA ZIBA • *Department of Biochemistry and Cell Biology, Max F. Perutz Laboratories, University of Vienna, Vienna, Austria*
KATARZYNA ZIENTARA-RYTTER • *Institute of Biochemistry and Biophysics, Polish Academy of Sciences, Warsaw, Poland*

Part I

Ubiquitin Conjugation and Deconjugation Analysis

Chapter 1

Approaches to Determine Protein Ubiquitination Residue Types

Qian Chen, Xiaoyuan Yang, and Qi Xie

Abstract

Ubiquitination is an important posttranslational modification in eukaryotic organisms and plays a central role in many signaling pathways in plants. Most ubiquitination typically occurs on substrate lysine residues, forming a covalent isopeptide bond. Some recent reports suggested ubiquitin can be attached to non-lysine sites such as serine/threonine, cysteine or the N-terminal methionine, via oxyester or thioester linkages, respectively. In the present protocol, we developed a convenient in vitro assay for investigating ubiquitination on Ser/Thr and Cys residues.

Key words Ubiquitination, Posttranslational modification, Serine/threonine, Cysteine, Hydrolysis

1 Introduction

Ubiquitination is an important posttranslational modification that controls many cellular processes. Many proteins involved in the ubiquitin system play crucial roles in signal transduction and biophysical processes. The effects of ubiquitination on its protein substrates are diverse and influence protein stability and activity, protein–protein interactions, and subcellular localization [1]. Degradation is the usual fate of polyubiquitinated proteins. Ubiquitination is catalyzed by ubiquitin-activating enzyme (E1), ubiquitin-conjugating enzyme (E2), and ubiquitin-ligase (E3), the action of which forms an isopeptide bond between the carboxyl group of the C-terminus of ubiquitin and the ε amino group of a lysine residue in substrate.

Somewhat surprisingly, many proteins are unstable and rapidly degraded when all lysine residues are mutated to arginine. The RING type E3 ligase mK3 targets the major histocompatibility complex (MHC) class I, an ER-associated degradation (ERAD) substrate, for degradation via ubiquitination of its cytosolic tail [2]. However, a lysine-deficient mutant was also ubiquitinated by mK3. In 2007, the same researcher found that the K-less heavy chain

L. Maria Lois and Rune Matthiesen (eds.), *Plant Proteostasis: Methods and Protocols,* Methods in Molecular Biology, vol. 1450,
DOI 10.1007/978-1-4939-3759-2_1, © Springer Science+Business Media New York 2016

(HC) of MHC class I could be degraded, while the KSCT-less HC was stable [3]. Ubiquitination was not influenced by reducing agents, which indicated that the modification was not occurring on a Cys residue, and closer inspection revealed that the MHC class I was ubiquitinated at Lys, Ser, and Thr sites [3]. Similarly, the ERAD substrate NS-1 is ubiquitinated by the HRD1 E3 ligase at Lys and Ser/Thr sites [4], and ubiquitination also occurs on the N-terminal amino of substrates [5] and Cys in substrates [6]. In plants, the $SCF^{TIR1/AFB}$ ubiquitin ligase substrate IAA1 was found to be ubiquitinated on Lys and Ser/Thr residues, which promoted rapid degradation [7]. Ubiquitination can therefore occur on multiple different amino acid residues, which introduces a great deal of complexity and flexibility.

Several methods have been established to identify ubiquitination sites, including mutagenesis followed by degradation assays and LC–MS/MS. However, mutagenesis is laborious if there are numerous Lys or Ser/Thr sites present. Although LC–MS/MS is highly efficient, it is expensive and false positives can be problematic. Biochemical approaches are convenient and can differentiate between ubiquitination on Ser/Thr or Cys in a shorter time. Ubiquitination of Ser/Thr results in a covalent oxyester bond that is sensitive to mild alkaline treatment, while the thioester bond between Cys and the C-terminus of ubiquitin is sensitive to reducing agents, and the isopeptide bond between Lys and ubiquitin is stable under both mild alkaline and reducing conditions [8, 9]. Detection of ubiquitination products following treatment with reducing agents or mild alkaline buffer can therefore determine the site(s) of ubiquitin attachment, and this method was tested in plants.

2 Materials

2.1 Extraction of Plant Proteins

1. 1 M dithiothreitol (DTT): dissolve 1.545 g DTT in 10 ml 0.01 M sodium acetate, filter using a 0.22 μM sterile membrane, aliquot and store at −20 °C (*see* **Note 1**).
2. 100 mM phenylmethanesulfonyl fluoride (PMSF): dissolve 0.174 g PMSF in 10 ml isopropanol, aliquot and store at −20 °C (*see* **Note 1**).
3. Native extraction buffer: 50 mM Tris–MES (pH 8.0), 0.5 M sucrose, 1 mM $MgCl_2$, 10 mM EDTA, autoclave and store at 4 °C. Add DTT to 1 mM, PMSF to 1 mM, and a protease inhibitor cocktail Complete Mini tablet (Roche) per 10 ml buffer immediately before use (*see* **Note 2**).

2.2 SDS-PAGE

1. Store 30% acrylamide/bis-acrylamide (29:1) solution (Genestar) at 4 °C.
2. Store *N*,*N*,*N*,*N*′-tetramethyl-ethylenediamine (TEMED; AMRESCO) at 4 °C.

3. 10% ammonium persulfate (APS): dissolve 1 g APS in 10 ml H_2O (10% in W/V) (*see* **Note 3**) and store at 4 °C.
4. 10% SDS: dissolve 100 g in 1 l H_2O and store at room temperature.
5. 1 M Tris–HCl (pH 6.8): dissolve 121.2 g Tris base in 800 ml H_2O, adjust pH to 6.8 with HCl, and adjust volume to 1 l with additional H_2O. Autoclave and store at room temperature.
6. 1.5 M Tris–HCl (pH 8.8): dissolve 181.6 g Tris base in 800 ml H_2O, adjust pH to 8.8 with HCl, and adjust volume to 1 l with additional H_2O. Autoclave and store at room temperature.
7. 10% SDS-PAGE separating gel (5 ml): 1.9 ml H_2O, 1.7 ml 30% acrylamide-bis-acrylamide (29:1), 1.3 ml 1.5 M Tris–HCl (pH 8.8), 0.05 ml 10% SDS, 0.05 ml 10% APS and 0.002 ml TEMED.
8. Stacking gel (3 ml): 2.1 ml H_2O, 0.5 ml 30% acrylamide-bis-acrylamide (29:1), 0.38 ml 1 M Tris–HCl (pH 6.8), 0.03 ml 10% SDS, 0.03 ml 10% APS and 0.002 ml TEMED.

2.3 Immunoblotting Components

1. Running buffer (1 l): dissolve 3.03 g Tris base, 14.4 g glycine and 1 g SDS in 1 l dH_2O.
2. Transfer buffer (1 l): dissolve 3.03 g Tris base and 14.4 g glycine in 800 ml dH_2O, and add 200 ml methanol.
3. 4× SDS loading buffer contained 0.25 M Tris–HCl (pH 6.8), 8% SDS, 40% glycerol, 0.005% bromophenol blue and 20% β-mercaptoethanol.
4. 10× PBS (1 l): 80 g NaCl, 2 g KCl, 35.8 g $Na_2HPO_4 \cdot 12H_2O$, 2.7 g KH_2PO_4, pH 7.4. Dilute to 1× PBS before use.
5. Blocking buffer: 5% skim milk powder in 1× PBS.
6. Antibody dilution buffer: 3% skim milk powder in 1× PBS.
7. Primary antibodies: specific for the epitopes or proteins of interest.
8. Secondary antibodies: goat anti-mouse or goat anti-rabbit (Proteinteach).
9. Horseradish peroxidase (HRP) substrate kit (Millipore).
10. X-ray film.

2.4 Immuno-precipitation

1. Store Protein G Dynabeads (Life Technologies) at 4 °C.
2. MG132 stock solution: MG132 is dissolved in DMSO and 10 mM stock. Aliquot it in small volume and stored at −80 °C (*see* **Note 4**).
3. A 4 °C cold room.
4. Thermo-mixer comfort (Eppendorf) equipped with a constant temperature setting.
5. Amicon Ultra-15 Centricons (Millipore; *see* **Note 5**).

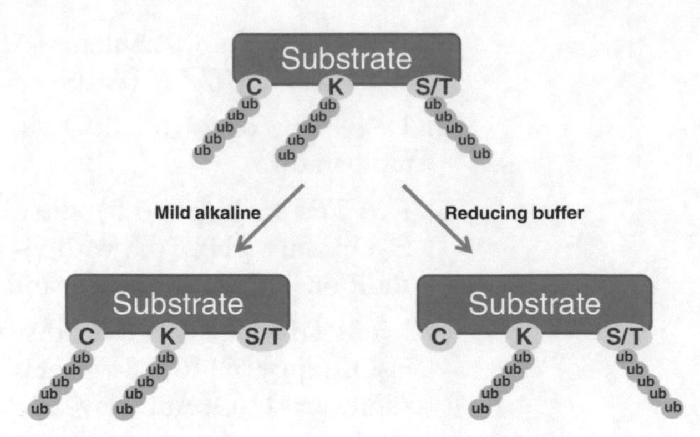

Fig. 1 The schematic diagram of ubiquitination occurring on different amino acids. Ubiquitination occurring on Ser/Thr residues is sensitive to mild alkaline (0.1 M NaOH) treatment, which also abolishes the signal from polyubiquitinated proteins. Ubiquitination occurring on Cys residues is sensitive to reducing agents, which also abolishes the signal from polyubiquitylated proteins

3 Methods

In this chapter, we describe the development of a detailed protocol for determining ubiquitination events on nonlysine residues. Substrate proteins are first obtained from transgenic plants or transient expression in *Nicotiana benthamiana* leaves as described previously [10]. Confirmation of ubiquitination is needed. Treatment with SDS followed by immunoprecipitation helps to exclude ubiquitinated interacting partners. Treatment with reducing agents or mild alkaline then determines if modification has occurred on Cys or Ser/Thr residues, respectively. All procedures should be carried out on ice unless otherwise specified. A schematic diagram is shown in Fig. 1.

3.1 Protein Expression and Extraction

1. Harvest transgenic plants or *N. benthamiana* leaves expressing substrates or control proteins and freeze immediately in liquid nitrogen.
2. Grind the material and resuspend in 1 ml native extraction buffer per 0.4 g powder (*see* **Note 6**).
3. Centrifuge at 14,000 × *g* for 6 min at 4 °C and the supernatant was transferred to a new tube. Repeat this centrifugation for four times and transfer the supernatant each time. Use supernatant in immunoprecipitation assay.

3.2 Immunoprecipitation

1. Filter supernatant using a 0.45 μM sterile membrane to remove debris (*see* **Note 7**).
2. Prepare Protein G Dynabeads by washing three times with ice-cold native buffer.

3. Antibody (10 μg antibody/ml native buffer) is incubated with Protein G Dynabeads in a centrifuge tube at 4 °C for 2 h. Wash the antibody three times with ice-cold 1× PBS and removes all liquid after the final wash.
4. Add 1 ml of supernatant and MG132 to a final concentration of 50 μM (*see* **Note 4**). Shake gently at room temperature for 30 min and at 4 °C for 2 h to bind substrate proteins (*see* **Note 8**).
5. Collect Dynabeads and wash five times with cold 1× PBS (150 mM NaCl was added to PBS; *see* **Note 9**). Use the immunoprecipitated mixture for in vivo ubiquitination assays and western blotting.

3.3 In Vivo Detection of Polyubiquitination

1. Immunoprecipitate substrates and control proteins using specific antibodies as described in Subheading 3.2.
2. Suspend beads with bound proteins in 40–60 μl 10 mM Tris–HCl (pH 6.8) containing 0.5% SDS, and heat at 95 °C for 5 min (*see* **Note 10**). This will remove bound proteins and disrupt protein–protein interactions.
3. Centrifuge at 14,000 × *g* for 30 s. Transfer supernatant to a new tube.
4. Dilute supernatant to 1 ml with 1× PBS.
5. Re-immunoprecipitate proteins for 2 h and collect beads as described above.
6. Elute samples using SDS sample buffer at 95 °C for 5 min and separate by SDS-PAGE. Western blotting will be conducted to detect polyubiquitinated substrates.

3.4 Detection of Polyubiquitination on Cys Residues

1. Immunoprecipitate substrate and control proteins as described in Subheading 3.2 and treat with SDS as described in Subheading 3.3.
2. Divide beads into two tubes and adjust volume to 30 μl with 1× PBS.
3. Add 10 μl nonreducing or reducing 4× SDS sample buffer, respectively, and boiled at 95 °C for 5 min.
4. Separate by SDS-PAGE and observe whether ubiquitination is decreased in reducing buffer by western blotting.

3.5 Detection of Polyubiquitination on Ser/Thr Residues

1. Immunoprecipitate substrate and control proteins as described in Subheading 3.2 and treat with SDS as described in Subheading 3.3.
2. Collect beads by centrifugation at 14,000 × *g* for 30 s, transfer supernatant to a new tube and adjust the final volume to 40 μl with 10 mM Tris–HCl (pH 6.8).

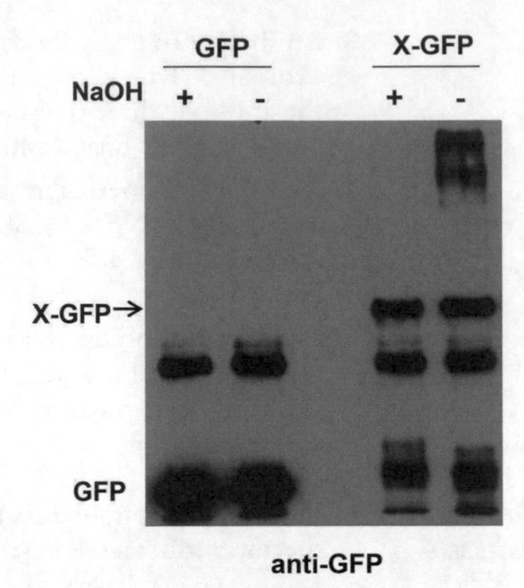

Fig. 2 The ubiquitination on Ser is sensitive to NaOH treatment. The protein X-GFP can be poly-ubiquitinated in vivo. X-GFP is immunoprecipitated and checked for ubiquitination after being treated with NaOH or 1× PBS as described in Subheading 3.5. The polyubiquitin of X-GFP is reduced by treatment with NaOH. GFP is used as a control in this assay

3. Divide supernatant into two equal aliquots and add NaOH to a final concentration of 0.1 M to one aliquot, and 1× PBS to the other aliquot to a final volume of 60 μl.
4. Incubate at 37 °C for 2 h.
5. Dilute to 4 ml with 10 mM Tris–HCl (pH 6.8) in Amicon Ultra-15 Centricons following hydrolysis and concentrate to a volume of 1 ml at 4 °C (*see* **Note 11**).
6. Test the pH using pH indicator paper, stop concentrating when the pH reaches 7–7.5, and repeat **step 5** if the pH is too high (*see* **Note 12**).
7. Re-immunoprecipitate proteins at 4 °C and elute in reducing buffer by boiling at 95 °C for 5 min. Separated by SDS-PAGE. An example was shown in Fig. 2.

3.6 Western Blotting

1. Separate samples by 10% SDS-PAGE at 160 V for 1 h.
2. Transfer to nitrocellulose membrane (in transfer buffer) at 100 V for 75 min.
3. Block membrane with blocking buffer for 1 h at room temperature or overnight at 4 °C.
4. Incubate membrane with primary antibody in antibody dilution buffer for 1 h at room temperature or overnight at 4 °C.
5. Remove antibody solution and wash membrane twice with 0.05% PBST for 15 min (*see* **Note 13**).

6. Incubate membrane with secondary antibody in antibody dilution buffer for 1 h at room temperature.
7. Wash membrane twice with 0.05% PBST for 15 min (*see* **Note 13**).
8. Detect the signal using a Millipore Chemiluminescence HRP Substrate Kit.

4 Notes

1. DTT (1 M) and PMSF (100 mM) stocks are unstable at room temperature or 4 °C and should be stored at −20 °C. Aliquot small volumes to avoid repeated freeze-thawing. Add to solutions immediately before use.
2. For membrane proteins, detergents such as NP-40 should be included to improve protein extraction. NP-40 should be added to native buffer just before use and the amount should be less than 1% in accordance with protein properties.
3. The "H_2O" used in this protocol is ultrapure water with an electrical resistivity of 18 MΩ cm at 25 °C.
4. For unstable substrate proteins, MG132 (or another proteasome inhibitor) should be used to prevent degradation. MG132 should not be freeze-thawed repeatedly and a working concentration of 50–100 μM is recommended.
5. An Amicon Centricon of less than one third of the molecular weight of the target protein should be used, and the centrifugal speed and centrifugal time should be in accordance with the manufacturer's instructions and with protein solubility.
6. The amount of buffer added can be adjusted according to the target protein expression level. The dilution ratio mentioned in the text was determined from empirical results.
7. This step is critical. Removing debris helps to minimize unrelated interacting proteins.
8. Ubiquitinated proteins can be unstable, rapidly degraded and hence difficult to detect. Incubating at room temperature with MG132 likely minimizes degradation and increases the amount of the ubiquitinated form.
9. Adding NaCl to wash buffer can reduce contaminating proteins during immunoprecipitation.
10. Interacting partner proteins may be coimmunoprecipitated. To avoid the false signal from ubiquitinated interacting proteins, beads should be boiled as described to disrupt protein–protein interactions.
11. The volume is dependent on protein solubility.

12. The pH is important and a high pH may affect separation by SDS-PAGE.
13. If a high background occurs during western blotting, add 0.1 % Tween 20 to PBST wash buffers. If the signal is too low, 1× PBS without Tween 20 should be used.

Acknowledgments

This project was supported by grants from the National Natural Science Foundation of China (grant number 91317308) and (grant number 31170781) and the Major State Basic Research Development Program of China [973 Program 2011CB915402] to Qi Xie.

References

1. Woelk T, Sigismund S, Penengo L, Polo S (2007) The ubiquitination code: a signalling problem. Cell Div 2:11
2. Wang XL, Connors R, Harris MR, Hansen TH, Lybarger L (2005) Requirements for the selective degradation of endoplasmic reticulum-resident major histocompatibility complex class I proteins by the viral immune evasion molecule mK3. J Virol 79:4099
3. Wang XL et al (2007) Ubiquitination of serine, threonine, or lysine residues on the cytoplasmic tail can induce ERAD of MHC-I by viral E3 ligase mK3. J Cell Biol 177:613
4. Shimizu Y, Okuda-Shimizu Y, Hendershot LM (2010) Ubiquitylation of an ERAD substrate occurs on multiple types of amino acids. Mol Cell 40:917
5. Ciechanover A, Ben-Saadon R (2004) N-terminal ubiquitination: more protein substrates join in. Trends Cell Biol 14:103
6. Cadwell K, Coscoy L (2005) Ubiquitination on nonlysine residues by a viral E3 ubiquitin ligase. Science 309:127
7. Gilkerson J, Kelley DR, Tam R, Estelle M, Callis J (2015) Lysine residues are not required for proteasome-mediated proteolysis of the auxin/indole acidic acid protein IAA1. Plant Physiol 168:708
8. Hershko A, Ciechanover A, Heller H, Haas AL, Rose IA (1980) Proposed role of ATP in protein breakdown: conjugation of protein with multiple chains of the polypeptide of ATP-dependent proteolysis. Proc Natl Acad Sci U S A 77:1783
9. Greene TW (1981) Protective groups in organic synthesis. Wiley, New York, NY, p 349
10. Liu LJ et al (2010) An efficient system to detect protein ubiquitination by agroinfiltration in Nicotiana benthamiana. Plant J 61:893

Chapter 2

Immunoprecipitation of Cullin-RING Ligases (CRLs) in *Arabidopsis thaliana* Seedlings

Anna Franciosini and Giovanna Serino

Abstract

CRL (Cullin-RING ubiquitin ligase) is the major class of plant E3 ubiquitin ligases. Immunoprecipitation-based methods are useful techniques for revealing interactions among Cullin-RING Ligase (CRL) subunits or between CRLs and other proteins, as well as for detecting poly-ubiquitin modifications of the CRLs themselves. Here, we describe two immunoprecipitation (IP) procedures suitable for CRLs in Arabidopsis: a procedure for IP analysis of CRL subunits and their interactors and a second procedure for in vivo ubiquitination analysis of the CRLs. Both protocols can be divided into two major steps: (1) preparation of cell extracts without disruption of protein interactions and (2) affinity purification of the protein complexes and subsequent detection. We provide a thorough description of all the steps, as well as advice on how to choose proper buffers for these analyses. We also suggest a series of negative controls that can be used to verify the specificity of the procedure.

Key words Cullin-RING ubiquitin ligase, Immunoprecipitation, Coimmunoprecipitation, Ubiquitin, Immunoblot

1 Introduction

Cullin-RING Ligases (CRLs) are the largest family of E3 ubiquitin ligases and recruit specific substrates for poly-ubiquitination [1]. Since their discovery in yeast almost 20 years ago [2, 3], CRLs have also been involved in almost all developmental and physiological plant processes [4]. CRLs are modular assemblies built on a backbone cullin subunit (CUL1, CUL3, and CUL4 in *Arabidopsis*) holding at their carboxy-terminal domain a RING-box protein (RBX1), which serves as a site for the interaction with the E2 ubiquitin conjugating enzyme, and at their amino-terminal domain a variable substrate receptor subunit, often connected via a bringing adaptor [5] (Fig. 1). Depending on the type of the cullin subunit, each recruiting an interchangeable substrate receptor, distinct subclasses of CRLs can be assembled, with different substrate specificity. Detection of protein–protein interactions among CRL

L. Maria Lois and Rune Matthiesen (eds.), *Plant Proteostasis: Methods and Protocols*, Methods in Molecular Biology, vol. 1450, DOI 10.1007/978-1-4939-3759-2_2, © Springer Science+Business Media New York 2016

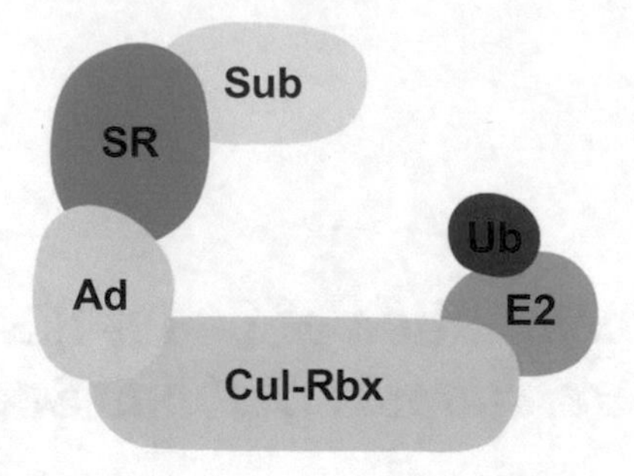

Fig. 1 Model of a CRL ubiquitin ligase. A typical CRL is composed of a cullin scaffold subunit, which interacts with RBX1 (Cul-Rbx), that in turn binds the E2 ubiquitin conjugating enzyme. Specific substrates are recruited by a variable substrate receptor (SR) anchored to the cullin through an adaptor (Ad)

subunits, as well as CRL subunit interaction with other proteins is therefore essential to provide insights on the individual cellular function of a given CRL.

CRLs activity is dynamically regulated. In absence of the substrate, CRLs can ubiquitinate their own substrate receptor, thus providing a mechanism to decrease the activity of a specific CRL when it is not necessary. CRL subunits can also be ubiquitinated by other E3 ubiquitin ligases [6]. Indeed, the turnover of several substrate receptors has been shown to be controlled by specific E3s. Therefore, determining the stability or the poly-ubiquitination status of a particular substrate receptor can offer a more complete overview on the biological role of a given CRL.

Two detailed step-by-step procedures are described here. The first one (*see* Subheading 3.1) illustrates how to immunoprecipitate a tagged CRL subunit to detect its direct or indirect interaction with another CRL subunit or other proteins of interest for which antibodies are available. Because the interaction between the protein of interest and its binding partner may be transient, a cross-linking step before protein extraction can be employed. Next, total proteins need to be extracted, and the composition of the grinding buffer may need to be adjusted (i.e., salt concentration, pH, amount of detergents), depending on the strength of the protein–protein interaction to be investigated. In addition, to enhance the overall yield of the immunoprecipitation (IP) and to increase the likelihood of immunoprecipitating interacting proteins, two classes of compounds could be added to the grinding buffer immediately before use: protease inhibitors, to avoid unwanted proteolysis during protein extraction, and phosphatase inhibitors, to preserve the phosphorylation state of immunoprecipitated proteins. Once proteins are extracted, the CRL complex is affinity purified by capturing the CRL subunit and its binding partners with a commercially available antibody immobilized on a solid support (beads). The CRL complex attached to the beads is then precipitated and isolated (IP sample) through centrifugation, while the unbounded proteins are washed out. Finally, the IP sample is analyzed by immmunoblotting using both an antibody against the tagged protein, to control that the CRL subunit has been correctly immunoprecipitated, and an antibody against other proteins to investigate their suspected interaction with the CRL of interest.

The second protocol (*see* Subheading 3.2) describes an IP-based procedure to examine whether a substrate receptor subunit of a given CRL is poly-ubiquitinated in vivo. The critical aspect of this experiment consists in preserving the integrity of the poly-ubiquitin chain conjugated to the protein of interest. Thus, before the protein extraction and the IP steps, it might be useful to incubate Arabidopsis seedlings with a proteasome inhibitor (such as MG132) in order to stabilize the poly-ubiquitinated proteins. In addition, it might be necessary to use a denaturing protein extraction buffer supplied with *N*-ethylmaleimide (NEM). NEM blocks a cysteine residue of the active site of the deubiquitinating enzymes, thus avoiding their unwanted activity. The second part of this protocol follows the same principles described for the first protocol: the CRL subunit is subjected to affinity purification using an antibody-coupled resin, and the samples are later analyzed by immunoblotting. The presence of an ubiquitin chain on the protein of interest can be observed by using an epitope tag antibody and an anti-ubiquitin antibody. This protocol may be also used to investigate whether a protein, which is not a component of CRL complexes, is covalently conjugated to an ubiquitin chain. The procedure can be used either with epitope tag antibodies or with native antibodies/affinity matrixes. For a protocol for antibody–resin coupling, please refer to [7].

2 Materials

2.1 Co-IP of CRLs

2.1.1 Plant Material and Growth

1. MS solid medium: 4.4 g/L Murashige and Skoog medium including Gamborg B5 vitamins, 10 g/L sucrose, 0.5 g/L MES, 0.8 % plant agar, pH adjusted to 5.7 with KOH.
2. DSP (dithiobis(succinimidyl propionate)) cross-linker stock solution: 10 mM dissolved in DMSO (*see* **Note 1**).
3. Cross-link reaction buffer: Phosphate Buffer Saline (PBS). Add 1 mM DSP before use.
4. Cross-link stop solution: 1 M Tris–HCl pH 7.5.

2.1.2 Total Protein Extraction

1. Grinding buffer A: 50 mM Tris–HCl pH 7.5, 50 mM $MgCl_2$, 150 mM NaCl, 0.1 % NP-40. Add 20 mM β-glycerophosphate, 20 mM NaF, 5 mM Na_3VO_4, phosphatase inhibitors, and

100 mM PMSF and 1× complete protease inhibitors (Roche; Sigma) immediately prior to use.

2. 2× Loading Buffer: 125 mM Tris–HCl pH 6.8, 5% β-mercaptoethanol, 4% (w/v) SDS, 10% w/v glycerol, 0.01% Bromophenol Blue. Store at 4 °C.
3. Liquid nitrogen.
4. Mortar and pestle.
5. Refrigerated centrifuge.

2.1.3 Immuno-precipitation

1. Washing buffer A: 50 mM Tris–HCl pH 7.5, 50 mM $MgCl_2$, 150 mM NaCl, 0.1% NP-40. Store at 4 °C.
2. Primary antibody against the protein to be pulled down.
3. 2× Loading Buffer: 125 mM Tris–HCl pH 6.8, 5% β-mercaptoethanol, 4% (w/v) SDS, 10% w/v glycerol, 0.01% Bromophenol Blue. Store at 4 °C.
4. Refrigerated centrifuge.
5. Rotator with 1.5 mL tube holders.

2.1.4 SDS-PAGE

1. Mini-PROTEAN TGX precast gel (Biorad), stored at 4 °C (*see* **Note 5**). The range of acrylamide concentration should be chosen depending on the predicted molecular weight of the proteins being separated.
2. Running Buffer 10×: 250 mM Tris, 1.92 M glycine, 1% SDS. Store a room temperature.
3. Prestained molecular mass marker.
4. Mini-PROTEAN precast gel cassette (Biorad) (*see* **Note 5**).
5. Power supply.

2.1.5 Immunoblotting and Detection

1. Transfer Buffer: 25 mM Tris, 192 mM glycine, 0.1% SDS, 20% methanol.
2. Methanol.
3. PVDF membrane cut slightly larger than the dimensions of the gel.
4. Filter paper cut slightly larger than dimensions of the gel.
5. Phosphate buffer saline with Tween-20 (PBS-T): 10 mM Na phosphate buffer pH 7.4, 150 mM NaCl, 0.05% Tween-20.
6. Blocking Buffer: 1% blocking reagent (Roche) dissolved in PBS-T.
7. Primary antibody against the immunoprecipitated protein.
8. Primary antibody against the coimmunoprecipitated protein.
9. HRP-conjugated secondary antibody.
10. Enhanced chemiluminescent (ECL) reagent.

11. X-ray films.
12. Mini Trans-Blot cell (Biorad) (*see* **Note 5**).
13. Power supply.
14. Shaker.
15. Image acquisition system (e.g., ChemiDoc, Biorad).

2.2 In Vivo Ubiquitination Analysis of CRLs

2.2.1 Plant Material and Growth

1. MS solid medium: 4.4 g/L Murashige and Skoog medium including Gamborg B5 vitamins, 10 g/L sucrose, 0.5 g/L MES, 0.8 % plant agar, pH adjusted to 5.7 with KOH.
2. MS liquid medium: 4.4 g/L Murashige and Skoog medium including Gamborg B5 vitamins, 10 g/L sucrose, 0.5 g/L MES, pH adjusted to 5.7 with KOH.
3. MG132 stock solution: 50 mM MG132 dissolved in DMSO. Store at −20 °C.

2.2.2 Total Protein Extraction

1. Grinding buffer B: 50 mM Tris–HCl pH 7.5, 150 mM NaCl, 1 % Triton X-100, 1 mM EDTA, 10% glycerol. Add 50 μM MG132, 10 mM NEM, 100 mM PMSF, and 1× Complete protease inhibitor cocktail (Roche) immediately prior to use.
2. 2× Loading Buffer: 125 mM Tris–HCl pH 6.8, 5 % β-mercaptoethanol, 4 % (w/v) SDS, 10 % w/v glycerol, 0.01 % Bromophenol Blue. Store at 4 °C.
3. Liquid nitrogen.
4. Mortar and pestle.
5. Refrigerated centrifuge.

2.2.3 Immuno-precipitation

1. Washing buffer B: 50 mM Tris–HCl pH 7.5, 150 mM NaCl, 1 % Triton X-100, 1 mM EDTA, 10 % glycerol. Store at 4 °C.
2. Primary antibody against the protein to be pulled down (*see* **Note 3**).
3. 2× Loading Buffer: 125 mM Tris pH 6.8, 5 % β-mercaptoethanol, 4 % (w/v) SDS, 10 % w/v glycerol, 0.01 % Bromophenol Blue. Store at 4 °C.
4. Refrigerated centrifuge.
5. Rotator with 1.5 mL tube holders.

2.2.4 SDS-PAGE

1. Gradient Mini-PROTEAN TGX precast gel (Biorad), stored at 4 °C (*see* **Note 5**).
2. Running Buffer 10×: 250 mM Tris, 1.92 M glycine, 1 % SDS. Store at room temperature.
3. Prestained molecular marker.
4. Mini-PROTEAN precast gel cassette (Biorad) (*see* **Note 5**).
5. Power supply.

2.2.5 Immunoblot and Detection

1. Transfer Buffer: 25 mM Tris, 192 mM glycine, 0.1% SDS, 20% methanol.
2. Methanol.
3. PVDF membrane cut slightly larger than the dimensions of the gel.
4. Filter paper cut slightly larger than dimensions of the gel.
5. Phosphate buffer saline with Tween-20 (PBS-T): 10 mM Na phosphate buffer pH 7.4, 150 mM NaCl, 0.05% Tween-20.
6. Blocking Buffer: 1% blocking reagent (Roche) dissolved in PBS-T.
7. Primary antibody against the immunoprecipitated protein.
8. Primary antibody against ubiquitin.
9. HRP-conjugated secondary antibody.
10. Enhanced chemiluminescent (ECL) reagent.
11. X-ray films.
12. Mini Trans-Blot cell (Biorad) (*see* **Note 5**).
13. Power supply.
14. Shaker.
15. Image acquisition system (e.g., ChemiDoc, Biorad).

3 Methods

3.1 IP of CRLs

3.1.1 Plant Material

1. Grow *Arabidopsis* seedlings on MS solid medium for 5–7 days at 22 °C.
2. Transfer 300–500 mg of seedlings in cross-linking reaction buffer (*see* **Notes 1** and **2**).
3. Incubate for 30′ at room temperature with gentle shaking.
4. Add the cross-linking stop solution to a final concentration of 10 mM and incubate for 15′ at room temperature.
5. Collect the seedlings in a 1.5 mL microcentrifuge tube and immediately freeze the sample in liquid nitrogen.

3.1.2 Total Protein Extraction

1. Transfer the frozen plant material in a mortar and grind them while keeping it frozen, until a fine powder is obtained. Collect the powder in a microcentrifuge tube and immediately add 300–500 μL of Grinding Buffer A. Vortex to homogenize the sample and then place the tube on ice.
2. Centrifuge the sample at 16,000 × *g* for 15′ at 4 °C, and transfer the supernatant in a new tube.
3. Remove a 20 μL aliquot to serve as a total extract control. Add 20 μL of 2× Loading Buffer and boil for 5′. Store at −20 °C for later analysis.

3.1.3 Immunoprecipitation

1. Equilibrate the antibody-coupled beads (*see* **Note 3**). Add 500 μL Grinding Buffer A to a 30 μL of beads. Centrifuge at 1500 × *g* for 4′ a room temperature and remove the supernatant.
2. Add the crude extract (from step 2 in Subheading 3.1.2) to the antibody-coupled beads.
3. Place the tube in a rotator and incubate with gentle agitation from 1 to 4 h at 4 °C (*see* **Note 4**).
4. Pellet the beads by centrifuging the tube at 1000 × *g* for 5′ at 4 °C. Add 1 mL of Washing Buffer A and incubate for 5′ with gentle agitation at 4 °C.
5. Repeat the washing (step 4) three times.
6. Pellet the beads at 1000 × *g* for 5′ at 4 °C and add 30 μL of 2× Loading Buffer. Boil for 5′.

3.1.4 SDS-PAGE

1. Prepare the Mini-PROTEAN TGX precast gel in the apparatus as indicated in the manufacturer's instruction (*see* **Note 5**). Fill the cassette with Running Buffer 1×.
2. Load on the gel the prestained molecular marker and an equal volume of the protein samples from step 3 in Subheading 3.1.2 (total extract) and from step 6 in Subheading 3.1.3 (immunoprecipitate) (*see* **Note 4**).
3. Connect the electrophoresis chamber to the power supply and run the gel from 100 to 200 V until the dye reaches the bottom of the gel.

3.1.5 Immunoblot and Detection

1. Before transfering the separated proteins from the gel to the membrane, activate the PVDF membrane in methanol for 10′ with gentle shaking. Transfer the PVDF membrane in Transfer Buffer to avoid its drying.
2. Prepare the gel sandwich with the filter papers, the gel, and the membrane in the Mini Trans-Blot (Biorad) cassette holder as indicated by the manufacturer's instruction. Fill the tank with Transfer Buffer, and connect the apparatus to the power supply.
3. Set the power supply at 100 V and run for 1 h.
4. After transfer, block membrane in 1% blocking reagent in PBS-T for 1 h at room temperature or at 4 °C overnight with gentle shaking.
5. Pour off the blocking solution and replace it with fresh 0.5–1% blocking solution containing the primary antibody.
6. Incubate for 3–6 h at room temperature or at 4 °C overnight with gentle shaking.

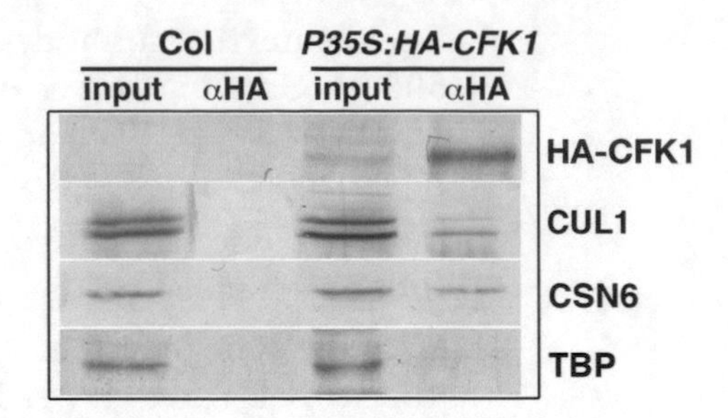

Fig. 2 Co-IP of the CRL substrate adaptor subunit CFK1 with CUL1 and CSN6. The Arabidopsis CRL substrate adaptor subunit CFK1 was fused to HA epitope to obtain plants expressing HA-CFK fusion protein [10]. Total protein extract from wild-type (Col-0) and HA-CFK1 expressing seedlings were immunoprecipitated using anti-HA affinity matrix followed by immunoblot to detect the interaction between CFK1 and other proteins. HA-CFK1 coimmunoprecipitates with CUL1, a subunit of the CRL complex, and with CSN6 protein. TBP (TATA Binding Protein) antibody was used as negative control of the interaction. "input" indicates the total extract [10]

7. Pour off the primary antibody and replace it with PBS-T. Wash with gentle shaking for 10′.
8. Repeat step 7 two more times.
9. Pour off the PBS-T and add the secondary antibody in 0.5–1 % blocking reagent in PBS-T. The secondary antibody is chosen based on the primary antibody used in the step 4.
10. Incubate at 1–2 h at room temperature or at 4 °C overnight with gentle shaking.
11. Pour off the secondary antibody and replace it with PBS-T. Wash with gentle shaking for 10′.
12. Repeat step 11 four more times.
13. Pour the PBS-T off the membrane and add ECL reagent on the blotted side of the membrane. Incubation time depends on the ECL reagent used.
14. Expose the membrane to the X-ray film or use an image acquisition system. The time of exposure may vary from experiment to experiment. Figure 2 represents an example of this procedure.

3.2 *In Vivo* Ubiquitination Analysis of CRLs

3.2.1 Plant Material and Growth

1. Grow Arabidopsis seedlings on MS solid medium for 5–7 days at 22 °C.
2. Transfer 300–500 mg of seedlings in MS liquid medium supplied with 50 μM MG132, and 300–500 mg in MS liquid medium with DMSO as negative control (*see* **Note 6**).
3. Incubate from 2 to 4 h in the *Arabidopsis* growth chamber.
4. Collect the seedlings in a 1.5 mL microcentrifuge tube and immediately freeze the sample in liquid nitrogen.

3.2.2 Total Protein Extraction

1. Transfer the plant material in a mortar and pestle, under liquid nitrogen, to a fine power. Collect the power in a microcentrifuge tube and immediately add 300–500 μL of Grinding Buffer B. Vortex to homogenize the sample, and then place the tube on ice.
2. Centrifuge the sample at 16,000 × *g* for 15′ at 4 °C, and transfer the supernatant in a new tube.

3.2.3 Immunoprecipitation

1. Equilibrate the antibody-coupled beads (*see* **Note 3**). Add 500 μL Grinding Buffer B to a 30 μL of beads. Centrifuge at 1500 × *g* for 4′ a room temperature and remove the supernatant.
2. Add the crude extract (from step 2 in Subheading 3.2.2) to the beads.
3. Put the tube in the tube rotator and incubate with gentle agitation from 1 to 4 h at 4 °C (*see* **Note 4**).
4. Pellet the beads in a centrifuge at 1000 × *g* for 5′ at 4 °C. Add 1 mL of Washing Buffer B and incubate for 5′ with gentle agitation at 4 °C.
5. Repeat the washing (step 4) three times.
6. Pellet the beads at 1000 × *g* for 5′ at 4 °C and add 30 μL of 2× Loading Buffer. Boil for 5′.

3.2.4 SDS-PAGE

1. Load on the gel the prestained molecular marker and samples from the step 6 in Subheading 3.2.3. Follow the procedure as described in Subheading 3.1.4.

3.2.5 Immunoblot and Detection

1. Follow the procedure as described in Subheading 3.1.5. For the in vivo ubiquitination analysis an antibody against the immunoprecipitated protein and an antibody against ubiquitin are used. A representative result of this procedure is shown in Fig. 3.

4 Notes

1. We use the chemical cross-linker DSP to covalently preserve the interactions among the CRL subunits. This step is not always required to detect protein–protein interaction and depends on the strength of the interaction. If this step is omitted, proceed directly to step 5 in Subheading 3.1.1.
2. The total protein extract (indicated as "input" in Fig. 2) serves as a positive control of the extraction. Extracts and immunoprecipitates from wild-type seedlings, not expressing the tagged protein, can be used as a negative control of the experiment. In addition, antibodies against proteins not supposed to

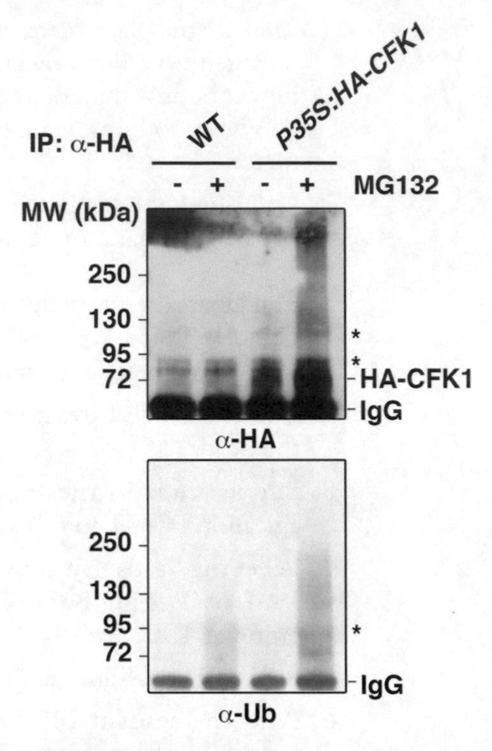

Fig. 3 In vivo ubiquitination analysis of the CRL substrate adaptor subunit CFK1. Wild-type and HA-CFK1 expressing seedlings were incubated with the proteasome inhibitor MG132. The crude extracts were prepared and immunoprecipitated with an anti-HA resin. The immunoprecipitated proteins were detected with anti-HA (top panel) and anti-ubiquitin (bottom panel) antibody. The increase in the higher molecular mass species in presence of MG132 recognized by the antibody against HA and against ubiquitin indicates that CFK1 is ubiquitinated in vivo [10]

interact can be employed. We suggest to use them in step 5 in Subheading 3.1.5 as a further negative control of the IP.

3. Both native antibodies or epitope antibodies can be used. Table 1 shows a list of epitope tags and their corresponding antibodies and matrices. We have successfully employed anti-HA agarose affinity gel from Sigma-Aldrich (Figs. 2 and 3), but other commercially available antibodies and resins (e.g., Covance, Roche) can be used. If a direct antibody against the protein to be immunoprecipitate is available, it can be coupled directly to protein A or protein G matrix and used for the IP. For this procedure, refer to other general IP protocols [7].
4. The incubation time might depends also on the antibody–resin that will be employed. Refer to the manufacturers' instruction to set up the IP time.
5. Here we provide the instructions for the SDS-PAGE based on the Mini-PROTEAN precast gels from Biorad, but other commercially systems or handcast gel can be employed [8, 9].

Table 1
List of epitope tags, commercially antibodies, and antibodies-conjugated matrices (and their corresponding catalog numbers) successfully used in Arabidopsis for IP protocols

TAG	Antibody	Matrix
HA	Sigma-Aldrich (H3663) Covance (MMS-101P) Santa Cruz (sc-805)	Sigma-Aldrich (E6779) Roche (11815016001)
c-myc	Sigma-Aldrich (M4439) Covance (MMS-164P) Santa Cruz (sc-40)	Covance (AFC-150P) Sigma-Aldrich (E6654)
FLAG	Sigma-Aldrich (F3165)	Sigma-Aldrich (A2220)
GFP	Sigma-Aldrich (G1544) Abcam (ab290)	ChromoTek (gta-20) Miltenyi Biotec (130-091-288)

6. Wild-type Arabidopsis seedlings, not expressing the tagged protein, can be used as negative control of the experiment. Because the MG132 proteasome inhibitor is dissolved in DMSO, immunoprecipitates from seedlings treated only with DMSO can be used as an additional negative control for the in vivo ubiquitination assay.

References

1. Lydeard JR, Schulman BA, Harper JW (2013) Building and remodelling Cullin-RING E3 ubiquitin ligases. EMBO Rep 14: 1050–1061
2. Bai C, Sen P, Hofmann K et al (1996) SKP1 connects cell cycle regulators to the ubiquitin proteolysis machinery through a novel motif, the F-box. Cell 86:263–274
3. Feldman RMR, Correll C, Kaplan K et al (1997) A complex of Cdc4p, Skp1p, and Cdc53p/Cullin catalyzes ubiquitination of the phosphorylated CDK inhibitor Sic1p. Cell 91:221–230
4. Choi C, Gray W, Mooney S et al (2014) Composition, roles, and regulation of cullin-based ubiquitin E3 ligases. Arabidopsis Book 12:e0175
5. Zheng N, Schulman B, Song L et al (2002) Structure of the Cul1–Rbx1–Skp1–F boxSkp2 SCF ubiquitin ligase complex. Nature 416: 703–709
6. Hua Z, Vierstra R (2011) The cullin-RING ubiquitin-protein ligases. Annu Rev Plant Biol 62:299–334
7. Serino G, Deng XW (2007) Protein coimmunoprecipitation in Arabidopsis. CSH protocols. Cold Spring Harbor Laboratory press, pdb. prot4683.
8. Smith BJ (1994) SDS Polyacrylamide gel electrophoresis of proteins. In: Walker JM (ed) Basic protein and peptide protocols, vol 32, Methods in molecular biology. Humana Press, Totowa, NJ, pp 23–34
9. Walker JM (1994) Gradient SDS polyacrylamide gel electrophoresis of proteins. In: Walker JM (ed) Basic protein and peptide protocols, vol 32, Methods in molecular biology. Humana Press, Totowa, NJ, pp 35–38
10. Franciosini A, Lombardi B, Iafrate S et al (2013) The Arabidopsis COP9 signalosome interacting F-box kelch 1 protein forms an SCF ubiquitin ligase and regulates hypocotyl elongation. Mol Plant 6:1616–1629

Chapter 3

Radioligand Binding Assays for Determining Dissociation Constants of Phytohormone Receptors

Antje Hellmuth and Luz Irina A. Calderón Villalobos

Abstract

In receptor–ligand interactions, dissociation constants provide a key parameter for characterizing binding. Here, we describe filter-based radioligand binding assays at equilibrium, either varying ligand concentrations up to receptor saturation or outcompeting ligand from its receptor with increasing concentrations of ligand analogue. Using the auxin coreceptor system, we illustrate how to use a saturation binding assay to determine the apparent dissociation constant (K_D') for the formation of a ternary TIR1–auxin–AUX/IAA complex. Also, we show how to determine the inhibitory constant (K_i) for auxin binding by the coreceptor complex via a competition binding assay. These assays can be applied broadly to characterize a one-site binding reaction of a hormone to its receptor.

Key words Radioligand binding, Hormone receptor, Binding affinities, Saturation binding, Competition binding

1 Introduction

Radioligand binding assays have been widely used in biochemistry and pharmacology to determine the binding affinities of ligands, such as small molecules, nucleic acids, and peptides to their receptors. Intermolecular forces, such as ionic bonds, hydrogen bonds, and van der Waals forces, define affinity binding. Affinity, as a measure of how "tightly" a ligand binds to its receptor, is described using the equilibrium dissociation constant K_D. Based on the law of mass action, K_D of a reversible binding reaction, Receptor (R) + Ligand $(L) \rightleftarrows$ Complex (RL) is defined as:

$$K_D = \frac{[R][L]}{[RL]}$$

at equilibrium at a given temperature. This means, if complex concentration at equilibrium is high and only little free ligand and free receptor is left, the K_D is a low value signifying high affinity

L. Maria Lois and Rune Matthiesen (eds.), *Plant Proteostasis: Methods and Protocols*, Methods in Molecular Biology, vol. 1450,
DOI 10.1007/978-1-4939-3759-2_3,

binding. In turn, if free ligand and receptor outweigh the complex at equilibrium, i.e., binding affinity is low, K_D will take on a high value.

For a typical saturation binding assay (Fig. 1a), the total receptor concentration $[R_{total}]$ is kept low and constant, while different concentrations of ligand $[L_{total}]$ are tested. The amount of total receptor used is a sum of free receptor concentration and complex concentration at equilibrium: $[R_{total}]=[R]+[RL]$. Also, $[R_{total}]$ is

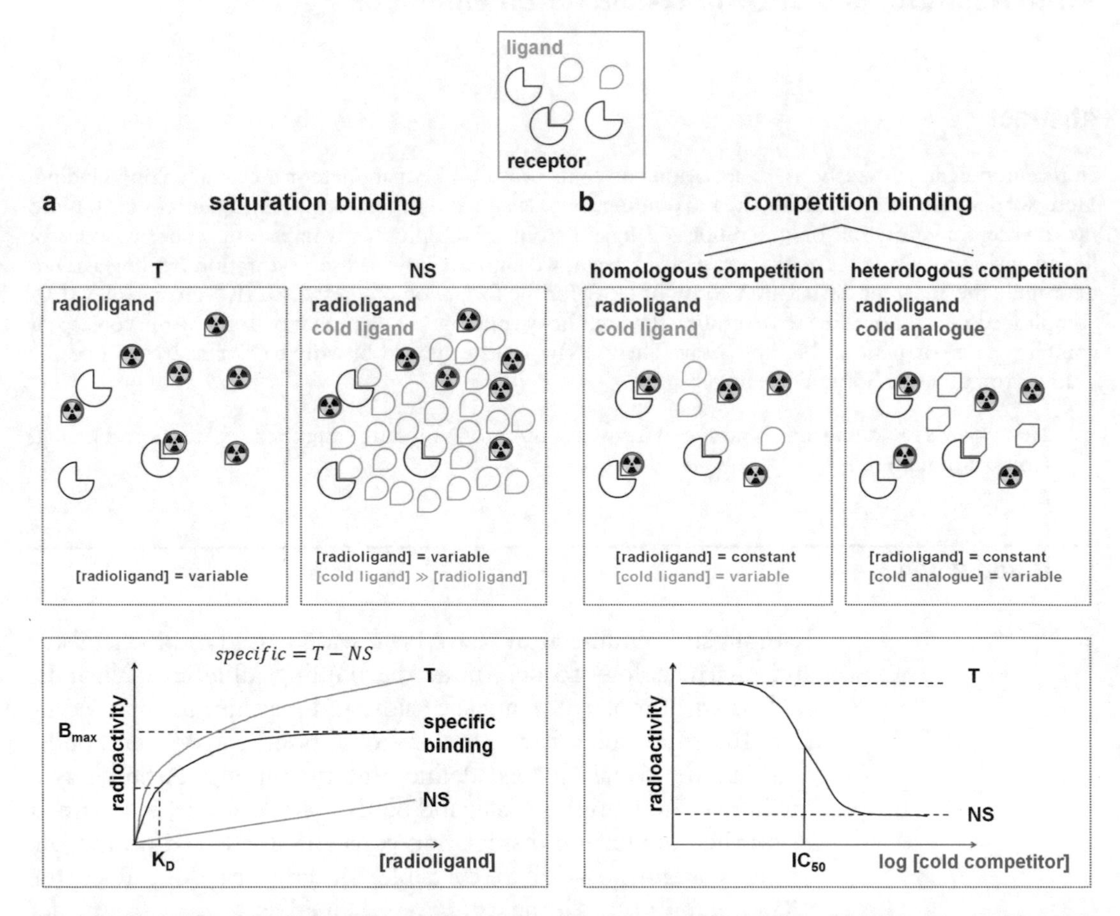

Fig. 1 Binding of ligand to its receptor can be assessed via two types of radioligand binding assays: Saturation and Competition Binding. (**a**) Saturation binding assays allow measuring of total (T) and nonspecific (NS) binding. To measure T and NS binding, receptor concentration is kept constant, while radioligand concentration is varied. NS samples are prepared with excess of unlabeled (cold) ligand. Specific binding results from subtracting NS from T, and will be plotted against radioligand concentration. Nonlinear regression allows determination of the dissociation constant K_D. (**b**) Competition binding assays can be performed as homologous (unlabeled competitor is identical with radioligand) or heterologous (unlabeled competitor is an analogue of the radioligand) assays. In both cases, receptor and radioligand concentrations are kept constant, while the concentration of unlabeled (cold) competitor is varied. Plotting the measured radioligand bound against logmolar concentration of cold competitor allows nonlinear regression to determine half-maximal inhibitory concentration IC_{50} of competitor

equivalent to maximum binding B_{max}, which will be determined in the experiment. Consequently, the equation for K_D can be rearranged to:

$$K_D = \frac{(B_{max} - [RL]) \cdot [L]}{[RL]} = \frac{B_{max} \cdot [L]}{[RL]} - [L]$$

$$[RL] = \frac{B_{max} \cdot [L]}{K_D + [L]}$$

Assuming only a small fraction of ligand is binding to the available receptor, so that $[L_{total}] \approx [L]$, the resulting equation is:

$$[RL] = \frac{B_{max} \cdot [L_{total}]}{K_D + [L_{total}]}$$

where $[L_{total}]$ is a known variable and B_{max} and K_D are specific constants we aim to determine. To quantify $[RL]$ as a function of the varied $[L_{total}]$, a radio isotope-labeled form of ligand (hereafter referred to as radioligand) is utilized, and the RL complex is immobilized on a glass fiber filter. Unbound radioligand is rapidly washed out with cold buffer to minimize disturbance of equilibrium. This is best accomplished using a suitable vacuum manifold or harvester. Take into account that if a given receptor–ligand complex exhibits a high dissociation rate, washing procedures can easily disturb the equilibrium, and might result in failure of capturing the true amount of the RL complex. In this case, consider alternative methods [1], e.g., surface plasmon resonance [2], isothermal titration calorimetry [3], fluorescence polarization [4], or thermophoresis [5]. With the assumption that radioligand can bind nonspecifically, for instance to nonreceptor sites on the protein, the filter paper, the tubes, etc., reactions need to be prepared in two sets: (1) containing only receptor and radioligand to obtain total (T) binding and (2) containing receptor and radioligand with the addition of excess of unlabeled ligand to obtain nonspecific (NS) binding. Subtraction of NS from T binding will result in specific binding values that are plotted against ligand concentration. Nonlinear regression with the equation derived above will yield a specific K_D (binding affinity) of a ligand for the receptor or vice versa.

Often, ligand analogues, i.e., compounds that bind to the same binding site in the receptor, are not available in a radiolabeled form. In this case, heterologous competition binding experiments can be carried out and allow for determination of inhibitory constant K_i, i.e., the K_D of binding of the analogue to the receptor. Note that beside heterologous competition assays, one can also perform homologous competition assays with the identical, unlabeled form of radioligand.

For a competition binding experiment (Fig. 1b), the total receptor concentration and radioligand concentration [*Radioligand*] are kept constant, while the concentration of the unlabeled competing analogue [*competitor*] is varied from zero to excess. A data curve for outcompetition of radioligand from the receptor binding site is obtained, and usually plotted against logmolar concentration of competitor. The curve follows the equation:

$$[RL] = \mathrm{NS} + \frac{\mathrm{T} - \mathrm{NS}}{1 + 10^{([\mathrm{competitor}] - \log \mathrm{IC}_{50})}},$$

where plateaus at both ends of the curve correspond to T and NS. As mentioned above, here too, the difference between T and NS binding gives specific binding. The concentration of competitor that is needed to reduce specific binding by half is referred to as half-maximal (or 50%) inhibitory concentration IC_{50}.

The inhibitory constant K_i can be determined from the IC_{50} via the Cheng–Prusoff equation [6]. This requires at least an estimation of K_D of radioligand for the receptor.

$$K_i = \frac{\mathrm{IC}_{50}}{1 + \frac{[\mathrm{Radioligand}]}{K_D}}$$

The methods of saturation and competition binding described here, further assume a one-site binding and no cooperativity. If there is more than one binding site to the receptor, or a more complex molecular mechanism, one can still apply one-site models, but has to refer to the apparent dissociation constant K_D' and interpret results appropriately. For more detailed information, we refer the reader to the vast literature resources on binding theory, for instance [7–12].

Various phytohormone receptor systems have been characterized using recombinant receptor and radiolabeled phytohormone or their analogues, including receptors for auxin [13], brassinosteroids [14, 15], jasmonic acid [16], salicylic acid [17, 18], gibberellins [19], and abscisic acid [20].

Our studies on the auxin sensing mechanism have included in vitro radioligand binding assays, which have served as powerful tool to understand how the ligand auxin might be bound by its receptor in vivo [13]. Auxin sensing requires the concerted action of F-box proteins TIR1/AFB1-5 and their targets for degradation, AUX/IAA transcriptional repressors [21]. TIR1/AFBs and AUX/IAAs act together to perceive auxin, and transmit the auxin signal, which in fact triggers changes in gene expression thereby modulating cell division, elongation, and differentiation [21–24]. Auxin occupies a pocket at the interface between TIR1 and AUX/IAAs proteins, so that auxin and AUX/IAA binding sites are spatially connected. Auxin ultimately acts as molecular glue, increasing TIR1–AUX/IAA affinity by cooperatively binding to both proteins [25]. Since an auxin receptor with full ligand-binding capability is constituted by a TIR1/AFB and an AUX/IAA protein, TIR1/AFB, and AUX/IAA together are referred to as a coreceptor system for auxin sensing [13]. In binding assays for assessing auxin sensing by the coreceptor system, all assumptions: equilibrium reached, no ligand depletion, one-site binding, no cooperativity are met. Yet, it needs to be taken into account that the precise molecular mechanism of binding hierarchy of auxin to the single receptor components is still unclear and, therefore, radioligand binding assays described here indicate an apparent K_D' for formation of a TIR1–auxin–AUX/IAA ternary complex.

2 Materials

2.1 Saturation and Competition Binding Reactions

1. Binding buffer: 50 mM Tris–HCl pH 8.0, 200 mM NaCl, 10% Glycerol, 0.1% Tween-20. Always prepare fresh, filter through a 0.45 μm pore size membrane, and cool at 4 °C. Freshly add 1 mM PMSF and Roche cOmplete EDTA-free protease inhibitor cocktail before use.
2. Receptor: Recombinantly expressed and purified protein. Proteins *a* and *b* are needed for a coreceptor system. In case of auxin sensing:
   (a) TIR1–ASK1 complex at ≥1 mg/mL, usually stored in 50 mM Tris–HCl pH 8.0, 200 mM NaCl, 10% Glycerol.
   (b) AUX/IAA at ≥0.2 mg/mL in binding buffer (*see* **Note 1**).
3. Radioligand: Tritiated indole-3-acetic acid [5-^{3}H] (hereafter ^{3}H-IAA), e.g., with specific activity 25 Ci/mmol at 1 mCi/mL concentration, dissolved in ethanol. For convenient pipetting, the concentration of the radioligand stock you acquire from the provider should be at least tenfold higher than the maximal concentration you will test in the assay. Store at −20 °C in the dark.
4. Unlabeled ligand or analogue (cold competitor): 100 mM indole-3-acetic acid (cold IAA stock) dissolved in absolute ethanol is used for saturation and homologous competition binding experiments. Other auxinic compounds (agonists and antagonists) at 100 mM concentration dissolved in absolute ethanol are used for heterologous competition binding experiments. Alternatively, cold competitor stock solutions can be prepared using DMSO as solvent.
5. 5 mL reaction tubes (75 × 12 mm) or any harvester-suitable tubes ≥1 mL.
6. Ice container holding sample racks for harvester.
7. Orbital shaker.

2.2 Quantifying Bound Radioligand

1. Wash buffer: 50 mM Tris–HCl pH 8.0, 200 mM NaCl. Filter wash buffer through a 0.45 μm pore size membrane and cool at 4 °C. Approximately 10 mL of wash buffer per sample should be calculated. Also include additional volume for priming and rinsing the harvester taking into account the internal volume of the device (*see* **Note 2**).
2. Filter paper buffer: Use wash buffer and add 0.5% (v/v) of a 50% (w/v) aqueous polyethylenimine (PEI) or polyaziridine solution. PEI is a cationic polymer and filters coated in PEI have increased binding.
3. Glass fiber filter paper, e.g., Whatman GF/B paper (fired).
4. Harvester or vacuum manifold for filter-binding assays, e.g., Brandel 24-sample system.
5. Forceps.
6. Scintillation liquid for universal application including aqueous samples or for glass fiber filters (e.g., National Diagnostics EcoScint, Zinsser Analytic Filtersafe) (*see* **Note 3**).
7. Liquid scintillation vials, e.g., Beckman Mini PolyQ Vials.
8. Liquid scintillation counter, e.g., Beckman LS6500.

2.3 Data Analysis

1. GraphPad Prism5 or higher. Alternatively, any comparable data analysis software that includes or can be configured to perform one-site saturation binding (hyperbola) and one-site competition IC_{50} regression.

3 Methods

Perform all procedures on ice or in a cold room.

3.1 Radioactivity Calculations

To calculate the remaining concentration of radioactivity c_{stock} in the ^{3}H-IAA stock solution, determine the decay time t passed since determination of the given specific activity (*see* **Note 4**).

1. Calculate remaining fraction of radioactivity after decay F using the following equation:

$$F = e^{\left(\frac{-\ln 2}{t_{0.5}} \cdot t\right)}$$

As we use ^{3}H-IAA here, $t_{0.5}$ is 4537 days.

2. Calculate the original concentration c_0 from the specific activity A and the radioligand concentration c_R (*see* **Note 4**), using the following equation:

$$c_0 = \frac{c_R}{A}$$

(*see* **Note 5**).

Table 1
Example of a sample set for saturation binding

	Triplicates			[radioligand] (nM)	Components
T	c_1^A	c_1^B	c_1^C	600	Receptor + radioligand
T	c_2^A	c_2^B	c_2^C	300	Receptor + radioligand
T	c_3^A	c_3^B	c_3^C	150	Receptor + radioligand
T	c_4^A	c_4^B	c_4^C	100	Receptor + radioligand
T	c_5^A	c_5^B	c_5^C	60	Receptor + radioligand
T	c_6^A	c_6^B	c_6^C	30	Receptor + radioligand
T	c_7^A	c_7^B	c_7^C	15	Receptor + radioligand
NS	c_1^A	c_1^B	c_1^C	600	Receptor + radioligand + excess unlabeled ligand
NS	c_2^A	c_2^B	c_2^C	300	Receptor + radioligand + excess unlabeled ligand
NS	c_3^A	c_3^B	c_3^C	150	Receptor + radioligand + excess unlabeled ligand
NS	c_4^A	c_4^B	c_4^C	100	Receptor + radioligand + excess unlabeled ligand
NS	c_5^A	c_5^B	c_5^C	60	Receptor + radioligand + excess unlabeled ligand
NS	c_6^A	c_6^B	c_6^C	30	Receptor + radioligand + excess unlabeled ligand
NS	c_7^A	c_7^B	c_7^C	15	Receptor + radioligand + excess unlabeled ligand

3. Calculate remaining concentration c_{stock} by multiplying the original concentration c_0 with remaining fraction of radioactivity after decay *F*.

$$c_{stock} = F \cdot c_0$$

3.2 Saturation Binding Experiment

1. Label tubes for reactions in triplicates for at least five different concentrations of ^{3}H-IAA. The concentrations should lie around the expected K_D and include high (i.e., up to tenfold K_D) concentrations to approximate B_{max}. Of each reaction you will have to prepare one set of tubes for measuring total binding and another set for measuring nonspecific binding. For example, you want to test seven concentrations of radioligand concentration c_1–c_7 in triplicates (A, B, and C). For every cn^A, cn^B, cn^C you will have to prepare a total binding reaction (T) and a nonspecific binding reaction (NS), resulting in number of samples $N=42$ samples (*see* Exemplary Sample Table 1).
2. To have 100 μL reactions that include: binding buffer, 10–15 nM TIR1–ASK1 complex and 1–5 μM AUX/IAA, prepare a $N+3$-times master mix and distribute the appropriate sample volume to individual tubes.

3. Using the 100 mM cold IAA stock, prepare a 25 mM predilution in binding buffer. For NS binding samples, add 4 μL of the 25 mM predilution and gently mix reaction, to obtain 1 mM cold IAA concentration per reaction (>1000× excess unlabeled ligand). Keep in mind though that the excess of unlabeled ligand is relative to the maximal radioligand concentration in your assay. For T binding samples, add equivalent volume of binding buffer.
4. Prepare a dilution series of ^{3}H-IAA stock that allows you to pipet the same amount of volume, e.g., 5 μL, per sample obtaining the desired final concentration of radioligand.
5. After all components have been added to the 100 μL reaction, incubate on ice or in a 4 °C cold chamber with gentle shaking. This incubation allows the binding reaction between ligand and receptor to reach equilibrium. Usually 30–60 min suffice, but incubation time may vary depending on binding kinetics. Note that K_D is temperature-dependent.

3.3 Competition Binding Experiment

1. Label tubes for reactions in triplicates for at least eight different cold competitor concentrations spanning approximately eight (8) orders of magnitude around the 50% inhibitory concentration IC_{50}. Include a sample without cold competitor for determining T binding, and another one with excess of cold competitor (10,000-fold K_D) for NS binding. T and NS are required for reliable regression. The concentration of radioligand should be $\geq K_D$, and give at least a 1000 cpm signal for T, to ensure a sufficient dynamic range.
2. Prepare 100 μL-reactions mixing binding buffer, 10–15 nM TIR1–ASK1 complex, 1–5 μM AUX/IAA and appropriate concentration of ^{3}H-IAA (see above). We recommend to prepare an $N+3$-times master mix. Distribute the appropriate sample volume to individual tubes.
3. Using the 100 mM cold competitor stock, prepare a 10 mM predilution in binding buffer. From that, prepare a dilution series of cold competitor that allows you to pipet the same amount of volume, e.g., 5 μL, per sample obtaining the desired final concentration of competitor. For T, add the equivalent volume of binding buffer.
4. After all components have been added to the 100 μL reaction, incubate on ice with gentle shaking. This incubation allows equilibrium formation. Usually 30–60 min suffice, but incubation time may vary depending on binding kinetics. Note that K_D is temperature-dependent.

3.4 Quantifying Bound Radioligand

1. Prepare the required number of glass fiber filter papers by soaking them in filter paper buffer at room temperature for 30 min.
2. Rinse the harvester with water and prime with wash buffer.
3. Insert a filter paper and aspirate the samples through the filter. Immediately wash the filter two times with 2 mL and once with 4 mL chilled wash buffer.
4. Carefully remove the filter paper with forceps from the harvester and transfer the filter discs corresponding to your samples to scintillation vials.
5. If due to the number of samples more than one filter paper is required, wash the harvester twice with 4 mL wash buffer, before inserting the next filter.
6. In addition, collect few filter discs corresponding to empty sample tubes to measure background signal.
7. Add 4 mL of scintillation liquid per vial, cap vial and ensure complete immersion of filter disc by shaking or vortexing vigorously.
8. Incubate for several hours or overnight at room temperature.
9. Perform scintillation counting for ^{3}H isotope, e.g., 1 min counting time.

3.5 Data Analysis

1. After obtaining counts per minute (cpm) values from scintillation counting, subtract background counts from all samples.

3.6 Analysis for Saturation Binding

1. Subtract cpm values for NS from cpm values for T to obtain specific binding values in cpm.
2. Plot mean cpm values for specific binding (y) against radioligand concentration (x).
3. Perform nonlinear regression using the following equation:

$$y = \frac{B_{\mathrm{max}} \cdot x}{K_{\mathrm{D}} + x}$$

4. Resulting hyperbola will approximate B_{max}.
5. K_{D} can be obtained by solving the resulting regression equation for x at $y - 0.5B_{\mathrm{max}}$.

3.7 Analysis for Competition Binding

1. Plot mean cpm values (y) against logarithm of cold competitor concentration (x).
2. Perform nonlinear regression using the following equation:

$$y = \mathrm{NS} + \frac{T - \mathrm{NS}}{1 + 10^{(x - \log \mathrm{IC}_{50})}}$$

3. From the regression you obtain $\log IC_{50}$. Calculate the K_i for binding of the cold competitor to the receptor according to Cheng–Prusoff [6]:

$$K_i = \frac{IC_{50}}{1 + \frac{[\text{Radioligand}]}{K_D}}$$

Be aware that as mentioned before one needs to have an estimate for a K_D for binding of the radioligand to the receptor. *See* **Note 6** for troubleshooting on data fitting.

4 Notes

1. We recommend using fresh, affinity-purified protein samples. Do not freeze, keep on ice, and use latest 5 days after purification.
2. We typically prepare ≥1.5 L of wash buffer for one set of 24 samples for a Brandel 24-sample harvester, 3 L are sufficient for up to four sets of 24 samples and so on.
3. The volume of scintillation liquid per sample varies depending on size of the filter discs obtained from the harvester system and the vials used in the liquid scintillation counter. We use 4 mL per filter discs obtained from a Brandel 24-sample harvester system in Beckmann PolyQ Mini Vials to completely immerse the filter disc.
4. Information on specific activity should be provided in a technical data sheet shipped with the radioligand.
5. With c_R given in mCi/mL and A given in mCi/mmol, the resulting c_0 will be in mM.
6. If fitting the data does not result in reliable curves and reasonable K_D values, consider the following:
   (a) Is the radioligand and/or unlabeled ligand or analogue still intact? Auxinic compounds, e.g., are highly photolabile and decompose over extended time even when stored at dark, or at −20 °C.
   (b) Do your receptor preparations contain enough active species? Ideally have an alternative assay at hand to check integrity and/or activity of your protein preparations.
   (c) Do the assumptions made apply to your system? If cooperativity or other special scenarios apply to your binding reaction, consult the appropriate models for fitting the data. If you are not sure if equilibrium has been reached, try to determine the half-life of the complex and extend incubation time accordingly if necessary.

Acknowledgments

This work was supported by the IPB core funding from the Leibniz Association and the Deutsche Forschungsgemeinschaft (DFG) through the research grant CA716/2-1 to L.I.A.C.V.

References

1. de Jong LAA, Uges DRA, Franke JP, Bischoff R (2005) Receptor–ligand binding assays: technologies and applications. J Chromatogr B 829:1–25
2. Nguyen H, Park J, Kang S, Kim M (2015) Surface plasmon resonance: a versatile technique for biosensor applications. Sensors 15: 10481–10510
3. Velazquez-Campoy A, Freire E (2006) Isothermal titration calorimetry to determine association constants for high-affinity ligands. Nat Protoc 1:186–191
4. Rossi AM, Taylor CW (2011) Analysis of protein-ligand interactions by fluorescence polarization. Nat Protoc 6:365–387
5. Jerabek-Willemsen M, André T, Wanner R, Roth HM, Duhr S, Baaske P, Breitsprecher D (2014) Microscale thermophoresis: interaction analysis and beyond. J Mol Struct 1077: 101–113
6. Cheng Y, Prusoff WH (1973) Relationship between the inhibition constant (K1) and the concentration of inhibitor which causes 50 per cent inhibition (I50) of an enzymatic reaction. Biochem Pharmacol 22:3099–3108
7. Bigott-Hennkens HM, Dannoon S, Lewis MR, Jurisson SS (2008) In vitro receptor binding assays: general methods and considerations. Q J Nucl Med Mol Imaging 52:245–253
8. Bylund DB, Toews ML (1993) Radioligand binding methods: practical guide and tips. Am J Physiol 265:L421–L429
9. Carter CM, Leighton-Davies JR, Charlton SJ (2007) Miniaturized receptor binding assays: complications arising from ligand depletion. J Biomol Screen 12:255–266
10. Hulme EC, Trevethick MA (2010) Ligand binding assays at equilibrium: validation and interpretation. Br J Pharmacol 161:1219–1237
11. Maguire JJ, Kuc RE, Davenport AP (2012) Radioligand binding assays and their analysis. Methods Mol Biol 897:31–77
12. Motulsky HJ, Neubig RR (2010) Analyzing binding data. Curr Protoc Neurosci Chapter 7: Unit 7.5
13. Calderón Villalobos LI, Lee S, De Oliveira C, Ivetac A, Brandt W, Armitage L, Sheard LB, Tan X, Parry G, Mao H et al (2012) A combinatorial TIR1/AFB-Aux/IAA co-receptor system for differential sensing of auxin. Nat Chem Biol 8:477–485
14. Wang ZY, Seto H, Fujioka S, Yoshida S, Chory J (2001) BRI1 is a critical component of a plasma-membrane receptor for plant steroids. Nature 410:380–383
15. Cano-Delgado A, Yin Y, Yu C, Vafeados D, Mora-Garcia S, Cheng JC, Nam KH, Li J, Chory J (2004) BRL1 and BRL3 are novel brassinosteroid receptors that function in vascular differentiation in Arabidopsis. Development 131:5341–5351
16. Sheard LB, Tan X, Mao H, Withers J, Ben-Nissan G, Hinds TR, Kobayashi Y, Hsu FF, Sharon M, Browse J et al (2010) Jasmonate perception by inositol-phosphate-potentiated COI1-JAZ co-receptor. Nature 468:400–405
17. Slaymaker DH, Navarre DA, Clark D, del Pozo O, Martin GB, Klessig DF (2002) The tobacco salicylic acid-binding protein 3 (SABP3) is the chloroplast carbonic anhydrase, which exhibits antioxidant activity and plays a role in the hypersensitive defense response. Proc Natl Acad Sci U S A 99:11640–11645
18. Fu ZQ, Yan S, Saleh A, Wang W, Ruble J, Oka N, Mohan R, Spoel SH, Tada Y, Zheng N, Dong X (2012) NPR3 and NPR4 are receptors for the immune signal salicylic acid in plants. Nature 486:228–232
19. Ueguchi-Tanaka M, Ashikari M, Nakajima M, Itoh H, Katoh E, Kobayashi M, Chow TY, Hsing YI, Kitano H, Yamaguchi I, Matsuoka M (2005) GIBBERELLIN INSENSITIVE DWARF1 encodes a soluble receptor for gibberellin. Nature 437:693–698
20. Soon FF, Suino-Powell KM, Li J, Yong EL, Xu HE, Melcher K (2012) Abscisic acid signaling: thermal stability shift assays as tool to analyze hormone perception and signal transduction. PLoS One 7:e47857
21. Calderón Villalobos LI, Tan X, Zheng N, Estelle M (2010) Auxin perception-structural

insights. Cold Spring Harb Perspect Biol 2: a005546

22. Dharmasiri N, Dharmasiri S, Estelle M (2005) The F-box protein TIR1 is an auxin receptor. Nature 435:441–445
23. Kepinski S, Leyser O (2005) The Arabidopsis F-box protein TIR1 is an auxin receptor. Nature 435:446–451
24. Chapman EJ, Estelle M (2009) Mechanism of auxin-regulated gene expression in plants. Annu Rev Genet 43:265–285
25. Tan X, Calderón Villalobos LI, Sharon M, Zheng C, Robinson CV, Estelle M, Zheng N (2007) Mechanism of auxin perception by the TIR1 ubiquitin ligase. Nature 446: 640–645

Chapter 4

Measuring the Enzyme Activity of Arabidopsis Deubiquitylating Enzymes

Kamila Kalinowska, Marie-Kristin Nagel, and Erika Isono

Abstract

Deubiquitylating enzymes, or DUBs, are important regulators of ubiquitin homeostasis and substrate stability, though the molecular mechanisms of most of the DUBs in plants are not yet understood. As different ubiquitin chain types are implicated in different biological pathways, it is important to analyze the enzyme characteristic for studying a DUB. Quantitative analysis of DUB activity is also important to determine enzyme kinetics and the influence of DUB binding proteins on the enzyme activity. Here, we show methods to analyze DUB activity using immunodetection, Coomassie Brilliant Blue staining, and fluorescence measurement that can be useful for understanding the basic characteristic of DUBs.

Key words Deubiquitinating enzymes, DUB assay, K48-linked ubiquitin chains, K63-linked ubiquitin chains

1 Introduction

Ubiquitylation is a reversible posttranslational modification that is key to various cellular processes in almost all physiological pathways of plants [1]. It must be strictly controlled and regulated at multiple steps during these processes. The attachment of ubiquitin to the target proteins is carried out by the sequential activities of the ubiquitin activating enzyme (E1), ubiquitin conjugating enzymes (E2s), and ubiquitin ligases (E3s) [2]. The ubiquitylation status of the substrate proteins is also controlled by the activity of deubiquitylating enzymes (DUBs: also deubiquitinating enzymes or deubiquitinases) that can deconjugate ubiquitin or ubiquitin-like proteins from their substrates [3].

The Arabidopsis genome codes for around 50 DUBs [4], though for most of these the exact molecular and biological functions are not yet understood. DUBs remove covalently attached ubiquitin molecules from substrates or hydrolyze the peptide bond between ubiquitin molecules. DUBs can play multiple roles in cellular processes. Firstly, they are essential for the posttranslational

L. Maria Lois and Rune Matthiesen (eds.), *Plant Proteostasis: Methods and Protocols*, Methods in Molecular Biology, vol. 1450, DOI 10.1007/978-1-4939-3759-2_4, © Springer Science+Business Media New York 2016

activation of ubiquitin molecules. Secondly, they are also responsible for the recycling of ubiquitin molecules by removing them from the substrates prior to degradation. Thirdly, DUBs can also actively regulate the stability of ubiquitylated proteins by deubiquitylating them before they are recognized by the degradation machinery. Finally, since the attachment of ubiquitin can affect the binding affinity of the ubiquitylated protein to other proteins, DUBs can also influence protein–protein interaction. Eukaryotes have five DUB families that can be classified according to the difference in their catalytic domains [5]: the ubiquitin-specific proteases (UBPs or USPs), the ubiquitin C-terminal hydrolases (UCHs), the ovarian tumor proteases (OTUs), the Machado–Joseph domain (MJD) or Josephine domain proteases, and the JAB1/MPN/MOV34 (JAMM) proteases. Except DUBs of the JAMM family that are zinc-dependent metalloproteases, all other DUBs are cysteine proteases.

Monoubiquitylation as well as seven different ubiquitin chain linkages (K6-, K11-, K27-, K29-, K33-, K48-, and K63-linkages) are found in vivo [6], indicating that all chain types can have biological significance. In addition, linear or mixed ubiquitin chains also have been shown to have important biological functions [7, 8]. Due to their distinct topology, different ubiquitin linkages can be recognized by different set of proteins and thereby can be involved in different pathways. With the now available information about chain-type specificity of human DUBs, highly specific DUBs can also be used as tools to identify ubiquitin chain types of a ubiquitylated protein [9].

Since substrate identification of DUBs is not trivial, identification of interactors of DUBs and analysis of the enzymatic characteristics are crucial to determine the pathway a given DUB might be involved. In vitro assay for studying DUB activities are therefore useful tools to analyze ubiquitin chain-type specificities of DUBs and also to examine whether interacting proteins can influence DUB activity. The availability of various types of commercial ubiquitin chains enables quantitative and reproducible assays with simple equipment. Fluorescence- or luminescence-based substrates also offer possibility of determining the enzyme kinetics. In this chapter, we describe immunoblot-, Coomassie Brilliant Blue (CBB) staining-, and fluorescence-based analysis of DUB activity. We show the example with the Arabidopsis DUB AMSH3, which is a conserved DUB implicated in intracellular protein trafficking [10, 11].

2 Materials

2.1 Recombinant GST-Tagged DUB Purification from Bacteria

1. Buffer A: 50 mM Tris–HCl, 100 mM NaCl, 10% (w/v) glycerol. Adjust the pH to 7.5, cool the buffer down to 4 °C overnight, then readjust the pH again to pH 7.5. Store at 4 °C.
2. Buffer A supplemented with 0.2% (v/v) Triton X-100 and 1× complete EDTA-free protease inhibitor (Roche), prepare directly before use.

3. Buffer A supplemented with 1 mM dithiotreitol (DTT), prepare directly before use.
4. Ultrasonic homogenizer.
5. Refrigerated centrifuge for 50 ml tubes.
6. Refrigerated table top centrifuge for 1.5 ml tubes.
7. GST purification matrix (e.g., Glutathione Sepharose 4B from GE Healthcare).
8. Mini-spin columns, e.g., Mini Bio-Spin Chromatography Columns (Bio-Rad).
   (a) 40 mM reduced glutathion, in case GST-fusion proteins will be eluted with the tag.
   (b) PreScission protease (GE Healthcare), in case the expression vector contains a PreScission protease recognition site (such as the pGEX-6P-series from GE Healthcare) (*see* **Note 1**).
9. Protein molecular weight standards.

2.2 Deubiquitylation Assay (DUB Assay)

1. DUB Assay Buffer: 50 mM Tris–HCl pH 7.2 (*see* **Note 2**), 25 mM KCl, 5 mM $MgCl_2$, 1 mM DTT. Prepare directly before use or store at –20 °C.
2. (a) Reaction substrate: In vitro ubiquitylated T7-Sic1^{PY} (Ubn-T7-Sic1^{PY}) prepared using the ubiquitylating enzymes E1(Uba1), E2 (Ubc4), and E3(Rsp5) [12].
   (b) Reaction substrate: Commercially available di- or polyubiquitin (Ub_{2-7}) chains (e.g., from Enzo Life Sciences) (*see* **Note 3**).
3. Heating block.
4. 4× NuPAGE SDS Sample Buffer: 564 mM Tris base, 416 mM Tris hydrochloride, 8% (w/v) SDS, 40% (w/v) glycerol, 2.04 mM EDTA, 0.88 mM SERVA Blue G250, 0.70 mM Phenol Red. Store at –20 °C or room temperature.

2.3 Gradient Gel Electrophoresis

1. NuPAGE 4–12% Bis-Tris gel (Thermo Fisher Scientific, *see* **Note 4**).
2. 20× MES SDS Running Buffer: 1 M MES, 1 M Tris base, 2% (w/v) SDS, 20 mM EDTA. Store the 20× stock at 4 °C. Use the 20× stock to prepare the 1× running buffer before use.
3. Prestained molecular mass marker.
4. Gel apparatus, e.g., XCellSureLock Mini-Cell Electrophoresis System for NuPAGE (Thermo Fisher Scientific).
5. Protein standard markers with known concentration, e.g., BenchMark Protein Ladder (Thermo Fisher Scientific).

2.4 Protein Transfer and Western Blotting

1. 20× NuPAGE Transfer Buffer: 0.5 M Bicine, 0.5 M Bis-Tris (free base), 20 mM EDTA. Store at 4 °C. Prepare 2× transfer buffer with 10% (v/v) methanol before use.

2. Horizontal shaker.
3. Semidry transfer apparatus.
4. Four filter papers (1.5 mm) cut in the size 0.5 cm larger than the protein gel.
5. PVDF membrane or nitrocellulose membrane cut in the size of the protein gel.
6. 100% methanol, in case a PVDF membrane is used.
7. Tris-buffered saline (TBS): 0.5 M Tris–HCl pH 7.5, 1.5 M NaCl, 10 mM $MgCl_2$. Store at room temperature.
8. Tris-buffered saline with Tween-20 (TBST): use 10× TBS stock to prepare 1× working solution. Add Tween-20 to 0.05% (v/v). Store at room temperature.
9. Blocking buffer: 10% (w/v) powdered milk in TBST. Prepare directly before use.
10. Monoclonal anti-ubiquitin (anti-Ub) antibody P4D1 (e.g., from Santa Cruz) (*see* **Note 5**).
11. Anti-mouse HRP-conjugated antibody.
12. Enhanced chemiluminescent (ECL) reagents.
13. Chemiluminescence detection apparatus, e.g., LAS4000 mini system (Fuji Film).

2.5 CBB Staining

1. CBB staining solution: 40% (v/v) ethanol, 7% (v/v) acetic acid, 0.25% (w/v) CBB.
2. Destaining solution: 40% (v/v) ethanol, 7% (v/v) acetic acid.

2.6 Fluorescence-Based DUB Assay

1. TAMRA DUB Buffer: 50 mM Tris–HCl pH 7.5, 100 mM NaCl, 0.1% (w/v) Pluronic F-127, 1 mM Tris(2-carboxyethyl)phosphine (TCEP). Prepare before use.
2. Reaction substrate: diubiquitin (K63-linked) FRET TAMRA Position 3 (from R&D Systems) (*see* **Note 6**).
3. Reaction plates, e.g., 96-well black plate.
4. Fluorescence plate reader (e.g., Synergy 2 Multi-Mode Microplate Reader from BioTek) with filters for excitation wavelength of 530 nm and emission wave length of 590 nm.

3 Methods

3.1 Recombinant DUB Purification from Bacteria

1. Cool down a refrigerated centrifuge for 50-ml tubes to 4 °C.
2. Add 20 ml Buffer A supplemented with 0.2% (w/v) Triton X-100 and 1× complete EDTA-free protease inhibitor to *E. coli* pellet from 250 ml culture and resuspend pellet.

3. Place the tube in a beaker filled with ice and water and sonicate the sample for 15 min with four cycles at 20% output. Repeat the sonication for another 15 min if necessary.
4. Centrifuge the postsonication solution at 15,000 × *g* for 10 min at 4 °C.
5. Transfer the supernatant into a new 50-ml tube and keep on ice.
6. Take 100 μl of Glutathione Sepharose 4B (75 μl bed volume) with a cut tip and transfer to a 1.5 ml tube (**Note 7**).
7. Add 1 ml buffer A to the beads and centrifuge them for 1 min at 800 × *g* at 4 °C. Remove supernatant. Repeat washing three times.
8. Add 500 μl buffer A to the beads. Transfer the beads with a cut tip to the 50-ml tube containing the protein supernatant.
9. Incubate the protein solution with beads for 2 h at 4 °C with rotation.
10. Centrifuge the protein solution for 3 min at 800 × *g* at 4 °C. Discard supernatant.
11. Add 20 ml buffer A and centrifuge again as above. Repeat washing three times.
12. Discard the washing buffer, leaving ca. 500 μl buffer in the tube. Using a pipette and a tip with a cut-off end, transfer the beads on a mini-spin column.
13. Add 500 μl buffer A containing 0.2% (v/v) Triton to the beads and centrifuge them for 5 s at 800 × *g* at 4 °C. Discard the flow-through. Repeat washing three times.
14. Wash the beads three times with 500 μl buffer A as above.
15. Elute the purified protein with 40 mM reduced glutathione by incubating 10 min at room temperature (*see* **Note 8**). If PreScission protease is used, add 200 μl buffer A containing 1 μl of PreScission protease to the beads and rotate for 16–20 h.
16. Take 6 μl of the purified protein, add 1.5 μl 5× Laemmli buffer and incubate for 5 min at 95 °C. Analyze the purity and concentration of the purified protein on a CBB-stained SDS-PAGE gel using proteins standards.

3.2 Deubiquitylation Assay

1. Set the temperature at 30 °C on a heating block (*see* **Note 9**).
2. Prepare individual reaction tubes for each time point for the experiment and aliquote 3-pmol, 2-pmol, or 8-pmol of recombinant DUB for Ubn-T7-Sic1^{PY} (Immunoblot), polyubiquitin chains (Immunoblot), or diubiquitin (CBB detection), respectively, in the DUB Assay Buffer to make a total volume of 10 μl. Preincubate the tubes for 5–10 min at 30 °C. If DUB inhibitors are to be tested, they can be added to the reaction mixture at this point (*see* **Note 10**).

3. While preincubating the reaction mixture, prepare the ubiquitin substrates to a concentration of 250 ng/µl in DUB Assay Buffer. Start the reaction by adding the following amount of substrates to the preincubated reaction mixture from **step 2**: (a) 500 ng Ubn-T7-Sic1^{PY}, (b) 250 ng polyubiquitin chains for immunodetection or (c) 1 µg diubiquitin for CBB detection. Incubate in the heating block at 30 °C for the desired amount of time.
4. Terminate the reaction by adding the 4× NuPAGE SDS Sample Buffer and place the tubes on ice.
5. Once all reactions are terminated, incubate samples at 80 °C for 10 min and let cool at room temperature.

3.3 Gradient Gel Electrophoresis

1. Prepare 800 ml 1× running buffer. Unpack the precast NuPAGE 4–12% Bis-Tris Gel, remove the comb and wash the sample loading pockets with distilled water.
2. Assemble the gel apparatus and the gel. Fill the apparatus with 1× running buffer.
3. Load on all of your reaction mixture in each lane.
4. Run the gel at 200 V for 35 min. For the polyubiquitin chains, continue with a NuPAGE transfer and immunoblotting (*see* Subheading 3.4). For diubiquitin, continue with CBB staining (*see* Subheading 3.5) (*see* **Note 11**).

3.4 Protein Transfer and Western Blotting

1. Disassemble the gel plates and incubate the gel for 15 min in the 2× NuPAGE transfer buffer containing 10% (v/v) methanol.
2. Soak the filter papers in the 2× NuPAGE semidry transfer buffer and assemble a stack in the following order (from bottom to top): Two filter papers, PVDF (washed for 30 s in 100% Methanol) or nitrocellulose membrane, gel and two filter papers. Eliminate any air bubbles by rolling a glass tube over the transfer package after adding each filter paper.
3. Transfer the protein to the membrane for 25 min at 15 V.
4. Optionally, boil the membrane in ultrapure water on a heating plate for 10 min (*see* **Note 12**).
5. After the transfer, place the membrane in the blocking buffer and incubate for 15–30 min at room temperature on a shaker.
6. Prepare a 1:1000 dilution of the primary anti-ubiquitin (P4D1) antibody in the blocking buffer.
7. Incubate the primary antibody with the membrane at room temperature with shaking for at least 1 h or overnight at 4 °C.
8. Remove the solution with the primary antibody. Wash the membrane for 15 min with TBST buffer. Repeat the step three times, using fresh TBST buffer each time.
9. Prepare a proper dilution of the secondary antibody in TBST (anti-mouse HRP-conjugated antibody).

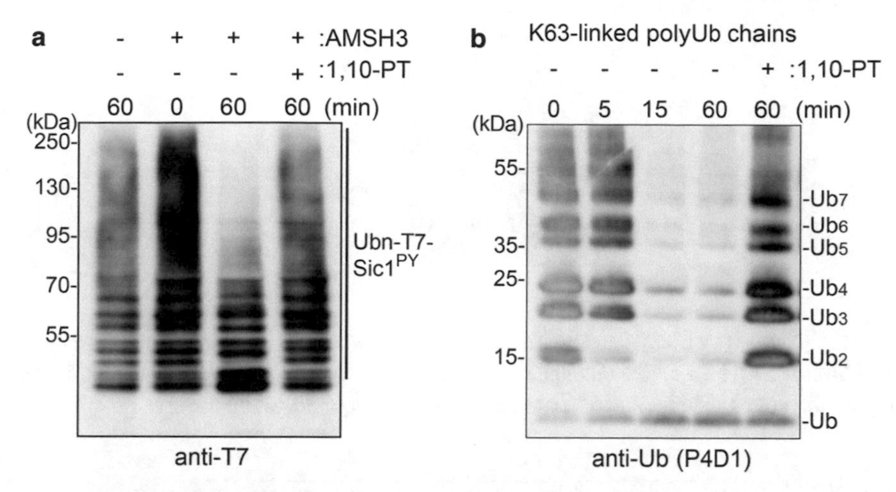

Fig. 1 DUB assay using immunoblot detection. (**a**) DUB assay of AMSH3 using Ubn-Sic1^{PY} as substrate. 1,10-PT was preincubated with AMSH3 for 10 min before the addition of the substrate. The reaction mixture was subjected to SDS-PAGE and immunoblotting with anti-ubiquitin antibody. (**b**) DUB assay using commercial K63-linked polyubiquitin chains. Reactions were terminated at the indicated time points. 1,10-PT was preincubated with AMSH3 for 10 min before the addition of the substrate

10. Incubate secondary antibody with the membrane at room temperature with shaking for at least 45 min or overnight at 4 °C.
11. Wash the membrane as in **step** 7.
12. Take out the membrane from the washing solution and remove excess liquid. Incubate the membrane with the ECL solution (600 μl are sufficient for a 6.5 × 8.0 cm membrane) for 5 min. Remove excess ECL solution with a paper towel and detect the chemiluminescence. Optimal exposure time varies between experiments. For a typical result of a DUB assay using polyubiquitin chains, *see* Fig. 1a, b.

3.5 CBB Staining

1. After electrophoresis, disassemble the gel plates. Incubate the gel in the destaining solution on a shaker for 15 min at room temperature, in order to remove excess SDS from the gel.
2. Discard the destaining solution. Pour the CBB-staining solution over the gel and gently shake for 60 min at room temperature.
3. Discard the staining solution. Wash the gel gently several times with tap water. Pour the destaining solution over the gel and incubate it on a shaker at room temperature until the gel background is reduced to a satisfactory extend. For a typical result of a DUB assay using diubiquitin, *see* Fig. 2.

3.6 Fluorescence-Based DUB Assay

1. Prepare dilutions of the DUB in the TAMRA DUB buffer, e.g., 0, 2.5, 5, and 10 nM to a total volume of 450 μl.
2. Pipet 100 μl of the reaction mixture into four wells of the reading plate (*see* **Note 13**).

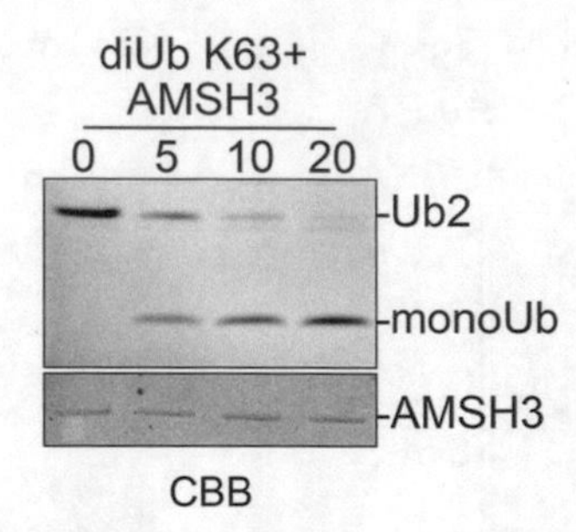

Fig. 2 DUB assay followed by Coomassie Brilliant Blue (CBB) staining. To the reaction tubes containing 8 pmol Arabidopsis AMSH3, 1 μg K63-linked diubiquitin was added and the reaction was conducted for 0, 5, 10, and 20 min. Degradation of diubiquitin and accumulation of monoubiquitin can be observed

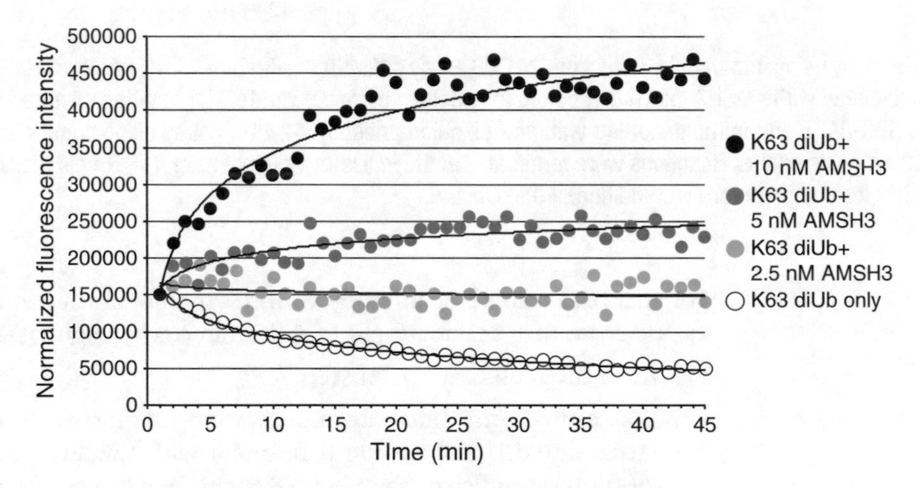

Fig. 3 Fluorescence-based DUB assay with diubiquitin FRET TAMRA. 96-well reaction plate containing 0, 2.5, 5, and 10 nM AMSH3. Reactions was started by addition of 0.2 μM diubiquitin (K63-linked) FRET TAMRA Position 3 and changes in the fluorescence were measured every minute over a time frame of 45 min

3. Dilute diubiquitin (K63-linked) FRET TAMRA Position 3 to a concentration of 10 μM in the TAMRA DUB buffer. Using a channel pipette, start the reaction by adding 2 μl of the substrate to have a final concentration of 0.2 μM in the assay.
4. Immediately close the fluorescence plate reader and start the reaction. Measure the changes in fluorescence every minute over a time period typically between 45- and 120 min (ex. 530 nm; em. 590 nm). A typical result of a TAMRA DUB assay is presented in Fig. 3 (*see* **Note 14**).

4 Notes

1. In some cases, fusion of large protein tags such as GST or MBP can affect enzyme activity. In these cases the tags should be cleaved off after purification using proteases such as Thrombin,

Factor Xa, or PreScission protease. PreScission protease has an advantage in that it is active at 4 °C. Moreover, since it is available as a GST-fusion protein, the protease remains on the beads. An untagged DUB can be detected on an immunoblot only when a specific antibody for the DUB is available. In case such antibody is not available, the presence of a tag allows detection of the DUB with a tag-specific antibody or otherwise the amount of the DUB in the reaction should be monitored by gel-staining.

2. The pH of the DUB Assay Buffer may need to be optimized for each DUB as different enzymes may have different optimal pH.
3. All major ubiquitin chain types (K6-, K11-, K27-, K29-, K33-, K48-, K63-linkages) as well as linear di- and tetraubiquitin are commercially available.
4. For Ubn-Sic1^{PY}, a standard 10% SDS-PAGE can also be used. Instead of a NuPAGE gel, self-made gradient gels or other commercially available gradient gels can also be used. However, we had the best experience with NuPAGE-gels for the detection of monoubiquitin by immunoblotting.
5. Other anti-ubiquitin antibodies can also be used.
6. FRET TAMRA diUb substrates are available with different fluorophore positions. It may be necessary to establish the most suitable substrate for your enzyme. We experienced that the attachment of the fluorophore to certain positions interfered with the DUB activity. UB-AMC is also a widely used substrate, but does not convey chain-type specificities.
7. The amount of the beads to be used depends on the volume of the cell culture, expression levels, and solubility of the recombinant protein as well as the binding capacity of the beads. Typically, for purification from a 250 ml culture, 50–100 μl bed volumes of beads are used.
8. Glutathione can be removed from the eluate by dialysis or with desalting columns.
9. Reaction temperature might have to be optimized depending on the origin of the DUB. Activity of some DUBs might be affected by temperature.
10. AMSH DUB activity can be inhibited by 10 mM EDTA, 1 mM 1,10-phenanthroline (1,10-PT), and 5 mM *N*-ethylmaleimide (NEM). Other DUB inhibitors that can be used for in vitro assays include 2 μM Ub-aldehyde or 250 μM *N*,*N*,*N*-tetrakis (2-pyridylmethyl) ethylenediamine (TPEN).
11. If the quality and amount of the purified DUB is high enough, CBB staining provides faster results than immunoblotting. Silver staining or fluorescent dyes can also be used. In case the results have to be quantified, direct staining of the gel is more precise over immunoblotting.

12. Boiling of the membrane after the transfer may enhance the detection of monoubiquitin, though in an in vitro DUB assay this step could be skipped.
13. Different plate types are suitable for different readers. The reaction volume depends on the plate type. Typically, we recommend a 50–100 μl reaction volume for a 96-well plate. When using smaller volumes, make sure that the reaction mix fully covers the whole area of the bottom of the well.
14. Using different concentrations of fluorescent substrates, Michaelis–Menten kinetics can be determined.

References

1. Vierstra RD (2012) The expanding universe of ubiquitin and ubiquitin-like modifiers. Plant Physiol 160(1):2–14. doi:10.1104/pp.112.200667
2. Hershko A, Ciechanover A (1998) The ubiquitin system. Annu Rev Biochem 67:425–479
3. Clague MJ, Coulson JM, Urbe S (2012) Cellular functions of the DUBs. J Cell Sci 125(Pt 2):277–286. doi:10.1242/jcs.090985
4. Isono E, Nagel MK (2014) Deubiquitylating enzymes and their emerging role in plant biology. Front Plant Sci 5:56. doi:10.3389/fpls.2014.00056
5. Komander D, Clague MJ, Urbe S (2009) Breaking the chains: structure and function of the deubiquitinases. Nat Rev Mol Cell Biol 10(8):550–563
6. Xu P, Duong DM, Seyfried NT, Cheng D, Xie Y, Robert J, Rush J, Hochstrasser M, Finley D, Peng J (2009) Quantitative proteomics reveals the function of unconventional ubiquitin chains in proteasomal degradation. Cell 137(1):133–145. doi:10.1016/j.cell.2009.01.041
7. Boname JM, Thomas M, Stagg HR, Xu P, Peng J, Lehner PJ (2010) Efficient internalization of MHC I requires lysine-11 and lysine-63 mixed linkage polyubiquitin chains. Traffic 11(2):210–220. doi:10.1111/j.1600-0854.2009.01011.x
8. Rahighi S, Ikeda F, Kawasaki M, Akutsu M, Suzuki N, Kato R, Kensche T, Uejima T, Bloor S, Komander D, Randow F, Wakatsuki S, Dikic I (2009) Specific recognition of linear ubiquitin chains by NEMO is important for NF-kappaB activation. Cell 136(6):1098–1109. doi:10.1016/j.cell.2009.03.007
9. Hospenthal MK, Mevissen TE, Komander D (2015) Deubiquitinase-based analysis of ubiquitin chain architecture using Ubiquitin Chain Restriction (UbiCRest). Nat Protoc 10(2):349–361. doi:10.1038/nprot.2015.018
10. McCullough J, Clague MJ, Urbe S (2004) AMSH is an endosome-associated ubiquitin isopeptidase. J Cell Biol 166(4):487–492
11. Isono E, Katsiarimpa A, Muller IK, Anzenberger F, Stierhof YD, Geldner N, Chory J, Schwechheimer C (2010) The deubiquitylating enzyme AMSH3 is required for intracellular trafficking and vacuole biogenesis in Arabidopsis thaliana. Plant Cell 22(6):1826–1837. doi:10.1105/tpc.110.075952, tpc.110.075952 [pii]
12. Saeki Y, Isono E, Toh-e A (2005) Preparation of ubiquitylated substrates by the PY motif-insertion method for monitoring 26S proteasome activity. Methods Enzymol 399:215–227. doi:10.1016/S0076-6879(05)99014-9, S0076-6879(05)99014-9 [pii]

Chapter 5

Fluorescent Reporters for Ubiquitin-Dependent Proteolysis in Plants

Katarzyna Zientara-Rytter and Agnieszka Sirko

Abstract

Ubiquitin is a small protein commonly used as a signal molecule which upon attachment to the proteins affects their function and their fate in the cells. For example, it can be used as a degradation marker by the cell. Ubiquitin plays a significant role in regulation of numerous cellular processes. Therefore, monitoring of ubiquitin-dependent proteolysis can provide important information. Here, we describe construction of YFP-based proteasome substrates containing modified ubiquitin and the protocol for their transient expression in plant cells for functional analysis of the ubiquitin/proteasome system. To facilitate further subcloning all plasmids generated by us are based on the Gateway® Cloning Technology and are compatible with the Gateway® destination vectors.

Key words Fluorescent reporters for ubiquitin, UFD, UPS

1 Introduction

There are two major degradation machineries in plants, ubiquitin–proteasome system (UPS) and autophagy pathway. UPS functions in two cellular compartments, the cytoplasm and the nucleus, and requires 26S protein complex (proteasome) for protein clearance. Contrary to UPS, autophagy operates only in the cytoplasm and involves vesicle transport to deliver various cellular components to be degraded in the vacuole. Both pathways may use a small highly conserved 76-aa protein, called ubiquitin (Ub), to mark substrates for degradation. Despite that ubiquitination is an important determinant of autophagy selectivity and that autophagy can take over degradation of the ubiquitin–proteasome pathway substrates when UPS is impaired [1] ubiquitination is not critical in autophagy and several ubiquitin-independent selective autophagy receptors has been already described [2]. In contrary, ubiquitination is crucial for tagging short-lived soluble proteins for degradation by UPS. It is worth to mention that it is UPS which is mainly responsible for regulated and progressive degradation of such intracellular proteins.

L. Maria Lois and Rune Matthiesen (eds.), *Plant Proteostasis: Methods and Protocols*, Methods in Molecular Biology, vol. 1450, DOI 10.1007/978-1-4939-3759-2_5, © Springer Science+Business Media New York 2016

Ubiquitination of proteins designated for UPS degradation occurs though a three-step sequential action of El (Ub-activating enzyme), E2 (Ub-conjugating enzyme), and E3 (Ub ligase) enzymes [3]. Proteins are usually modified by more than one Ub molecule. The sequential Ub ligation to another Ub molecule previously attached to the protein causes elongation of the Ub chain (polyubiquitination). The polyubiquitinated proteins, especially those marked by a K48 chain, are main substrates for proteasomal degradation. In eukaryotes, the N-end rule pathway and the Ub-fusion degradation (UFD) pathway are a part of the UPS and they regulate half-life of many proteins. The ubiquitination pathways for the N-end rule and UFD pathways have been mapped in detail. It is known that both pathways do not overlap and possess diverse degradation signals, as well as require different E2–4 factors, chaperons or ubiquitin-binding subunits involved in proteasomal targeting of each of the substrates. The UFD substrates are ubiquitinated within the N-terminal ubiquitin moiety. The UFD pathway recognizes these "nonremovable" N-terminal Ub moiety as a primary degron, whereas N-end rule substrates before ubiquitination on the substrate itself require cleavage of the N-terminal ubiquitin by isopeptidases.

Monitoring of Ub-dependent proteolysis is very important because of the requirement for Ub in both protein degradation pathways (UPS and autophagy) and its significant role in regulation of numerous cellular processes. As a matter of fact, it has been already shown that the production of GFP-based proteasome substrates by fusion to the N-terminus of GFP-specific degrons for N-rule or for UFD pathways successfully allow to perform functional analysis of the UPS and to monitor the cross-talk between UPS and autophagy in mammals and yeast [4, 5]. Here, we report the protocol for obtaining similar fluorescent reporters for ubiquitin-dependent proteolysis in plants. To investigate the functionality of these fluorescent reporters in plants, the plant expression cassettes encoding the stable (Ub-M-YFP) and the UFD substrate (Ub^{G76V}-YFP) fusion proteins were created under the 35S promoter in the appropriate binary plasmids. As expected, microscopy and western blot analysis of transiently transformed *N. benthamiana* epidermis expressing these Ub-X-YFP fusions showed stability of Ub-M-YFP while the expression of Ub^{G76V}-YFP resulted in low fluorescent intensity confirmed by western blot.

2 Materials

2.1 PCR and Plasmid Recombination

Prepare all solutions using ultra-pure sterile water freshly before use and keep them on ice. Grow *Escherichia coli* strains at 37 °C and *Agrobacterium tumefaciens* at 28 °C. Shake liquid cultures at 150–300 rpm in a rotary shaker.

1. Standard *E. coli* strains used for cloning and plasmid amplification or for recombinant protein expression should be grown in conventional bacterial growth media LB (10 g/l tryptone, 5 g/l yeast extract, and 10 g/l NaCl with pH adjusted to 7 with NaOH and additionally 15 g/l agar for LB plates) and SOB (20 g/l tryptone, 5 g/l yeast extract, and 0.5 g/l NaCl, 0.186 g/l KCl with pH adjusted to 7 by adding NaOH) [6].
2. Add antibiotics to the media: kanamycin (stock solution: kan, 50 mg/ml in water, final concentration: kan, 50 mg/l) for pENTR/D-TOPO vector carrying bacteria, chloramphenicol (stock solution: can, 30 mg/ml in ethanol, final concentration: cam, 30 mg/l) plus streptomycin (stock solution: sp, 50 mg/ml in water, final concentration: sp, 50 mg/l) for bacteria containing donor vector such as pH7YWG2 and streptomycin (sp, 50 mg/l) for bacteria carrying destination (binary) plasmid.
3. *A. tumefaciens* strain LBA4404 should be grown in YEB medium (5 g/l beef extract, 1 g/l yeast extract, 5 g/l peptone, 5 g/l sucrose, 0.5 g/l MgCl2, and additionally 15 g/l agar for YEB plates) [7] with rifampicin (stock solution: rif, 30 mg/ml in chloroform, final concentration: rif, 30 mg/l) plus streptomycin (sp, 75 mg/l).
4. Use primers listed below for amplifying cDNA for tobacco *Ub* (*see* **Note 1**):

   Forward (F): 5′-CACCATGCAGATCTTCGTGAAGACATTGAC-3′

   Reverse (R1): 5′-CTTACCCATACCACCACGGAGACGGAGGAC-3′

   Reverse (R2): 5′-CTTACCAAC**AAC**ACCACGGAGACGGAGGAC-3′

   The reverse primer R1 is designed to amplify full length ubiquitin with additional linker at the C-terminus coding MGK tripeptide (underlined).

   The reverse primer (R2) is designed to generate G76V substitution in Ub (bold) and to add additional linker (VGK tripeptide) at the C-terminus (underlined).
5. pENTR/D-TOPO cloning kit (Life Technologies, cat. number K2400-20).
6. LR recombination kit (Life Technologies, cat. numbers 11791-043 or 11791-020).
7. PCR reagents including polymerase and dNTPs (for example, Life Technologies Pfu polymerase, cat. number EP0502 and Life Technologies dNTP set, cat. number 10297-018).
8. The pH7YWG2 binary plasmid can be ordered on line from http://www.vib.be/en/research/services/Pages/Gateway-Services.aspx.

2.2 Transient Transformation

1. Cultivate *Nicotiana benthamiana* plants in soil in a growth chamber under the conditions of 60% relative humidity, with a day/night regime of 18 h light 300 μmol photons/m²/s at 21 °C and 6 h dark at 18 °C, or in a greenhouse till they fully expanded leaves achieved about 5–6 cm in diameter approx. 4–5 weeks (*see* **Note 2**).
2. Prepare four 10-ml syringes without needles, two microscopic cover slides and cover slips.

2.3 SDS-PAGE and Western Blot

1. Use either the precast SDS-PAGE gels (for example, 12% Mini-PROTEAN TGX, Bio-Rad, cat. number 456-1041) or the fresh 12% SDS-PAGE gels prepared according to the published protocols [6].
2. Extraction Buffer: 50 mM Tris–HCl, pH 8.0, 1 mM EDTA, 0.05% β-mercaptoethanol, 0.0005% PMSF.
3. Bio-Rad protein assay kit (cat number 500-0002).
4. 4× Laemli Sample Buffer: 200 mM Tris–HCl pH 6.8, 8% SDS, 40% glycerol, 50 mM DTT, 0.02% bromophenol blue. Store at 4 °C.
5. 1× Running Buffer: 25 mM Tris base, 192 mM glycine, 0.1% SDS, pH 8.3.
6. 1× Transfer Buffer: 25 mM Tris base, 0.192 M glycine.
7. Blocking Solution: 5% dried milk in PBS. Store at 4 °C.
8. 1× PBS: 137 mM NaCl, 2.7 mM KCl, 10 mM Na_2HPO_4, 1.8 mM KH_2PO_4.

3 Methods

3.1 Plasmid Creation

1. To amplify cDNA encoding ubiquitin from organism of choice (in this case tobacco) prepare fresh cDNA. Extract total plant RNA from frozen powdered material using, for example, the cold phenol method [8] (*see* **Note 3**) and subsequently use purified total RNA as templates for reverse transcription-polymerase reaction as described in **Note 4**.
2. Prepare 50 μl PCR mix solutions on ice. Mix the appropriate primers (F and R1 in one reaction; F and R2 in the second reaction) at 0.1–0.5 μM each with 1× PCR reaction buffer containing ~0.2 mM dNTPs and 1 U of Pfu polymerase. Finally add about 200 ng of plant cDNA to amplify and incorporate appropriate mutations into *Ub* cDNA.
3. Use following PCR reaction parameters: initial denaturation at 94–95 °C for 2 min, 30 cycles: 30 s at 94 °C (denaturation), 30 s at 62 °C (an annealing temperature) and elongation at 72 °C for 30 s, and a final 5-min finishing elongation at 72 °C.

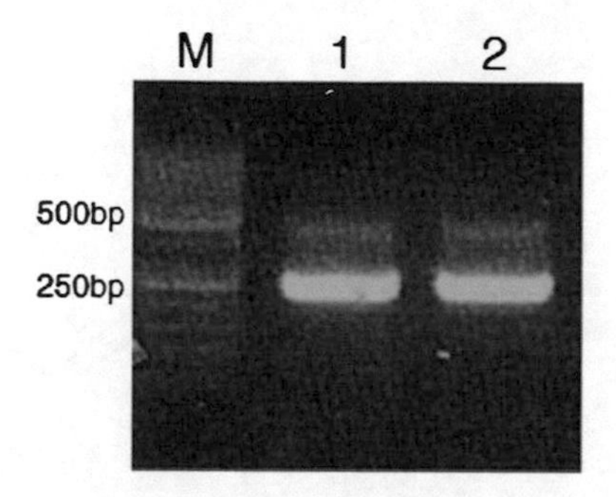

Fig. 1 Electrophoretogram (2 % agarose) of PCR products. PCR products derived from F + R1 or F + R2 primer sets used to amplify Ub-M and to Ub^{G76V}, respectively, were separated using a 2 % agarose gel in TAE buffer. *Lane M*: GeneRuler 50 bp DNA Ladder from Life Technologies, *lanes 1* and *2*: Ub-M and to Ub^{G76V}, respectively

4. Resolve PCR products on agarose gels (2 % in 1× TAE buffer) according to the standard procedure [6]. To visualize DNA bands in UV light ethidium bromide can be added to the gel to 0.2 μg/ml prior to pouring.
5. Excise the appropriate bands from the gel and purify using any gel extraction kit (Fig. 1).
6. Independently clone both PCR products into the pENTR/D--TOPO vector according to manufacturer's protocol and select positive colonies by sequencing.
7. Finally, create destination vector by LR recombination reaction using pH7YWG2 plasmid according to the manufacturer's procedure (*see* **Note 5**).
8. Analyze colonies by sequencing to verify proper insert incorporation into the open reading frame. The linker sequence between Ub and fluorescent protein should be as it is described on Fig. 2.
9. Transform independently *A. tumefaciens* by each of two destination plasmids using either the high-voltage electroporation or the heat shock method (*see* **Note 6**). After transformation, add 0.5 ml of YEB, incubate cells for 3–4 h at 28 °C in a water bath, spread mixture on the appropriate selective plates and incubate at 28 °C for 3–5 days.

3.2 Transient N. benthamiana Leaf Transformation

1. Cultivate *N. benthamiana* plants till fully expanded leaves achieve about 5–6 cm in diameter.
2. When young *N. benthamiana* plants are ready prepare 3 ml fresh overnight cultures of *A. tumefaciens* strains containing the appropriate destination (binary) plasmids.
3. Spin cells down, and wash twice in sterile ddH_2O to remove medium.

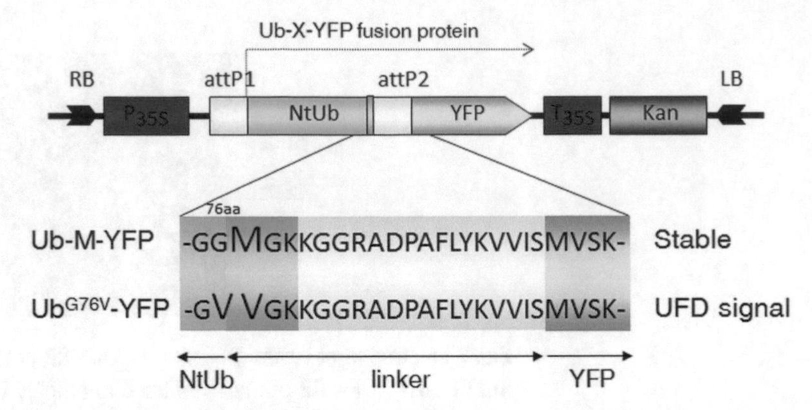

Fig. 2 Schematic representation of the T-DNA cassettes containing Ub-X-YFP fusion proteins. The amino acid sequence of the ubiquitin is shown in *blue*, the linker sequence is shown in *green* (inserted amino acids during PCR reaction) and in *gray* (residues corresponding to attP2 region) and the YFP sequence is highlighted in *yellow*. The glycine residue in position −1 of Ub^{G76V}-YFP is substituted by valine

4. Re-suspend each cell suspensions in 5–10 ml ddH_2O to obtain a final cell density of about $1–6 \times 10^8$ cells/ml (OD_{600} from 0.1 to 0.6).
5. Inoculate one half of the leaf (the lower epidermis) with the prepared suspension of *A. tumefaciens* containing the plasmid enabling expression of the stable fluorescent protein (Ub-M-YFP) and the second half of the leaf with the second suspension (for expression of the unstable fluorescent protein Ub^{G76V}-YFP) using needle-less syringes by placing the syringe against the underside of the leaves and gentle pressing. Leave half a centimeter borders near the leaf nerve to prevent the transfer of bacteria (Fig. 3).
6. After 2–3 days of incubation harvest the leaf for transgene expression analysis by confocal microscope and western blot.

3.3 Microscopy Observation

1. On the third day post-agroinfiltration, cut the scrape (for example, 1 × 1 cm square) from the center of each leaf half with a sharp scalpel. Collect the rest of plant material for western blot analysis.
2. Immediately place each scrap into the separate 10-ml syringe without needle, cover by your finger (protected by laboratory gloves) the bottom opening of the syringe and fill it with 5 ml of water. Remove your finger from the bottom of the syringe and pull the air. One more time use your finger to cover the bottom of the syringe and pull the plunger to create a vacuum. You should see the air bubbles coming from the leaf scrap. Release plunger. The leaf scrap should become dark green due to flooding the intercellular air spaces with water.

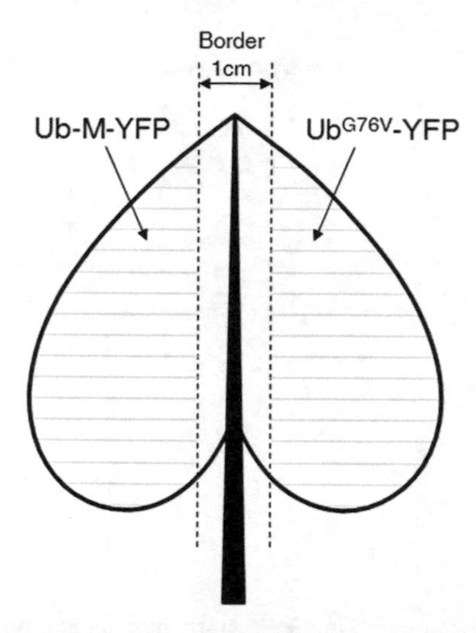

Fig. 3 Scheme of the leaf lower epidermis division into two halves for two simultaneous agroinfiltrations. The safe border preventing bacterial penetration of second leaf half is marked

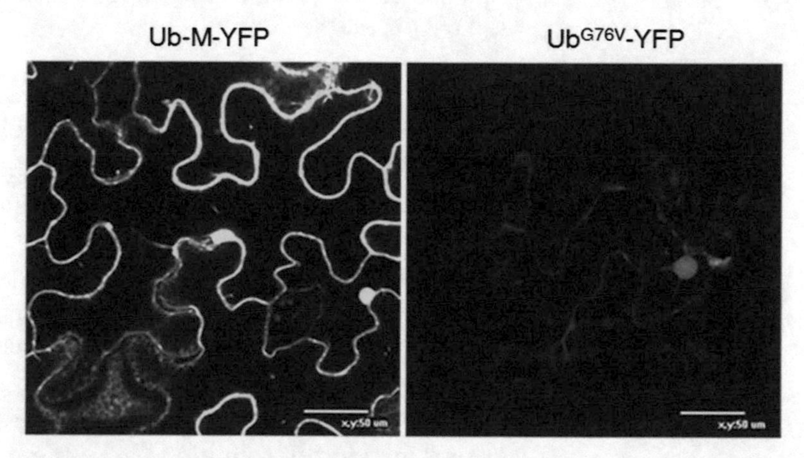

Fig. 4 Representative images of the expression level of the Ub-X-YFP fusion proteins in *N. benthamiana*. Stability of the YFP chimeras was analyzed 3d post-agroinfiltration using fluorescent confocal microscope by determining intensity of the YFP fluorescence under the same set of the parameters

3. Take out leaf scraps from the syringe and prepare microscopic slides (*see* **Note 7**).
4. Analyze each scrap in the fluorescent confocal microscope under the same set of the parameters (Fig. 4).

3.4 Western Blot Analysis

1. Third day post-agroinfiltration collect each half of the leaf tissue into the Eppendorf tube (approximately 100 mg) for western blot analysis. Freeze the plant material in liquid nitrogen.
2. Grind tissue in 100 μl of Extraction Buffer to extract proteins.

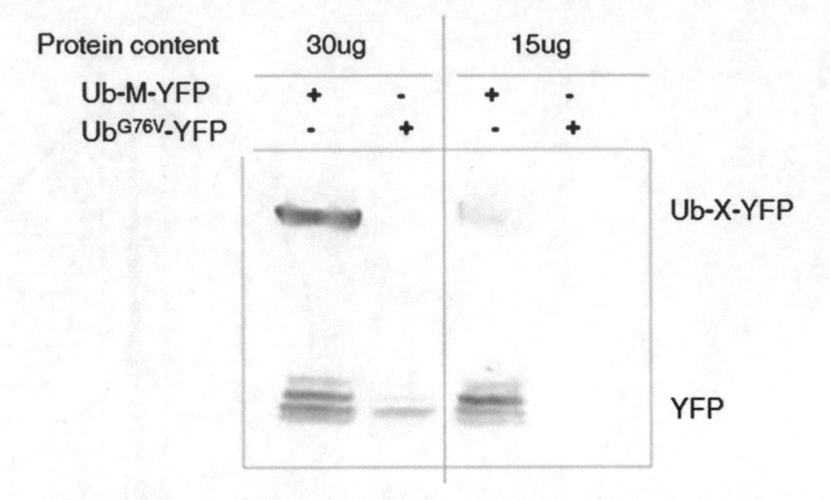

Fig. 5 Western blot indicating the amount of the Ub-X-YFP fusion proteins. Expression of the YFP chimeras was analyzed 3d post-agroinfiltration by western blot with an anti-GFP antibody. The uncleaved precursors and polypeptides with sizes corresponding to the cleaved degraded products are indicated as Ub-X-YFP and YFP, respectively

3. Shake the tubes in the thermomixer for 5 min and spin 10 min in a microcentrifuge at 10 rpm in a cold room.
4. Collect the supernatants in the new Eppendorf tubes and determine the concentration of the recombinant proteins using a standard method, for example, with the Bio-Rad protein assay kit.
5. Take a volume with a given protein amount (usually 15–30 μg), add 1/3 of the volume of 4× Laemli Sample Buffer and denature by boiling for 5 min. Subsequently cool on ice and spin briefly.
6. Load samples into 12% SDS-polyacrylamide gel and run the gel in reducing conditions [6] using conditions recommended by the manufacturer of the electrophoresis apparatus.
7. Perform the western blot analysis according to a standard procedure [6] using anti-GFP antibody (Fig. 5).

4 Notes

1. To create the substrate for N-end rule pathway which will be compatible with described above fluorescent reporters one can introduce in reverse primer R1 a single point mutation to form one more linker (RGK tripeptide) at the C-terminus. The lysine residues in position 3, 4, and 15 in the linker between ubiquitin and the fluorescent protein (downstream of the ubiquitin moiety) will be potential ubiquitination sites *for* ubiquitin ligase of the N-end rule pathway.

2. *N. benthamiana* plants can be alternatively cultivated hydroponically in hydroponic containers filled by 0.5× Hoagland's nutrient solution. Hoagland's media should be buffered with 2 mM (0.39 g/l) MES (2-[*N*-Morpholino]ethanesulfonic acid) and adjusted to pH 5.5 with 1 M KOH before autoclaving (for 15 min at 121 °C).
3. Add 1 ml TRI REAGENT® (Sigma-Aldrich) to 50–100 mg of homogenized plant tissue, vortex, and keep 5 min at room temperature. After 10 min centrifugation at 12,000 g transfer the supernatant to a new microfuge tube and add 100 μl of bromochloropropane, vortex 15 s, and keep 2–15 min at room temperature. Spin 12,000 g for 15 min at 4 °C, collect the upper phase (~600 μl) to a new microfuge tube and add 500 μl of isopropanol for RNA precipitation. Then vortexed the samples and after 2–15 min on ice, spin 12,000 g for 8 min (cold or room temperature), discard the supernatant, add 1 ml of 70% ethanol and mix (RNA wash). Spin 7500 g for 5 min to remove the ethanol and leave the pelleted RNA to air dry for 2–3 min. Finally, add 10–20 μl of TE pH 8 and resuspend RNA by pipetting. Alternatively, for total RNA isolation use DirectZol™ RNA MiniPrep (ZYMO RESEARCH, #R2052) followed by DNAseI treatment according to the manufacturer's protocols. To check the quality of RNA isolated from plant tissues, electrophoresis in 1.2% agarose gels (conditions as described for DNA) can be performed.
4. Each of 20-μl reverse transcription reaction contained 5 μg of total RNA, 2 pmol of specific antisense primer, 1 mM dNTPs mix, 10 mM DTT, and 1 μl of PowerScript™ Reverse Transcriptase (BD Biosciences Clontech) in the buffer supplied by the manufacturer. The RNA and primers were preheated to 70 °C for 10 min and snap-cooled in ice water before adding the remaining components. The RT reactions were carried out for 1 h at 42 °C and were terminated by heating to 70 °C for 15 min. Then 1-μl aliquots of the reaction mixtures were used for PCR, with specific primer pairs designed for the selected cDNAs.
5. Use other donor vectors (for example, pH7CWG2, or pSITE-1NB, pSITE-2NA, pSITE-2NB, pSITE-4NA, or pSITE-4NB [9] which can be ordered on line from The Arabidopsis Information Resource (TAIR) web page (www.arabidopsis.org)) to create other stable and unstable fluorescent proteins.
6. Preparation of electrocompetent bacteria may be performed according to Dower et al. [10]. The same protocol can be used for *E. coli* and *A. tumefaciens*.
7. Use a toothpick and the Vaseline to draw the window along the edges of cover glass to prevent water evaporation and tissue crushing.

Acknowledgments

This work was supported by the Ministry of Science and Higher Education in frame of the project No W16/7.PR/2011 realized in the years 2011–2015 and the National Science Centre (grant No 2012/05/N/NZ1/00699 realized in the years 2013–2014). KZ-R was also supported by the European Union Mazovia Fellowship "Development of science—development of the region—scholarship support for Mazovias' Ph.D.'s".

References

1. Pal R, Palmieri M, Sardiello M, Rodney GG (2014) Impaired-UPS can be compensated by activation of autophagy in neurodegenerative diseases. Biophys J 106:670a
2. Veljanovski V, Batoko H (2014) Selective autophagy of non-ubiquitylated targets in plants: looking for cognate receptor/adaptor proteins. Front Plant Sci 5:308
3. Lilienbaum A (2013) Relationship between the proteasomal system and autophagy. Int J Biochem Mol Biol 4:1–26
4. Dantuma NP, Lindsten K, Glas R, Jellne M, Masucci MG (2000) Short-lived green fluorescent proteins for quantifying ubiquitin/proteasome-dependent proteolysis in living cells. Nat Biotechnol 18:538–543
5. Lindsten K, Dantuma NP (2003) Monitoring the ubiquitin/proteasome system in conformational diseases. Ageing Res Rev 2:433–449
6. Sambrook J, Frisch EF, Maniattis T (1989) Molecular cloning: a laboratory manual. Cold Spring Harbor Laboratory Press, Cold Spring Harbor, NY
7. Vervliet G, Holsters M, Teuchy H, Van Montagu M, Schell J (1975) Characterization of different plaque-forming and defective temperate phages in Agrobacterium. J Gen Virol 26:33–48
8. Chomczynski P, Sacchi N (1987) Single-step method of RNA isolation by acid guanidinium thiocyanate-phenol-chloroform extraction. Anal Biochem 162:156–159
9. Chakrabarty R, Banerjee R, Chung SM, Farman M, Citovsky V, Hogenhout SA, Tzfira T, Goodin M (2007) PSITE vectors for stable integration or transient expression of autofluorescent protein fusions in plants: probing Nicotiana benthamiana-virus interactions. Mol Plant Microbe Interact 20:740–750
10. Dower WJ, Miller JF, Ragsdale CW (1988) High efficiency transformation of E. coli by high voltage electroporation. Nucleic Acids Res 16:6127–6145

Chapter 6

Generation of Artificial N-end Rule Substrate Proteins In Vivo and In Vitro

Christin Naumann, Augustin C. Mot, and Nico Dissmeyer

Abstract

In order to determine the stability of a protein or protein fragment dependent on its N-terminal amino acid, and therefore relate its half-life to the N-end rule pathway of targeted protein degradation (NERD), non-Methionine (Met) amino acids need to be exposed at their amino terminal in most cases. Per definition, at this position, destabilizing residues are generally unlikely to occur without further posttranslational modification of immature (pre-)proproteins. Moreover, almost exclusively, stabilizing, or not per se destabilizing residues are N-terminally exposed upon Met excision by Met aminopeptidases. To date, there exist two prominent protocols to study the impact of destabilizing residues at the N-terminal of a given protein by selectively exposing the amino acid residue to be tested. Such proteins can be used to study NERD substrate candidates and analyze NERD enzymatic components. Namely, the well-established ubiquitin fusion technique (UFT) is used in vivo or in cell-free transcription/translation systems in vitro to produce a desired N-terminal residue in a protein of interest, whereas the proteolytic cleavage of recombinant fusion proteins by tobacco etch virus (TEV) protease is used in vitro to purify proteins with distinct N-termini. Here, we discuss how to accomplish in vivo and in vitro expression and modification of NERD substrate proteins that may be used as stability tester or activity reporter proteins and to characterize potential NERD substrates.

The methods to generate artificial substrates via UFT or TEV cleavage are described here and can be used either in vivo in the context of stably transformed plants and cell culture expressing chimeric constructs or in vitro in cell-free systems such as rabbit reticulocyte lysate as well as after expression and purification of recombinant proteins from various hosts.

Key words N-end rule pathway, Ubiquitin fusion technique, TEV protease, N-terminomics, Protease, Degradomics

1 Introduction

The abundance and activity of all cellular proteins, the proteome, have to be strictly regulated to ensure their proper function. Proteostasis control is accomplished on transcriptional, translational, and posttranslational levels. One of these protein quality control checkpoints is the ubiquitin-proteasome system (UPS), a part of the cellular protein modification machinery utilizing the small protein modifier ubiquitin (Ub), which can lead to the

L. Maria Lois and Rune Matthiesen (eds.), *Plant Proteostasis: Methods and Protocols*, Methods in Molecular Biology, vol. 1450, DOI 10.1007/978-1-4939-3759-2_6, © Springer Science+Business Media New York 2016

degradation of, e.g., misfolded proteins or of those which function is either not any longer needed for cell viability or even may cause cytotoxic effects.

The N-end rule degradation pathway (NERD) of targeted proteolysis is a specialized part of the UPS, *see* Fig. 1a. It links the half-life of a protein to its N-terminal amino acid and is built up in a hierarchical way comprising—for some substrates—a multi-step biochemical reaction cascade involving several highly specific

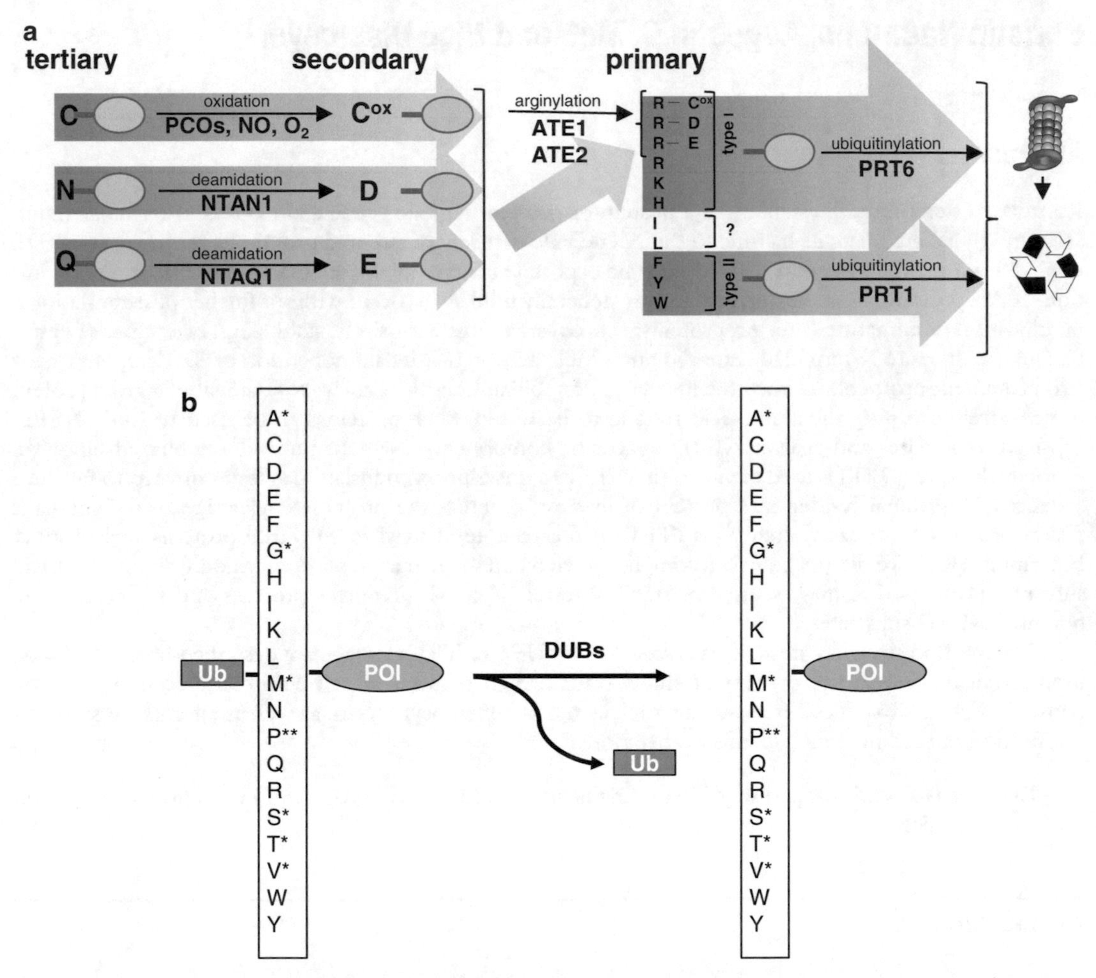

Fig. 1 N-end rule degradation pathway (NERD) of targeted proteolysis and Ubiquitin fusion technique (UFT). (**a**) NERD in plants. Substrates bearing an N-terminal primary, secondary, or tertiary destabilizing residue can be recognized and enzymatically modified by amidases (NTAs), arginyl transferases (ATEs), and NERD E3 Ub ligases (PRTs). Cys can be nonenzymatically oxidized by reactive oxygen species (1–4); or enzymatically by plant cysteine oxidases (PCOs; [15]), (**b**) UFT and amino acids possible to engineer at the N-terminus. UFT allows in vivo generation of needed N-termini. Ub-POI fusion proteins expressed are deubiquitinated via DUBs. After Ub removal amino acid in position 1 is exposed. Single-letter abbreviations for amino acid residues are as follows: A, Ala; C, Cys; D, Asp; E, Glu; F, Phe; G, Gly; H, His; I, Ile; K, Lys; L, Leu; M, Met; N, Asn; P, Pro; Q, Gln; R, Arg; S, Ser; T, Thr; V, Val; W, Trp; and Y, Tyr. *: amino acids not per se destabilizing, **: Ub-Pro is slowly processed by DUBs [25]

protein modifying enzymes. NERD function is conserved among all kingdoms albeit involving various ways to accomplish proteolysis, i.e., either enzymes of the eukaryotic UPS or bacterial or organellar proteases [1–5]. A protein bearing a destabilizing N-terminal such as basic, bulky or hydrophobic side chains, in this context known as N-degrons, which may occur as possible protein cleavage products can be recognized via NERD E3 Ub protein ligases—the so called N-recognins—followed by polyubiquitination and degradation by the 26S proteasome. Both steps are regulated in yeast and animals by functional homologs of the Ubr1 E3 Ub protein ligase [2–4]. Proteins bearing a tertiary destabilizing residue (Cys, Asn, Gln) can be modified in an enzymatic (deamidation of Asn and Gln by N-terminal amidases (NTAs) or oxidation of Cys) or nonenzymatic (oxidation of Cys by reactive oxygen species (ROS) such as nitric oxide (NO)) reaction to become secondary destabilizing residues (oxidized Cys [Cys^{ox}], Asp, Glu). These amino acids can be recognized via tRNA-arginyltransferases (ATEs) that attach an Arg residue as a primary destabilizing side chain to their substrates (Fig. 1a).

Known NERD substrates in yeast and animals are proteins and peptide fragments with mainly regulatory functions but also related to human diseases such as Alzheimer's and Parkinson's [6]. In plants, only five proteins, i.e., members of the group 7 ethylene response factors (ERFs), are associated with NERD-mediated degradation [7–9]. These targets are transcription factors involved in water–stress response. In their specific case, the proteins start with Met followed by Cys (MC-starting proteins). Met is rapidly cleaved off by MAPs if the second amino acid is stabilizing, e.g., Gly or not per se destabilizing, Cys. Under normoxic conditions, the Cys is oxidized and the proteins are degraded via NERD. Under hypoxic conditions, the absence of oxygen leads to their stabilization. Other proteins are discussed as NERD substrates but not verified until now.

In *Arabidopsis*, the two Ubr1 functional homologs PROTEOLYSIS (PRT) 1 [10, 11] and 6 [12], the two arginyltransferases ATE1 and 2 [13, 14], as well as plant cysteine oxidases (PCOs) [15] are known NERD components but their physiological role and molecular mechanisms are largely unknown. Additionally, PRT1 is—as a plant pioneer protein—completely unrelated to known Ubr enzymes [6]. Plant NERD seems to be mechanistically diverse due to nonhomologous ATEs and the mild phenotype of plant *prt* mutants hints toward the existence of further important NERD enzymes showing lower sequence similarities with known homologs.

To further relate plant NERD to biological functions, it is pivotal to identify substrates of the different enzymatic components and the entire degradation pathway.

1.1 Importance of Creating Proteins Comprising Defined N-Termini

To analyze the stability of NERD substrate candidate proteins, it is inevitable to use methods that allow creation of distinct non-Met N-termini of choice for studies both in vivo and in vitro. The need of specific and exposed N-termini as a starting point for protein stability assays and testing of hypotheses becomes clear from Fig. 1a, where the questionable residues are ordered according to their possible influence on protein stability. Generation of the mentioned non-Met N-termini is not possible applying regular translation as all open reading frames, regardless of transcribed and translated in vivo or in vitro, need to commence with a start codon. Therefore, freshly synthesized proteins will necessarily contain the initiator Met residue. In vivo, in most cases, cotranslational N-terminal Met excision (NME) by MAPs leads to the cleavage of the initiator Met residue if the second amino acid is not a primary, secondary or tertiary destabilizing one according to NERD or another Met. One exception is Cys which may be presented at the N-terminal after NME [16, 17]. However, Met-Cys-starting proteins are highly underrepresented, e.g., in the *Arabidopsis* proteome. Particularly, in vivo and in vitro studies demonstrated that MAPs remove the initiator Met only if the second residue has a small radius of gyration of the side chain. Thus, bulky amino acids do not allow Met removal [18].

NME occurs at the Met adjacent to residues at position two of the nascent protein which are classified as stabilizing residues—or as not per se destabilizing residues—such as Ala, Gly, Val, Ser, and Thr but also Pro and, as mentioned above, Cys. Met is retained at the N-terminal if the second residue is another Met [19–22]. We recently confirmed this Met excision "dogma" and the underrepresentation of charged and hydrophobic amino acids at the N-terminal also for *Arabidopsis* [23].

If proteins comprise N-terminal signal or transit peptides, they can be cleaved off after transport into the desired compartment, e.g. ER, nucleus, mitochondria, or chloroplasts. Many proteins, especially zymogenes, i.e., precursors of enzymes, are also translated as (pre-)proproteins or (pre-)propeptides bearing an N-terminal sequence which is cleaved off autocatalytically or by a protease. One example is the formation of active trypsin [24].

To characterize the stability of a protein of interest (POI) dependent on the N-terminal, it has to be made sure that the protein expressed in vivo or in vitro is bearing the N-terminal amino acid in question to which the half-life is to be correlated to. There exist two ways to artificially circumvent this problem for experimental purposes and produce test proteins in a way that they can be metabolized by NERD, i.e., expression of chimeric proteins via the so-called ubiquitin fusion technique (UFT) and, cleavage after recombinant production by the catalytic domain of the Nuclear Inclusion a (NIa) protein encoded by the tobacco etch virus (TEV) to generate distinct N-termini of proteins for further characterization. In the following, we explain three experimental strategies how to use UFT in vivo, UFT in vitro, and TEV cleavage to create test proteins in this context.

1.2 Ubiquitin Fusion Technique

UFT was originally established in *S. cerevisiae* to study protein stability according to a distinct N-terminal [25]. It was extensively used to characterize NERD substrates and enzymatic components [26, 27]. UFT is based on the natural occurrence and processing of the small protein Ub which is synthesized as a polymer [28] followed by cleavage into single Ub moieties after its last residue Gly76 by deubiquitinating enzymes (DUBs) [29, 30]. UFT was initially used to express Ub-X-β-galactosidase constructs in *S. cerevisiae* and *E. coli*. Only in the yeast, the fusion was cleaved by DUBs as the bacteria are lacking this machinery and the UPS [25]. UFT allows "automatic" cleavage if DUBs are present and it works for all amino acids in position X albeit the cleavage of Ub-Pro is not very efficient [25, 27, 30]. It is also possible to use UFT in Ub-lacking prokaryotes but it is necessary to cotransform, e.g., the yeast DUB Ubp1 which is then able to remove Ub from fusion proteins in the bacterial host [31].

We use this cotranslational cleavage mechanism to separate a Ub-X-POI fusion protein, with X being the wanted N-terminal amino acid. Also in this artificial context, the fusion is cotranslationally cleaved after Gly76 of Ub and the resulting C-terminal fragment of the fusion protein showing the engineered N-terminal (Fig. 1b).

1.3 UFT In Vivo

UFT allowed screening for mutants with impaired proteasomal degradation of proteins bearing a distinct N-terminus. In *Arabidopsis*, the E3 Ub ligase PRT1 was identified by a forward genetics screen based on Ub-F-DHFR reporter constructs. The F-construct was found to be stabilized in *prt1* mutant plants [10, 32]. Also the second identified plant NERD E3 ligase was validated using UFT [12]. Here, the Ub reference technique (URT) was used where the Ub is N-terminally tagged with a second reporter protein to follow fusion protein cleavage with an internal reference DHFR-Ub^{K48R} [27]. DHFR-Ub^{K48R} contains a mutated Ub to prevent its recognition as degradation signal of the Ub fusion degradation pathway (UFD). In our experiments, we can use the wild type version of Ub as we use it without a reference. In *prt6* mutant plants harboring DHFR-Ub^{K48R}-X-β-galactosidase (X = M, L, F, R), the R-β-gal was stabilized whereas in the wild type this construct is highly instable [12].

We usually express Ub-X-POI-YFP (X = D, R, G) to study stabilization of the POI according to it's N-terminus in the in vivo system of plant protoplasts. This system allows to transiently transform plant material including cell types derived from mutant lines and study, e.g., protein localization, movement or interaction [33, 34].

mCherry is cotransformed as a transformation control and the ratio of protoplasts showing a YFP and an mCherry signal in comparison to protoplasts transformed only with mCherry is calculated. In parallel we analyze the protein abundance via western blotting.

1.4 UFT In Vitro

In order to study protein half-life in vitro, UFT can be used in mammalian cell culture [35], *Xenopus* oocyte extracts [36], and commercially available reticulocyte lysate [37]. The latter allows to easily study stability effects in vitro in an eukaryotic background comprising a functional proteasome. In the plant field, reticulocyte lysate independent of UFT was applied to study protein degradation of MC-starting proteins of the group 7 ethylene response factors (ERFs; [7]). Here, UFT is not required as MAPs cleave off the first Met and the following Cys is enzymatically oxidized by PCOs [15] or nonenzymatically by ROS [4]. The proteins can then be degraded via NERD. These ERFs are the only described NERD substrates in plants so far. The same reticulocyte lysate system was used to follow the degradation of RGS4 and 5 as substrates of mammalian NERD [37]. Here, the process was monitored with biotin-labeled proteins.

In the protocol outlined below, UFT fusion proteins are expressed under control of the T7 promoter and their stability was monitored via SDS-PAGE followed by western blot and immunostaining after cycloheximide treatment.

1.5 TEV Cleavage

Recombinant protein production is often accomplished by expressing tagged proteins or entire fusion proteins comprising larger accompanying non-POI protein moieties which serve to enhance expression and solubility or facilitate further downstreaming, e.g., enrichment, cleavage, and purification, of the fusion itself or its separated parts. Highly sequence-specific proteases enable the cleavage of such fusion protein tags or fusion protein partners from the actual target POI. To precisely remove an N-terminal tag from a fusion protein is the key for controlled exposure of a desired N-terminal of a given POI, e.g., in the context of studying the impact of recognition by N-recognins and other enzymatic NERD components. Among such highly specific and versatile proteases, TEV protease, i.e., the catalytic domain of the Nuclear Inclusion a (NIa) protein encoded by the TEV, gained popularity both in in vitro and in vivo applications due to its high sequence specificity, low enzymatic promiscuity, non-toxicity, relatively high catalytic turnover, insensitivity to many proteinase inhibitors and ease of removal after its action [38, 39] (*see* **Note 1**). In our studies, we use the recognition sequence of TEV as a linker between an N-terminal octahistidine-MBP (maltose binding protein, $(His)_8$-MBP) which used both as an affinity and purification tag and the X-POI, where X stands again for the N-terminal amino acid to be engineered. TEV is highly sequence-specific to its canonical recognition site ENLYFQ|X, with X being the relevant P1′ amino acid and therefore the new N-terminally exposed residue of the POI [40]. TEV displays optimal performance on the peptide sequence ENLYFQ|S/G [41], where Q and S or G corresponds to the P1 and P1′ residues. P1′ is typically known to be G or S but can be successfully replaced by all known amino acids with no or little activity loss except for proline which drastically hindered the TEV recognition ability [40, 42, 43]. Recently, new proteases suitable as candidates for tag removal were described having much higher activity even at 0 °C with a broader optimal range for buffer and/or salt conditions [44].

In our case, we use a construct containing an N-terminal $(His)_8$-MBP fused to a Phe- or Gly-starting and C-terminally His-tagged substrate sequence as POI. This sequence is based on *E. coli* lacZ, known as eK (extension containing lysines/Ks) and has extensively been used to generate artificial NERD substrates to characterize this proteolytic pathway [25, 45]. $(His)_8$-MBP and eK are separated by a linker which contains a TEV cleavage site, as illustrated in Fig. 3a, b. The $(His)_8$-MBP tag facilitates purification and is easily removed by a TEV digest, allowing eK substrate generation and purification for X-eK-His (Fig. 4a, b). The F-eK-His is an excellently working artificial NERD substrate and was successfully tested in in vitro ubiquitination with G-eK-His as a negative control (Fig. 4c).

Examples for engineering various P1′ sites and substituting amino acids by replacing the G/S within the canonical TEV recognition site are Asp or Phe [42, 43, 46, 47], and Cys [48]. A new version of TEV protease almost completely lacks specificity for the amino acid at position P1′ and allows an even broader variety of N-terminals to be exposed via cleavage [49].

2 Materials

2.1 General Equipment

1. Climatized greenhouse or growth chambers.
2. Fridge or cold room for TEV preparation and cleavage of fusion proteins.
3. Cooling microcentrifuge or regular microcentrifuge in cold room.
4. 1.5-mL microcentrifuge tubes.
5. Flat-tip forceps and razor blade.
6. Dewar container with swimmer, liquid nitrogen.
7. Ice bucket, wet ice.
8. Vortex mixer.
9. Graduated cylinders and containers for reagent preparation and storage.
10. Pipets accurately delivering 2.5, 20, 200, and 1000 μL.

11. Soil mixture for *Arabidopsis* cultivation, selection and propagation: steamed (pasteurized for min. 3 h at 90 °C) soil mixture of Einheitserde Classic Kokos (45% (w/w) white peat, 20% (w/w) clay, 15% (w/w) block peat, 20% (w/w) coco fibers; cat. no. 10-00800-40, Einheitserdewerke Patzer, Gebr. Patzer); 25% (w/w) Vermiculite (grain size 2–3 mm; cat. no. 29.060220, Gärtnereibedarf Kamlott), 300–400 g/m^3 soil substrate of Exemptor (100 g/kg thiacloprid, cat. no. 802288, Hermann Meyer).
12. Confocal laser scanning microscope with 514 and 587 nm excitation wavelength, emission filters for 509 and 610 nm.

2.2 Cloning

1. cDNA from *Arabidopsis thaliana* (L.) Heynh., ecotype Columbia-0 (Col-0).
2. Standard cloning equipment.
3. Site-specific oligonucleotides (*see* Table 1).
4. Proofreading polymerase such as Pfu.
5. Gateway BP and LR kits (Invitrogen).
6. Gateway-compatible Entry and Destination vectors (*see* Table 2).
7. DH5α or equivalent cloning hosts (Invitrogen).
8. *E. coli* BL21-CodonPlus(DE3)-RIL (Stratagene/Agilent) or equivalent bacterial expression hosts.
9. Gel extraction Kit (Thermo Scientific, K0513).
10. DNA Maxi-Prep Kit (Macherey & Nagel NucleoBond, PC 500).

Table 1
Oligonucleotide primers used for PCR

Primer name	Sequence (5′–3′)	Annealing temperature (°C)
Sequences for UFT constructs		
ss_attB1_Ub	GGGGACAAGTTTGTACAAAAAAGCAG GCTTAGCCGCCACCATGCAGATCTTCGTCAAG	51
ss_bridge_Ub_attB1	ACCATGCAGATCTTCGTCAAGACGTTAAC	56
as_Ub_POI	[NNN]*n*CCCACCTCTAAGTCTTAAGACAAGATG	Depending on GOI
ss_Ub_POI	GTGGG[NNN]*n*	Depending on GOI
as_POI	[NNN]*n*	Depending on GOI
Sequences for TEV constructs		
ss_adapter_tev	GGGGACAAGTTTGTACAAAAAAGCA GGCTTAGAAAACCTGTATTTTCAG	48
ss_tev_X_POI	GCTTAGAAAACCTGTATTTTCAG<u>XXX</u>[NNN]*n*	Depending on GOI
as_attB2_POI	GGGGACCACTTTGTACAAGAAAGCTGGGTA[NNN]*n*	Depending on GOI

NNN: bases specific for annealing on DNA sequence encoding POI, XXX: codon for N-terminal amino acid residue of interest, GOI: gene of interest encoding POI

Table 2
Vectors used

Vector	Description	Source or reference
pDONR201	Gateway donor vector for single Gateway recombination (attP1/P2)	Invitrogen
pVP16	Gateway destination vector (attR1/R2) containing a 8×His:MBP coding sequence 5′ of the Gateway cassette leading to an N-terminal 8×His:MBP double affinity tag under control of Pro_{T5} and a Lac-Operon for protein induction (*see* **Note 11**). bla resistance in bacteria; Gateway recombinational cloning into this vector removes ccdB and cat	Kind gift of Russell L. Wrobel, Protein Structure Initiative (PSI) at the Center for Eukaryotic Structural Genomics (CESG), University of Wisconsin-Madison [54]
pOLENTE	Gateway.compatible destination vector (attR1/R2) for coupled transcription/translation, based on pTNT (Promega). bla resistance in bacteria; Gateway recombinational cloning into this vector removes ccdB and cat	Details on the cloning will be published elsewhere
pUBC-YFP	Gateway destination vector (attR1/R2) containing a YFP coding sequence 3′ of the Gateway cassette leading to a C-terminal YFP. Fusion protein is under control of the *Arabidopsis* Ubiquitin-10 (At4g05320) promotor. bla resistance in bacteria; Gateway recombinational cloning into this vector removes ccdB and cat	[55]
pRK793	pRK793 overproduces the TEV catalytic domain as an MBP fusion protein that cleaves itself in vivo to yield a TEV protease catalytic domain with an N-terminal His-tag and a C-terminal polyarginine tag, plasmid based on pMal-C2 (New England Biolabs).	pRK793 was a gift from David Waugh (Addgene plasmid # 8827; [53])
pVP16-tev-POI	Gateway expression vector (attB1/B2) based on pVP16 comprising a primer-born TEV recognition site (ENLYFQ-X) at the junction to the POI	This work

bla: β-lactamase, ccdB: cell death cassette, cat: chloramphenicol acetyltransferase

2.3 Protoplast Isolation (See Table 3)

1. 5 M sodium chloride (NaCl).
2. 1 M potassium chloride (KCl).
3. 0.1 M calcium chloride ($CaCl_2$).
4. 0.8 M mannitol (sterile-filtered).
5. 0.2 M MES (4-morpholineethanesulfonic acid; pH 5.7).
6. Haemocytometer.
7. Pre-cut 1000 μL pipet tips.
8. Macerozyme R10 (SERVA).
9. Cellulase R10 (SERVA).

Table 3
Buffers and solutions used for protoplast isolation and transformation

All buffers except W5 (storable for 1 week at 4 °C) have to be prepared freshly.

Buffer W5

Component	Stock conc.	Final conc.	50 mL	100 mL	200 mL	400 mL	500 mL	
NaCl	5 M	154 mM	1.54	3.08	6.16	12.32	15.4	mL
$CaCl_2$	1 M	125 mM	6.25	12.5	25	50	62.5	mL
KCl	0.1 M	5 mM	2.5	5	10	20	25	mL
MES pH 5.7	0.2 M	2 mM	0.5	1	2	4	5	mL
H_2O			39.21	78.42	156.84	313.68	392.1	mL

Buffer WI

Component	Stock conc.	Final conc.	5 mL	10 mL	15 mL	20 mL	30 mL	40 mL	
Mannitol	0.8 M	0.5 mM	3.15	6.3	9.45	12.6	18.9	25.2	mL
KCl	0.1 M	20 mM	1	2	3	4	6	8	mL
MES pH 5.7	0.2 M	4 mM	0.1	0.2	0.3	0.4	0.6	0.8	mL
H_2O			0.75	1.5	2.25	3	4.5	6	mL

Buffer MMG

Component	Stock conc.	Final conc.	5 mL	10 mL	15 mL	20 mL	30 mL	40 mL	
Mannitol	0.8 M	0.4 M	2.5	5	7.5	10	15	20	mL
$MgCl_2$	0.15 M	15 mM	0.5	1	1.5	2	3	4	mL
MES pH 5.7	0.2 M	4 mM	0.1	0.2	0.3	0.4	0.6	0.8	mL
H_2O			1.9	3.8	5.7	7.6	11.4	15.2	mL

PEG solution

Component	Stock conc.	Final conc.	5 mL	10 mL	15 mL	20 mL	30 mL	40 mL	
Mannitol	0.8 M	0.2 M	1.25	2.5	3.75	5	7.5	10	mL
$CaCl_2$	1 M	0.1 M	0.5	1	1.5	2	3	4	mL
PEG	Solid	40%	2	4	6	8	12	16	g
H_2O			1.5	3	4.5	6	9	12	mL

Enzyme solution

Component	Stock conc.	Final conc.	5 mL	10 mL	15 mL	20 mL	30 mL	40 mL	
Mannitol	0.8 M	0.4 M	2.5	5	7.5	10	15	20	mL
KCl	0.1 M	20 mM	1	2	3	4	6	8	mL
MES pH 5.7	0.2 M	20 mM	0.5	1	1.5	2	3	4	mL
H_2O			0.95	1.9	2.85	3.8	5.7	7.6	mL
Cellulose R10		1.50%	75	150	225	300	450	600	mg
Macerozyme R10		0.40%	20	40	60	80	120	160	mg
$CaCl_2$	1 M	10 mM	50	100	150	200	300	400	μL
BSA	0.1 g/mL	1 mg/mL	50	100	150	200	300	400	μL

10. BSA (Roth).
11. Sheet of regular printer paper.
12. Desiccator.
13. Black cloth (for desiccator).
14. Nylon mesh (100 μm mesh size).
15. Cell culture tubes (polystyrene, sterile, with screw cap; Greiner).

2.4 PEG-Mediated Transformation of DNA into Protoplasts (See Table 3)

1. 150 mM magnesium chloride ($MgCl_2$).
2. Polyethylene glycol (PEG) 4000.
3. 1 M potassium chloride (KCl).
4. 0.1 M calcium chloride ($CaCl_2$).
5. 0.8 M mannitol (sterile-filtered).
6. 0.2 M MES (4-morpholineethanesulfonic acid; pH 5.7).

2.5 Analysis of Protein Expression in Protoplasts

1. 1× SDS-loading buffer: 50 mM Tris–Cl (pH 6.8), 50 mM DTT, 1 % (v/v) SDS, 10 % (v/v) glycerol, 0.01 % (v/v) bromophenol blue.
2. Pre-cut 1000 μL pipet tips.

2.6 Protein Expression in a Cell-Free System

1. TNT T7 Coupled Reticulocyte Lysate System (Promega, L4610).
2. 0.5 M cycloheximide (Sigma).
3. 1× SDS loading buffer as in Subheading 2.5, **item 1**.

2.7 Protein Expression in Bacteria

1. LB medium (Roth).
2. 500 mL Erlenmeyer flasks.
3. 1 M IPTG (isopropyl β-D-1-thiogalactopyranoside).
4. 100 mM PMSF (phenylmethanesulfonylfluoride).
5. Ni-buffer: 100 mM Tris–Cl pH 8, 300 mM NaCl, 0.25 % (v/v) Tween, 10 % (v/v) glycerol.
6. Ni-elution buffer: 100 mM Tris–Cl pH 8, 300 mM NaCl, 0.25 % (v/v) Tween, 10 % (v/v) glycerol, 200 mM imidazole.
7. Amylose buffer: 20 mM Tris–Cl pH 7.4, 200 mM NaCl, 1 mM EDTA.
8. Amylose elution buffer: 20 mM Tris–Cl pH 7.4, 200 mM NaCl, 1 mM EDTA, 10 mM maltose.
9. Ni-NTA agarose (nickel-charged resin; Qiagen).
10. Amylose agarose (amylose resin; NEB).
11. Polypropylene columns (5 mL; Qiagen).
12. 40 g/mL lysozyme (Sigma) in Ni-buffer.
13. Imidazole (Merck).
14. Maltose (Roth).
15. Appropriate antibiotics (in our case, a stock solution of 50 mg/mL carbenicillin).

2.8 TEV Cleavage

1. Home-made TEV protease (*see* **Note 2**).
2. TEV reaction buffer (50 mM Tris–HCl pH 8, 0.5 mM EDTA, 1 mM DTT).

3. Ni-buffer: 100 mM Tris–HCl pH 8, 300 mM NaCl, 0.25% (v/v) Tween, 10% (v/v) glycerol.
4. Ni-elution buffer: 100 mM Tris–HCl pH 8, 300 mM NaCl, 0.25% (v/v) Tween, 10% (v/v) glycerol, 200 mM imidazole.
5. Amylose buffer: 20 mM Tris–HCl pH 7.4, 200 mM NaCl, 1 mM EDTA.
6. Amylose-elution buffer: 20 mM Tris–Cl pH 7.4, 200 mM NaCl, 1 mM EDTA, 10 mM maltose.
7. Ni-NTA agarose (nickel-charged resin; Qiagen).
8. Amylose resin (NEB).
9. Polypropylene columns (5 mL Qiagen).
10. Imidazole (Merck).
11. Maltose (Roth).
12. Amicon Ultra-15 (Merck Millipore; 30 and 10 kDa cut-off).

2.9 Protein Concentration Determination

1. Protein Assay Dye Reagent Concentrate (Bio-Rad).
2. 1 mg/mL bovine serum albumine (BSA).
3. Microplate reader.
4. Spectrophotometer set to 595 nm.

2.10 SDS-Polyacrylamide Gel Electrophoresis (SDS-PAGE)

1. SDS-PAGE equipment and power supply, we preferentially use small gel systems such as a BioRAD.
2. 30% acrylamide mix (30% Acrylamide/*N*,*N*′-Methylenebisacrylamide solution, ratio 37.5:1 in water, *see* **Note 3**).
3. 1.5 M Tris–HCl pH 8.8 (for separating gel).
4. 1 M Tris–HCl pH 6.8 (for stacking gel).
5. 10% (w/v) Ammonium persulfate (APS, immediately freeze upon preparation in single use aliquots, store at −20 °C).
6. 10% (w/v) sodium dodecyl sulfate (sodium lauryl sulfate, SDS).
7. *N*,*N*,*N*′,*N*′-tetramethyl-ethylenediamine (TEMED, *see* **Note 4**).
8. Isopropanol (*see* **Note 5**).
9. Gel loading tips, extended length.
10. Protein gel running buffer 10×; for 1 L: 30.2 g Tris base, 144.2 g glycine, and 10 g SDS in water.
11. Prestained molecular weight markers such as Precision Plus Protein Standard All Blue (Bio-Rad).
12. 5× SDS sample buffer: 0.25 M Tris-HCl pH 6.8, 50% glycerol, 5% SDS, 0.05% bromophenol blue, 0.25 M DTT [50]. For 1× SDS Sample buffer add 10 μL of 5× SDS sample buffer to 40 μL of water.

13. Prepare gel volumes according to the sizes of your gel system, all percentages in (v/v), modified from [51]:
    (a) For a 12% separating gel add 33% water, 40% acrylamide mix (30%), 25% 1.5 M Tris–Cl pH 8.8, 1% SDS (10% (w/v)) and APS (10% (w/v)), and 0.04% (v/v) TEMED.
    (b) For a 5% stacking gel add 68% water, 17% acrylamide mix (30%), 12.5% 1.5 M Tris–Cl pH 8.8, 1% SDS (10% (w/v)) and APS (10% (w/v)), and 0.1% (v/v) TEMED.

2.11 Western Transfer of Proteins and Detection

1. Semi-dry blot apparatus (such as BioRAD Trans-Blot SD Semi-Dry Transfer Cell).
2. Power supply.
3. Filter paper (Whatman).
4. PVDF membrane (GE Healthcare).
5. methanol for activation of PVDF membrane.
6. 3% (w/v) dry milk in TBST.
7. 5% (w/v) dry milk in TBST.
8. Tris-buffered saline (TBS) 10× for 1 L: solve 87.66 g of NaCl and 12.11 g of Tris-base in water; adjust to pH 7.5.
9. TBST for 1 L: add 100 mL of 10× TBS (for 150 mM NaCl and 10 mM Tris–Cl pH 8.0), 10 mL of 10% (v/v) Tween 20 (for 0.1% (v/v)) to water.
10. 10× semi-dry transfer buffer: for 1 L, add 58 g of Tris base (for 47 mM) and 29 g of glycin (for 50 mM) in water. 1× semi-dry transfer buffer: for 1 L, add 100 mL of 10× buffer and 200 mL of methanol to 1 L with water.
11. Saran wrap or plastic disposal bags to wrap membranes during ECL detection.
12. Enhanced chemiluminescent reagents for chemiluminescent imaging (ECL, SuperSignal West Femto, Pierce cat. no. 1858415).
13. BioMax Light Film for chemiluminescent imaging (Hartenstein).
14. Autoradiography cassette, film, and film developing unit.
15. Antibodies used for protoplast, reticulocyte lysate and recombinant protein work are listed in Table 4.

3 Methods

3.1 General Considerations

Here we outline how we use UFT and TEV fusions to study protein stability and degradation. Time considerations for UFT in protoplasts are 2–3 days and for UFT in reticulocyte lysate

Table 4
Antibodies used for protoplast, reticulocyte lysate and recombinant protein work

Antigen	Species, type	Name	Supplier	Cat. No.	Conditions	Range of use		
						Protoplast lysate	Reticulocyte lysate	Recombinant protein
1° antibodies								
HA tag	Mouse, monoclonal	HA.11	Covance or HISS	MMS-101	1:1000 dilution in TBST 4% milk	Yes	Yes	Yes
HA tag	Rabbit, polyclonal	HA-probe (Y-11)	Santa Cruz Biotechnology	sc-805	Western blot 1:200 dilution in TBST 3% milk	No	Not tested	Yes
His tag	Rabbit, polyclonal	His-probe (H-15)	Santa Cruz Biotechnology	sc-803	Western blot 1:200 dilution in TBST 3% milk	No	Not tested	Yes
His tag	Mouse, monoclonal	Anti-His antibody	GE Healthcare	27-4710-01	Western blot 1:1000 dilution in TBST 3% milk	Yes	Yes	Yes
Green fluorescent protein (GFP)	Rabbit, polyclonal	GFP (FL)	Santa Cruz Biotechnology	sc-8334	1:1000 dilution in TBST 4% milk	Yes	Not tested	Yes
Green fluorescent protein (GFP)	Mouse, monoclonal	GFP (B-2)	Santa Cruz Biotechnology	sc-9996	Western blot 1:1000 dilution in TBST 3% milk	Not tested	Not tested	Yes
Ubiquitin	Mouse, monoclonal	Ub (P4D1)	Santa Cruz Biotechnology	sc-8017	Western blot 1:1000 dilution in TBST 3% milk	No	Not tested	Yes
2° antibodies								
Mouse	Goat, IgG	HRP-conjugated antibody Immuno Pure Peroxidase	Pierce	31430	Western blot 1:5000 dilution in TBST 3% milk	Yes	Yes	Yes
Mouse	Goat, IgG	Anti-mouse IgG-HRP	Santa Cruz Biotechnology	sc-2005	Western blot 1:5000 dilution in TBST 3% milk	Not tested	Not tested	Yes
Rabbit	Goat, IgG	HRP-conjugated antibody Immuno Pure Peroxidase	Pierce	1858415 NCI8415	Western blot 1:5000 dilution in TBST 3% milk	Yes	Yes	Yes
Rabbit	Goat, IgG	Anti-rabbit IgG-HRP	Santa Cruz Biotechnology	sc-2004	Western blot 1:5000 dilution in TBST 3% milk	Not tested	Not tested	Yes

1–2 days, each starting from either transformation or transcription/translation. After stratification of 4–5 days at 4 °C in the dark, *Arabidopsis* seeds were germinated and plants grown under standard short day (8/16 h light/dark) greenhouse conditions.

3.2 Cloning

We applied classical PCR cloning strategies to obtain the chimeric DNA fusions containing *Arabidopsis* GOIs. In brief, primers containing the Gateway attB1 and attB2 sites (Table 1) were used for flanking both the 5′- and 3′-ends of each final fusion construct. This protocol was based on subsequent two-step fusion PCR and is explained in the following.

ORFs to express fusion proteins for UFT were cloned in a two step PCR. Ub was cloned using primer pair (ss_bridge_Ub_attB1/as_Ub_POI) from vector template DNA originally based on a synthetic human Ub gene (pRTUB8; [32, 52]).

In parallel, the GOI was cloned with an overhang to Ub with primers (ss_Ub_POI/as_attB2_POI) from cDNA (Fig. 2).

The fragments were purified via gel extraction and afterwards fused using primers (ss_attB1-Ub/as_attB2_POI) in a second PCR reaction (56 °C annealing temperature, 1 min extension time).

For attB1-tev-POI-attB2 fragments, the GOI was cloned from cDNA using primer ss_tev_X_POI as sense primer (X = N-terminal amino acid after TEV-cleavage of fusion protein). To complete the attB1 site, a second PCR with ss_adapter_TEV as sense primer was done (Fig. 3). The obtained fragments were recombined into pDONR201 vector using Gateway BP Clonase enzyme mix (Invitrogen) and analyzed via restriction digest and sequencing, primers can be found in Table 1. The insert of the resulting Entry vector was recombined into the respective destination vector (pVP16 for tev-constructs, pUBC-YFP for protoplasts, pOLENTE for the cell-free system) (Figs. 2 and 3). The isolated Expression vector was used to the respective experiment.

3.3 Protoplast Isolation

Preparation of protoplasts was done as described previously [33] and we only highlight some differences in our protocol.

All solutions except W5 need to be prepared freshly according to the needed volumes. W5 can be stored in the fridge. Solubilisation of PEG requires often more than 2 h, therefore, it is necessary to start with it in the beginning of the preparations.

1. Enzyme solution should be prepared as follows: Add MES, mannitol and KCl and preheat the solution to 55 °C. Now add the enzymes and keep the temperature for 10 min at 55 °C. Then, cool on ice and add $CaCl_2$ and BSA. Filter the solution through a 45 μm acetate filter into a Petri dish or similar.
2. Choose well-expanded leaves from 5- to 8-week-old plants grown in a stress-free environment under standardized conditions.

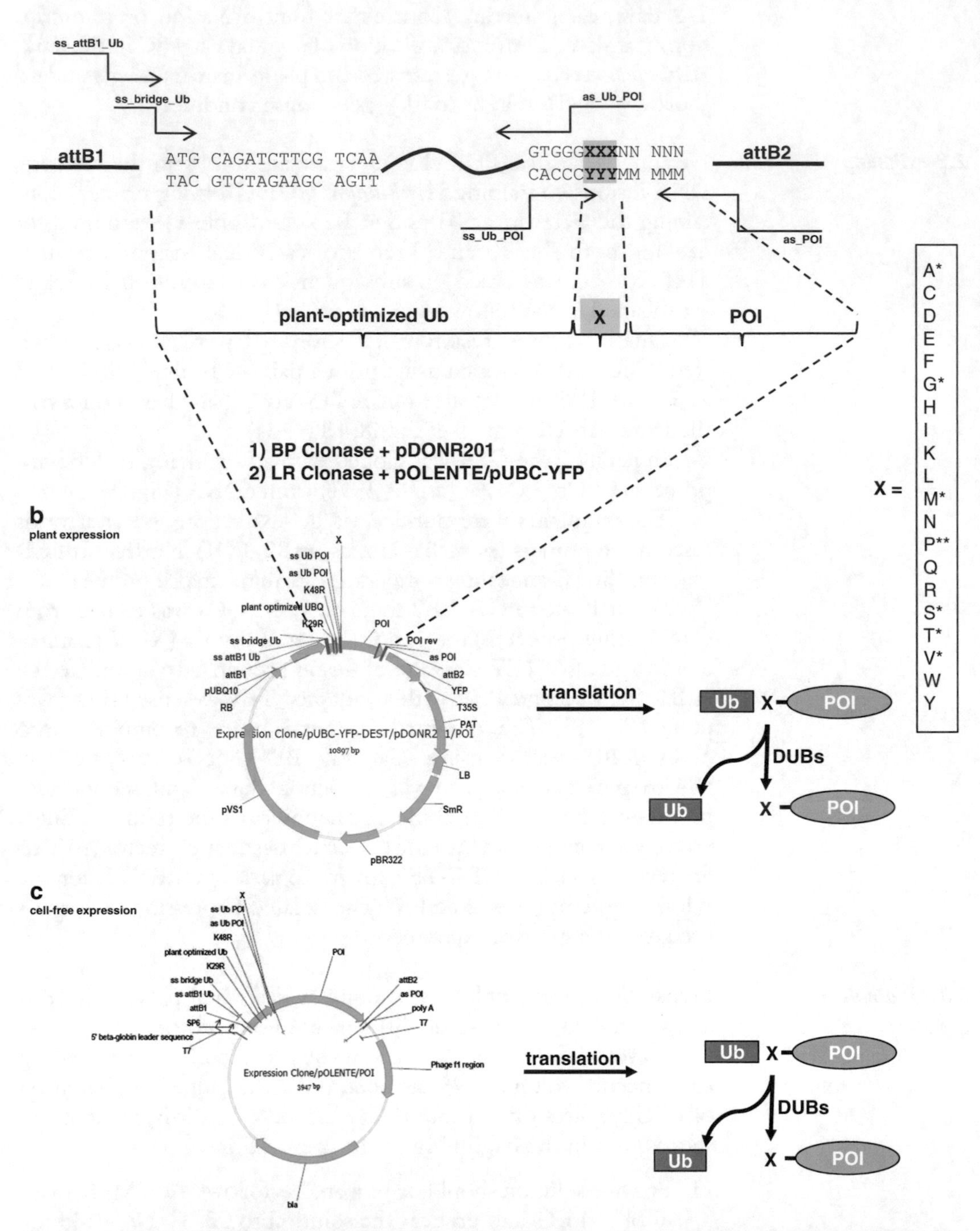

Fig. 2 Preparation of UFT constructs. (**a**) General cloning strategy. POI and Ub are subcloned from template DNA, PCR reactions are done as described in Subheading 3. The PCR fusion attB1-Ub-X-POI-attB2 is recombined into pDONR201 using Gateway BP clonase, followed by LR reaction into the designated destination vector. (**b**, **c**) Expression vectors and formation of X-POIs. (**b**) Cloning strategy for in vivo UFT. pUBC-YFP-Ub-X-POI for plant expression contains a Ub-POI-YFP fusion under control of a Ub promotor. After expression in planta, Ub is removed by DUBs and X is exposed as N-terminus. (**c**) Cloning strategy for in vitro UFT. pOLENTE-Ub-X-POI for in vitro expression in reticulocytes contains Ub-X-POI under control of the T7 promotor. After translation, Ub is removed by DUBs and X is exposed as N-terminal. *: amino acids not per se destabilizing, **: Ub-Pro is slowly processed by DUBs

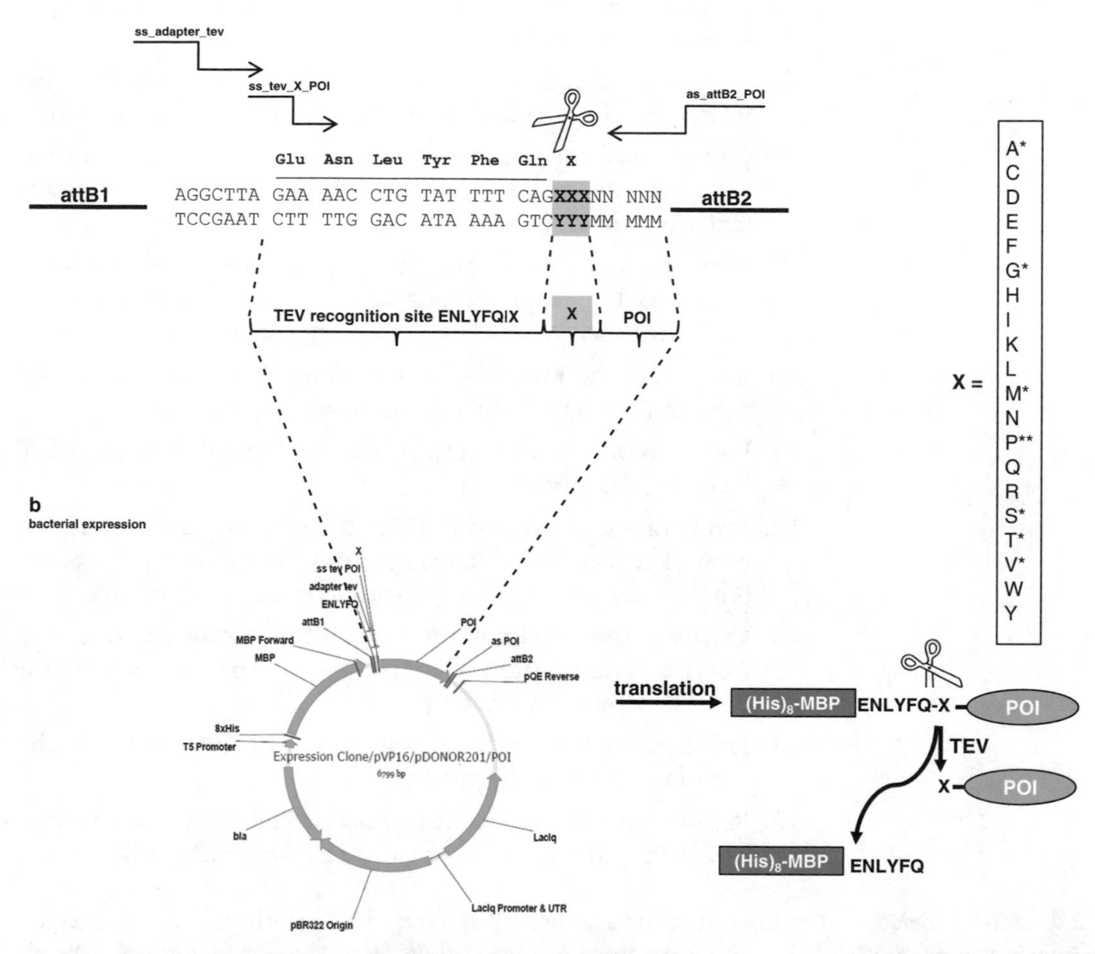

Fig. 3 Preparation of TEV cleavable fusion proteins. (**a**) General cloning strategy. GOI is cloned from cDNA with annotated primers containing the TEV recognition sequence ENLYFQ (tev) directly followed by X-POI. PCR reactions are done as described in Subheading 3. The PCR fusion attB1-tev-X-POI-attB2 is recombined into pDONR201 using Gateway BP clonase, followed by LR reaction into pVP16. (**b**) Expression vector and formation of X-POIs. pVP16-tev-POI is expressed in bacteria to obtain N-terminally tagged $(His)_8$-MBP-tev-POI. After purification, the $(His)_8$-MBP tag is removed by TEV cleavage as described in Subheading 3. *: amino acids not per se destabilizing, **: Ub-Pro is slowly processed by DUBs

3. For cell wall lysis, leaves can be cut individually or two to three leaves piled up. Prepare a clean printer paper sheet or similar to cut the leaves on.
4. Remove the top part and leaf stalk with a razor blade. Cut <0.5 mm leaf strips from the middle part of a leaf.
5. As soon as a leaf is cut into strips, transfer them into the prepared enzyme solution (around nine leaves in 3 mL of enzyme

solution in a Petri dish (3.5 cm)). Dip them completely into the solution by using a pair of flat-tip forceps.

6. Vacuum-infiltrate leaf strips for 30 min in the dark using a desiccator (covered with a black cloth).
7. Continue the digestion, in the dark for at least 3 h at RT. Best worked 20–22 °C, which are stable in an air-conditioned room.
8. Gently shake the enzyme solution to release the protoplasts. The solution should turn green and at least half of the leaf strips become transparent.
9. Filter the suspension through a nylon mesh (100 μm mesh size) into 12 mL pre-cooled cell culture tubes (polystyrene, sterile, with screw cap, Greiner). Keep tubes on ice.
10. Centrifuge the protoplast suspension for 1 min at 200 g at 4 °C and remove as much supernatant as possible.
11. Wash protoplasts with 2 mL of W5. Resuspend them by gently inverting the tubes.
12. Invert tube and take up 8 μL of the suspension to determine protoplast concentration using a haemocytometer (Always cut the tip ends when pipetting protoplasts to avoid damage).
13. Calculate the required volume of MMG solution to have a working concentration of 2×10^5 pp/mL (pp: protoplasts) for transformation (*see* **Note 6**).
14. Leave protoplasts on ice for 40 min. They will settle on the bottom of the tube by gravity.
15. Remove supernatant from protoplast pellet and do a second wash with 2 mL of W5. Let them rest for another 40 min.

3.4 PEG-Mediated Transformation of DNA into Protoplasts

1. During the sedimentation (**step 15** in Subheading 3.3), prepare either microcentrifuge tubes (for transformation of 100 or 200 μL of protoplast suspension, e.g., for fluorescence microscopy) or cell culture tubes (for transformation of 300–800 μL of protoplast suspension, e.g., for western blot analysis).
2. Add the required amount of plasmid DNA for transformation (10 μg of plasmid DNA/100 μL of protoplast suspension). Release the DNA at the bottom of the tubes to make sure successful transfer of the sample. For western blotting we normally use 300 μL of protoplast suspension.
3. Remove supernatant from the protoplast pellet and resuspend in the calculated volume of MMG solution at room temperature (20–22 °C) to get a concentration of 2×10^5 pp/mL. Mix gently by inverting.
4. Add protoplasts to plasmid DNA and mix gently by briefly inverting the tubes.

5. Add 1.1 protoplast suspension volumes of PEG solution to the tube and mix gently by inverting the tubes. Do one to two tubes at a time.
6. Incubate at room temperature (20–22 °C) for 5–10 min.
7. Add 4.4 protoplast suspension volumes of W5 to stop the transformation process. Mix by gently inverting the tubes.
8. Centrifuge for 1 min at 200 × *g* (4 °C) and remove as much supernatant as possible.
9. Add 1 protoplast suspension volumes of W1, mix by gently inverting the tubes.
10. Place tubes horizontally and incubate in the dark at room temperature (20–22 °C) over night.

3.5 Analysis of Protein Expression in Protoplasts via Fluorescence Microscopy

1. Carefully transfer 300 μL of protoplast suspension to a fresh 1.5 mL microcentrifuge tube using a cut 1000 μL pipet tip.
2. After overnight expression, YFP abundance can be checked using a laser scanning microscope. YFP has an excitation wavelength of 514 nm and an emission wavelength of 532 nm. mCherry is excited by 587 nm and emits a signal at 610 nm.
3. To calculate the stability of the POI dependent on its N-terminus, in each experiment, the number of transformed protoplasts was counted with both fluorescence signals (YFP and transformation control mCherry) in relation to just mCherry expressing protoplasts.

3.6 Analysis of Protein Expression in Protoplasts by Western Blotting

1. Spin down shortly at maximum speed, discard supernatant and add 12 μL of 1× SDS loading buffer.
2. Heat for 2–10 min at 96 °C.
3. Load to a 12% SDS-PAGE gel. In order to load the entire sample, use combs for broad pockets, e.g., a ten-sample comb, for SDS-PAGE and western blotting, see below.

3.7 Protein Expression in a Cell-Free System

To express proteins in a cell-free system, we use the TNT T7 Coupled Reticulocyte Lysate system (Promega) close to the manufacturer's instructions with few deviations.

1. We prepare only a quarter of the standard reaction volume for two time points and a master mix of DNA solution (111.11 μg/mL in DNAse-free water).
2. 5 μL of each sample are mixed with 32 μL of 1× SDS-loading buffer and frozen in liquid nitrogen to stop translation and possible degradation.
3. When all samples are taken, they are heated at 65 °C for 15–30 min to avoid protein degradation during storage.

4. The entire sample is carefully loaded to the SDS-PAGE for western blotting. Clotting of the reticulocyte lysate makes a complete sample transfer to the SDS-PAGE very difficult, however, that can be avoided by following this procedure.

3.8 Protein Expression in Bacteria

To define a specific N-terminal of a recombinant protein, we express them as a chimeric fusion with an N-terminal $(His)_8$-MBP-tag followed by a TEV-cleavage site as outlined in detail above and in Figs. 3 and 4. The fusion protein is purified in two steps, first, via

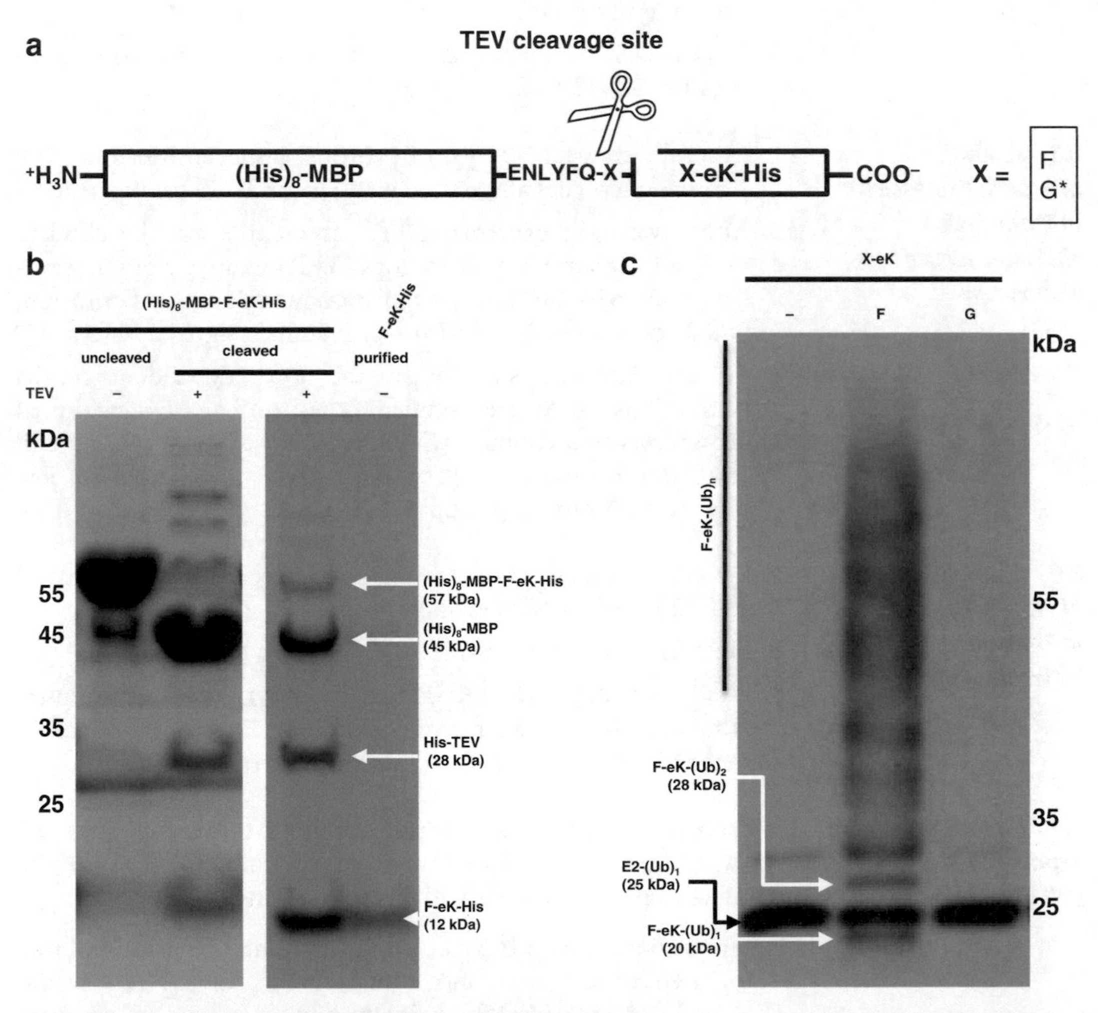

Fig. 4 TEV cleavage and ubiquitination of artificial NERD substrates. (**a**) Fusion protein used to generate desired N-termini at the eK-His artificial substrate using TEV protease. (**b**) Western blot (α-His, sc-803) after SDS-PAGE monitoring the efficiency of the TEV cleavage of the $(His)_8$-MBP-F-eK-His fusion protein and purified artificial substrate F-eK, (1) purified fusion protein, (2) and (3) reaction mixture after TEV cleavage of the fusion protein, and (4) final F-eK-His substrate after further purification first via amylose resin, second via Ni-NTA agarose, then via centrifugal filter units to remove $(His)_8$-MBP, uncleaved fusion protein and the His-tagged TEV protease. (**c**) In vitro ubiquitination assay of the purified F-eK, with G-eK as negative control on an anti-Ub western blot (α-Ub, sc-8017). *: not per se destabilizing

His tag directly followed by purification using the MBP tag. The expression volume is dependent on the amount of protein which is needed. We usually use 200 mL of cultures in 500 mL Erlenmeyer flasks. The expression vectors pVP16-tev-POI are transformed into *E. coli* BL21-CodonPlus(DE3)-RIL.

1. 4 mL of overnight preculture are prepared and 1 mL added to the main expression culture containing 200 mL of LB medium (1:200 inoculation) supplemented with appropriate antibiotics, in our case Carbenicillin at 50 μg/mL.
2. Grow the culture until OD_{600} of 0.3 (*see* **Note 7**).
3. Induce the expression by adding 200 μL of 1 M IPTG to 200 mL of main culture, shift the cultures to 20 °C and express for 16–18 h.
4. Harvest the bacteria by centrifugation (3500 × *g*, 4 °C, 10 min) and resuspend the pellet in 10 mL of Ni-buffer. The pellet can be stored at −20 °C for months.
5. To break down the cell walls, add 250 μL of lysozyme solution for 1 h on ice followed by cell disruption by French press. Immediately add 100 μL of 100 mM PMSF (for 1 mM PMSF final concentration) to inhibit proteases.
6. Centrifuge the lysate at 18,500 × *g* for 25 min, load the clear supernatant to the Ni-NTA agarose (nickel-charged resin) column of 1 mL bed volume equilibrated with Ni-buffer. In parallel, equilibrate an amylose column (1 mL bed volume) with amylose buffer for the second purification step. To get a better binding to the column, we recommend letting the supernatant flow through the column for three times.
7. Wash the column with 15 mL of Ni-buffer.
8. Equilibrated the amylose column with amylose buffer.
9. Elute the proteins with 5 mL of Ni-elution buffer containing 200 mM imidaziole and directly load the eluate to the amylose column which was previously equilibrated in the same buffer. Again, to achieve better binding let it pass three times over the column.
10. Wash column with 15 mL of amylose buffer.
11. Elute proteins in five fractions of 500 μL with amylose-elution buffer containing 10 mM maltose.
12. To prove that the protein was purified properly, we recommend taking samples from each step and monitoring the purity via SDS-PAGE or western blot, both procedures are described below.
13. The material is ready for TEV-mediated digestion and purification.

3.9 TEV Cleavage

1. Determine the OD_{280} of the eluate of the amylose column and of the TEV stock solution (*see* **Note 8**).
2. Mix fusion protein solution with TEV protease solution in a 100:1–50:1 ratio of OD_{280} and incubate for 16–18 h at 4 °C.
3. The reaction mix is loaded on centrifugal filter units to replace amylose-buffer containing maltose by amylose buffer lacking with maltose. The reaction mix is reduced by centrifugation to a volume of 1 mL and then diluted into 15–20 mL of amylose buffer without maltose. The reaction mixture can further be dialyzed against 3 L of amylose buffer for 6–16 h to remove the maltose.
4. Load the mix to an amylose column equilibrated with amylose buffer, here again to increase binding of MBP tag and uncleaved fusion protein, let it pass the column three times.
5. Harvest the flow through and elute the $(His)_8$-MBP tag and uncleaved fusion protein with amylose-elution buffer containing 10 mM maltose to clear the column and re-equilibrated.
6. Repeat the procedure three times to get a pure X-POI.
7. The flow through can be purified further by using a Ni-NTA column again, then also the His-tagged TEV protease is eliminated from the protein mix together with desired X-POI, in our case F/G-eK-His (Fig. 4a, b).
8. Due to the size difference, we recommend to use centrifugal filter units to separate and concentrate the proteins (*see* **Note 9**). Using 30 kDa cut-off centrifugal filters will retain the TEV protease but allow the F-eK-His to pass-through and this can be finally concentrated using 10 kDa cut-off centrifugal filters (*see* **Note 10**).
9. Check protein concentration and purity on SDS-PAGE and western blot.

3.10 Protein Concentration Determination

Use your favorite method to determine the overall protein content of the cleared supernatant and follow the manufacturer's instructions. A decent BSA calibration curve is received when a protein standard such as BSA (stock is 1 mg/mL) is added in steps of 0, 2, 4, 6, 8, 10, 12, 16, 20, 24 μL + 2 μL extraction buffer to each tube. The concentrations usually represent the concentrations present in the supernatants to be tested. The linear range of the assay for BSA is 0.2–0.9 mg/mL.

1. Prepare as many cuvettes as needed with protein assay components (800 μL of water, 200 μL of protein assay solution).
2. Add 2 μL of each sample to the sample cuvettes.
3. Invert cuvettes wearing gloves. Make sure not to carry-over any protein from one assay to the other by wiping the gloves with paper towels.

4. Incubate at room temperature for at least 5 min. Absorbance will increase over time; samples should incubate at room temperature for no more than 1 h.
5. Measure the absorbance at 595 nm.
6. Plot the standard and the sample values to extrapolate the protein concentrations.

3.11 SDS-Polyacrylamide Gel Electrophoresis (SDS-PAGE)

We recommend the use of mini gels (e.g., from Bio-Rad) for optimal separation of the protein at a thickness of 1.5 mm, the gel is also relatively robust for the subsequent handling.

1. Place the fully assembled SDS-PAGE gel apparatus onto the bench and rinse carefully all the gel pockets that should be used. It helps to run some 2× sample buffer on top of the pockets to visualize where improperly polymerized or clogging polyacrylamide is remaining.
2. Load the molecular weight marker in one pocket. Also load those wells that remain empty and are directly adjacent to the sample wells with SDS-PAGE sample buffer. We use 200-μL gel-loading tips with extended length to load the samples onto a gel. This helps to avoid spill-over.
3. Load the samples carefully into the gel.
4. Run the gel according to the manufacturer's instructions. We set the power supply initially to 85 V and after 15 min increase the voltage to 135 V for 1 h.
5. Switch off the power supply and disassemble the apparatus.
6. Cut between the stacking and the separating gel to remove and discard the stacking gel and cut the remaining part vertically if not all the pockets of the gel have been loaded to remove and discard the empty lanes. Also cut off the dye front and dispose it.
7. Rinse the gel briefly with semi-dry buffer.

3.12 Western Transfer of Proteins and Detection

In the following part, the proteins are blotted to PVDF, probed with the antibodies and detected via chemoluminescence. This allows detection of proteins in extracts of protoplasts, cell-free systems and after TEV cleavage.

1. For each blot, prepare the required number of sheets of Whatman filter paper (five sheets of 0.34 mm thick blotting paper (Whatman 3MM Chr), three sheets of 0.92 mm thick paper (Whatman 17Chr), or two sheets of 1.2 mm thick paper (Whatman GB005)) and one sheet of PVDF membrane of the appropriate size, slightly larger than the gel.
2. Activate the membrane for 2–5 min in methanol.
3. In a clean, fat-free tray, allow the cut gels and filter papers to wet by capillary action and equilibrate the membrane.

4. Assemble the lower part of the blotting sandwich: start with the bottom of gel cassette that will face the anode, filter papers, membrane, gel, filter paper.
5. Make sure to keep all the layers moist and take precautions not to include air bubbles in the setup. Roll out the air in all the layers using a rinsed test tube or glass rod.
6. Activate the power supply and transfer at roughly 1 mA/cm^2 gel size (height × width × 0.65 = mA/gel) for 1–2 h.
7. Once the transfer is complete, open blotting machine, carefully take the membrane into 10 mL of blocking solution (e.g., 5 % (w/v) dry milk in TBST) for 1 h at room temperature or overnight at 4 °C on a rocking platform.
8. Discard the buffer and rinse the membrane quickly in TBST prior to addition of the primary antibody (Table 4).
9. Incubate with primary antibody for 1 h at room temperature while rocking.
10. Remove the primary antibody and wash the membrane three times for 5 min each with 20 mL of TBST.
11. Freshly prepare the secondary horseradish-conjugated antibody (Table 4) and add to the membrane for 1 h at room temperature on a rocking platform.
12. Discard the secondary antibody and wash the membrane three times for 10 min each with TBST.
13. After the final wash, briefly dry the membrane on filter paper and label properly if not done previously.
14. Line the X-ray film cassette with Saran wrap or a plastic bag and position the blot into a wrap or plastic pocket to separate the membrane to be soaked in ECL detection reagent from the film.
15. Mix the ECL reagents according to the instructions, here in a ratio of 1:1, and immediately spread it over the membrane. Ensure even coverage.
16. Squeeze out excess liquid, blot with tissue paper and move towards the dark room with safe light conditions.
17. Expose the first film for a suitable exposure time, typically 30 s, and determine optimal exposure time later on.

4 Notes

1. TEV works here because other proteases such as thrombin and PreScission are characterized by an invariant P1′ residue and therefore may only be used to generate a small set of predefined freshly formed N-termini: thrombin cleaves preferentially

between Arg and Gly of LVPR|GS, PreScission protease between Gln and Gly of LEVLFQ|GP. Theoretically, also enterokinase (cleaves preferentially after Lys at DDDDK|X and at other basic residues, but not at the site if followed by Pro) and factor Xa (cleaves preferentially after Arg at IG/DGR|X and at other basic residues, but not at the site if followed by Pro or Arg) should work to generate artificial NERD substrates.

2. Home-made TEV is produced as follows: *E. coli* BL21(DE3)-RIL cells containing pRK793 (Table 2) are grown at 37 °C in LB containing 100 μg/mM ampicillin or 50 μg/mM carbenicillin (for pRK793) and 30 μg/mL chloramphenicol (for pRIL of the *E. coli* strain). When the cells reach mid log phase (OD_{600} of approx. 0.5), IPTG is added to a final concentration of 1 mM and the temperature is reduced to 30 °C. After 4 h of induction, the cells are collected by centrifugation. The protocol for TEV purification is as follows [53]: Dissolve the cell pellet in 10 mL of lysis buffer (50 mM NaH_2PO_4 (pH 8.0), 100 mM NaCl, 10% (v/v) glycerol, 25 mM imidazole) per 1 g of wet cell paste. Lyse the cells with a Manton-Gaulin homogenizer at 10,000–10,500 psi for three passes. 5% polyethelene imine (adjusted to pH 7.9 with HCl) are added to a final concentration of 0.1%, mixed by inversion and immediately centrifuged at 15,000 g for 30 min. The supernatant is run through a Ni-NTA column equilibrated with lysis buffer, using approx. 2 mL of resin per gram of wet cell paste. The column is washed with 7 volumes of lysis buffer and the TEV protease eluted with a linear gradient of lysis buffer to elution buffer (50 mM NaH_2PO_4 (pH 8.0), 100 mM NaCl, 10% (v/v) glycerol, 200 mM imidazole) in 10 column volumes overall. The appropriate fractions (tested by SDS-PAGE for presence of TEV) are pooled and EDTA and DTT added to a final concentration of 1 mM each. The TEV is concentrated using Amicon Ultra-15 (Merck Millipore) with 20 kDa cut-off. Concentrate the protease to approx. 1 mg/mL and flash freeze in liquid nitrogen. Store at −80 °C and test aliquots for efficiency before using.

3. Acrylamide is a neurotoxin and carcinogenic in the unpolymerized form. Even the polymerized gel contains still traces of it and thus, it has to be handled very carefully and spills and contamination avoided. Any waste has to be disposed off accordingly.

4. TEMED is best stored at room temperature in a desiccator. Buy small bottles as it may decline in quality after opening and thus, gels will take longer to polymerize.

5. To prevent bubbles at the border of the separating gel, it is overlaid with either isopropanol or water-saturated isobutanol. For the latter, shake equal volumes of water and isobutanol in

a glass bottle and allow separation. Use the top layer. Store at room temperature.

6. Calculation of required MMG volume: $V = (x \times 10^4 \times 2) / (2 \times 10^5) = x/10$ with x = average of number of protoplasts counted in four squares.
7. Consider that different strains behave different in growth speed.
8. The TEV protease catalytic domain is expressed from pRK793 (Table 2) with an N-terminal His-tag and a C-terminal polyarginine tag. The His-tag can be used to eliminate TEV from X-POI after cleavage.
9. Due to its apparent molecular volume and protein shape, TEV does not pass through 30 kDa cut-off centrifugal filter units, even though its size is 28 kDa. Thus, filter units of 10 kDa should be considered.
10. Amicon Ultra 15 centrifugal filters have a maximal loading volume of 15 mL.
11. The pVP16 Gateway bacterial (*E. coli*) expression vector was designed as a $(His)_8$-MBP fusion tag system to overcome the low solubility of recombinant eukaryotic proteins and to provide a generic Ni-IMAC purification strategy. The backbone is derived from pQE80 (Qiagen) to express an N-terminal fusion protein consisting of $(His)_8$-MBP and a linker region containing the TEV protease recognition site contiguous with the first residue of the target protein. TEV protease cleavage site in pVP16 is "partial" and not functional. Therefore, the full TEV cleavage site must be added directly 5′ to the first codon of the insert by PCR, which allows amino acids encoded by the attB1 site to be cut off after protein expression.

Acknowledgments

The authors thank Carolin Mai for critical reading and helpful comments on the manuscript and Andreas Bachmair for sharing pRTUB8 containing the plant codon-optimized synthetic human Ub gene. This work was supported by a grant for setting up the junior research group of the *ScienceCampus Halle—Plant-based Bioeconomy* to N.D., by a Ph.D. fellowship of the Landesgraduiertenförderung Sachsen-Anhalt awarded to C.N., and a grant of the Leibniz-DAAD Research Fellowship Programme by the Leibniz Association and the German Academic Exchange Service (DAAD) to A.C.M. and N.D. Financial support came from the Leibniz Association, the state of Saxony Anhalt, the Deutsche Forschungsgemeinschaft (DFG) Graduate Training Center GRK1026 "*Conformational Transitions in Macromolecular*

Interactions" at Halle, and the Leibniz Institute of Plant Biochemistry (IPB) at Halle, Germany. To complete work on this project, a Short Term Scientific Mission (STSM) of the European Cooperation in Science and Technology (COST) was granted to A.C.M. and N.D. by the COST Action BM1307—"*European network to integrate research on intracellular proteolysis pathways in health and disease (PROTEOSTASIS)*".

References

1. Graciet E, Wellmer F (2010) The plant N-end rule pathway: structure and functions. Trends Plant Sci 15:447–453
2. Varshavsky A (2011) The N-end rule pathway and regulation by proteolysis. Protein Sci 8:1298–1345
3. Tasaki T, Sriram SM, Park KS, Kwon YT (2012) The N-end rule pathway. Annu Rev Biochem 81:261–289
4. Gibbs DJ, Bacardit J, Bachmair A, Holdsworth MJ (2014) The eukaryotic N-end rule pathway: conserved mechanisms and diverse functions. Trends Cell Biol 24(10):603–611
5. Dougan DA, Truscott KN, Zeth K (2010) The bacterial N-end rule pathway: expect the unexpected. Mol Microbiol 76(3):545–558
6. Brower CS, Piatkov KI, Varshavsky A (2013) Neurodegeneration-associated protein fragments as short-lived substrates of the N-end rule pathway. Mol Cell 50(2):161–171
7. Gibbs DJ, Lee SC, Isa NM, Gramuglia S, Fukao T, Bassel GW, Correia CS, Corbineau F, Theodoulou FL, Bailey-Serres J, Holdsworth MJ (2011) Homeostatic response to hypoxia is regulated by the N-end rule pathway in plants. Nature 479:415–418
8. Licausi F, Kosmacz M, Weits DA, Giuntoli B, Giorgi FM, Voesenek LA, Perata P, van Dongen JT (2011) Oxygen sensing in plants is mediated by an N-end rule pathway for protein destabilization. Nature 479:419–422
9. Mendiondo GM, Gibbs DJ, Szurman-Zubrzycka M, Korn A, Marquez J, Szarejko I, Maluszynski M, King J, Axcell B, Smart K, Corbineau F, Holdsworth MJ (2016) Enhanced waterlogging tolerance in barley by manipulation of expression of the N-end rule pathway E3 ligase PROTEOLYSIS6. Plant Biotechnol J 14:40. doi:10.1111/pbi.12334
10. Potuschak T, Stary S, Schlögelhofer P, Becker F, Nejinskaia V, Bachmair A (1998) PRT1 of Arabidopsis thaliana encodes a component of the plant N-end rule pathway. Proc Natl Acad Sci U S A 95(14):7904–7908
11. Stary S, Yin X-J, Potuschak T, Schlögelhofer P, Nizhynska V, Bachmair A (2003) PRT1 of Arabidopsis is a ubiquitin protein ligase of the plant N-end rule pathway with specificity for aromatic amino-terminal residues. Plant Physiol 133(3):1360–1366
12. Garzón M, Eifler K, Faust A, Scheel H, Hofmann K, Koncz C, Yephremov A, Bachmair A (2007) PRT6/At5g02310 encodes an Arabidopsis ubiquitin ligase of the N-end rule pathway with arginine specificity and is not the CER3 locus. FEBS Lett 581(17):189–196
13. Yoshida S, Ito M, Callis J, Nishida I, Watanabe A (2002) A delayed leaf senescence mutant is defective in arginyl-tRNA:protein arginyltransferase, a component of the N-end rule pathway in Arabidopsis. Plant J 32(1):129–137
14. Graciet E, Walter F, Ó'Maoiléidigh DS, Pollmann S, Meyerowitz EM, Varshavsky A, Wellmer F (2009) The N-end rule pathway controls multiple functions during Arabidopsis shoot and leaf development. Proc Natl Acad Sci U S A 106(32):13618–13623
15. Weits DA, Giuntoli B, Kosmacz M, Parlanti S, Hubberten HM, Riegler H, Hoefgen R, Perata P, van Dongen JT, Licausi F (2014) Plant cysteine oxidases control the oxygen-dependent branch of the N-end-rule pathway. Nat Commun 5:3425. doi:10.1038/ncomms4425
16. Sherman F, Stewart JW, Tsunasawa S (1985) Methionine or not methionine at the beginning of a protein. Bioessays 3:27–31
17. Meinnel T, Serero A, Giglione C (2006) Impact of the N-terminal amino acid on targeted protein degradation. Biol Chem 387(7):839–851
18. Bienvenut WV, Sumpton D, Martinez A, Lilla S, Espagne C, Meinnel T, Giglione C (2012) Comparative large scale characterization of plant versus mammal proteins reveals similar and idiosyncratic N-α-acetylation features. Mol Cell Proteomics 11:M111.015131
19. Giglione C, Serero A, Pierre M, Boisson B, Meinnel T (2000) Identification of eukaryotic peptide deformylases reveals universality of

N-terminal protein processing mechanisms. EMBO J 19(21):5916–5929

20. Giglione C, Vallon O, Meinnel T (2003) Control of protein life-span by N-terminal methionine excision. EMBO J 22(1):13–23
21. Ross S, Giglione C, Pierre M, Espagne C, Meinnel T (2005) Functional and developmental impact of cytosolic protein N-terminal methionine excision in Arabidopsis. Plant Physiol 137:623–637
22. Frottin F, Martinez A, Peynot P, Mitra S, Holz RC, Giglione C, Meinnel T (2006) The proteomics of N-terminal methionine cleavage. Mol Cell Proteomics 12:2336–2349
23. Venne AS, Solari FA, Faden F, Pareti T, Dissmeyer N, Zahedi RP (2015) An improved workflow for quantitative N-terminal ChaFRADIC to study proteolytic events in Arabidopsis thaliana seedlings. Proteomics 15:2458
24. Whitcomb DC, Lowe ME (2007) Human pancreatic digestive enzymes. Dig Dis Sci 52(1):1–17
25. Bachmair A, Finley D, Varshavsky A (1986) In vivo half-life of a protein is a function of its amino-terminal residue. Science 234(4773): 179–186
26. Varshavsky A (2000) Ubiquitin fusion technique and its descendants. Methods Enzymol 327:578–593
27. Varshavsky A (2005) Ubiquitin fusion technique and related methods. Methods Enzymol 399:777–799
28. Finley D, Ozkaynak E, Varshavsky A (1987) The yeast polyubiquitin gene is essential for resistance to high temperatures, starvation, and other stresses. Cell 48(6):1035–1046
29. Baker RT (1996) Protein expression using ubiquitin fusion and cleavage. Curr Opin Biotechnol 7(5):541–546
30. Gilchrist CA, Gray DA, Baker RT (1997) A ubiquitin-specific protease that efficiently cleaves the ubiquitin-proline bond. J Biol Chem 272(51):32280–32285
31. Piatkov K, Graciet E, Varshavsky A (2013) Ubiquitin reference technique and its use in ubiquitin-lacking prokaryotes. PLoS One 8(6): e67952
32. Bachmair A, Becker F, Schell J (1993) Use of a reporter transgene to generate arabidopsis mutants in ubiquitin-dependent protein degradation. Proc Natl Acad Sci U S A 90(2):418–421
33. Yoo S, Cho Y, Sheen J (2007) Arabidopsis mesophyll protoplasts: a versatile cell system for transient gene expression analysis. Nat Protoc 2(7):1565–1572
34. Maldonado-Bonilla L, Eschen-Lippold L, Gago-Zachert S, Tabassum N, Bauer N, Scheel D, Lee J (2014) The Arabidopsis tandem zinc finger 9 protein binds RNA and mediates pathogen-associated molecular pattern-triggered immune responses. Plant Cell Physiol 55(2):412–425
35. Kwon Y, Kashina A, Davydov I, Hu R, An JY, Seo JW (2002) An essential role of N-terminal arginylation in cardiovascular development. Science 297(5578):96–99
36. Sheng J, Kumagai A, Dunphy WG, Varshavsky A (2002) Dissection of c-MOS degron. EMBO J 21(22):6061–6071
37. Lee MJ, Tasaki T, Moroi K, An JY, Kimura S, Davydov IV, Kwon YT (2005) RGS4 and RGS5 are in vivo substrates of the N-end rule pathway. Proc Natl Acad Sci U S A 102(42): 15030–15035
38. Carrington JC, Dougherty WG (1988) A viral cleavage site cassette: identification of amino acid sequences required for tobacco etch virus polyprotein processing. Proc Natl Acad Sci U S A 85:3391–3395
39. Parks TD, Leuther KK, Howard ED, Johnston SA, Dougherty WG (1994) Release of proteins and peptides from fusion proteins using a recombinant plant virus proteinase. Anal Biochem 216:413–417
40. Kapust RB, Tozser J, Copeland TD, Waugh DS (2002) The P1′ specificity of tobacco etch virus protease. Biochem Biophys Res Commun 294:949–955
41. Phan J, Zdanov A, Evdokimov AG, Tropea JE, Peters HK, Kapust RB, Li M, Wlodawer A, Waugh DS (2002) Structural basis for the substrate specificity of tobacco etch virus protease. J Biol Chem 277:50564–50572
42. Taxis C, Stier G, Spadaccini R, Knop M (2009) Efficient protein depletion by genetically controlled deprotection of a dormant N-degron. Mol Syst Biol 5:267
43. Taxis C, Knop M (2012) TIPI: TEV protease-mediated induction of protein instability. Methods Mol Biol 832:611–626
44. Frey S, Görlich D (2014) A new set of highly efficient, tag-cleaving proteases for purifying recombinant proteins. J Chromatogr A 1337: 95–105
45. Bachmair A, Varshavsky A (1989) The degradation signal in a short-lived protein. Cell 56:1019–1032
46. Jungbluth M, Renicke C, Taxis C (2010) Targeted protein depletion in Saccharomyces cerevisiae by activation of a bidirectional degron. BMC Syst Biol 4:176

47. Verhoeven KD, Altstadt OC, Savinov SN (2012) Intracellular detection and evolution of site-specific proteases using a genetic selection system. Appl Biochem Biotechnol 166:1340–1354
48. Yi L, Sun H, Itzen A, Triola G, Waldmann H, Goody RS, Wu YW (2011) One-pot dual-labeling of a protein by two chemoselective reactions. Angew Chem Int Ed Engl 50:8287–8290
49. Renicke C, Spadaccini R, Taxis C (2013) A tobacco etch virus protease with increased substrate tolerance at the P1′ position. PLoS One 8:e67915
50. Laemmli UK (1970) Cleavage of structural proteins during the assembly of the head of bacteriophage T4. Nature 227:680–685
51. Harlow E, Lane D (1988) Antibodies: a laboratory manual. Cold Spring Harbor Laboratory, Cold Spring Harbor, NY
52. Bachmair A, Becker F, Masterson RV, Schell J (1990) Perturbation of the ubiquitin system causes leaf curling, vascular tissue alterations and necrotic lesions in a higher plant. EMBO J 9:4543–4549
53. Kapust RB, Tozser J, Fox JD, Anderson DE, Cherry S, Copeland TD, Waugh DS (2001) Tobacco etch virus protease: mechanism of autolysis and rational design of stable mutants with wild-type catalytic proficiency. Protein Eng 14(12):993–1000
54. Thao S, Zhao Q, Kimball T, Steffen E, Blommel PG, Riters M, Newman CS, Fox BG, Wrobel RL (2004) Results from high-throughput DNA cloning of Arabidopsis thaliana target genes using site-specific recombination. J Struct Funct Genomics 5(4):267–276
55. Grefen C, Donald N, Hashimoto K, Kudla J, Schumacher K, Blatt MR (2010) A ubiquitin-10 promoter-based vector set for fluorescent protein tagging facilitates temporal stability and native protein distribution in transient and stable expression studies. Plant J 64(2):355–365

Chapter 7

Peptide Arrays for Binding Studies of E3 Ubiquitin Ligases

Maria Klecker and Nico Dissmeyer

Abstract

The automated SPOT (synthetic peptide arrays on membrane support technique) synthesis technology has entrenched as a rapid and robust method to generate peptide libraries on cellulose membrane supports. The synthesis method is based on conventional Fmoc chemistry building up peptides with free N-terminal amino acids starting at their cellulose-coupled C-termini. Several hundreds of peptide sequences can be assembled with this technique on one membrane comprising a strong binding potential due to high local peptide concentrations. Peptide orientation on SPOT membranes qualifies this array type for assaying substrate specificities of N-recognins, the recognition elements of the N-end rule pathway of targeted protein degradation (NERD). Pioneer studies described binding capability of mammalian and yeast enzymes depending on a peptide's N-terminus. SPOT arrays have been successfully used to describe substrate specificity of N-recognins which are the recognition elements of the N-end rule pathway of targeted protein degradation (NERD). Here, we describe the implementation of SPOT binding assays with focus on the identification of N-recognin substrates, applicable also for plant NERD enzymes.

Key words SPOT assay, N-end rule, Ubiquitin ligase, Protein-protein interaction, ResPep SL, Substrate screen, Peptide library

1 Introduction

1.1 Peptide Arrays for Protein Interaction Studies

Screening of protein activity and elucidating binding parameters are at the basis of enzyme characterization and functional assignment. However, the identification of substrates, their enzyme-binding sites or of potent inhibitors may imply extensive work of cloning, mutant generation, protein expression, and purification when performed on full-length recombinant proteins. Another drawback of many rather classical protein–protein interaction assays such as co-immunoprecipitation or pull-downs is the often intrinsically low binding affinity between enzyme and substrate. For this reason, the use of synthetic peptide libraries on membrane supports with high local peptide concentrations at each "spot" has become increasingly popular for protein interaction studies since the full automation of SPOT synthesis with robots was launched [1].

L. Maria Lois and Rune Matthiesen (eds.), *Plant Proteostasis: Methods and Protocols*, Methods in Molecular Biology, vol. 1450, DOI 10.1007/978-1-4939-3759-2_7,

Publications involving SPOT assays for mapping of epitopes, kinase, and phosphatase interaction sites or for protease substrate identification run into the hundreds [1–4]. In the last decade, the method was further seized for the study of enzymes which are functionally comprised by the pathway of N-end rule degradation (NERD) [5–8]. This brings about a new type of SPOT design which contains peptides derived only from the very N-termini of substrate candidates and opens up the technique for the application on enzymes such as E3 ubiquitin ligases.

1.2 SPOT Assays on N-End Rule Enzymes

The N-end rule of protein degradation relates the in vivo half-life of a protein to the nature of its N-terminal amino acid [9, 10]. In eukaryotes, the machinery behind involves a variety of enzyme classes including Met-aminopeptidases, N-terminal acetyltransferases, amidases, cysteine oxidases, Arg-tRNA transferases, and E3 ubiquitin ligases, all of which appear to select their substrates chiefly based on their very N-terminal residues.

On a SPOT membrane, the lengths of the arrayed peptides usually limit their ability to adopt the correct conformation. Furthermore, in vivo peptide binding may be dependent on the cellular environment (membranes, scaffold proteins) which differs from the situation on the SPOT membrane. However, according to the current understanding, the ability of a given N-terminus to act as a NERD degradation signal, i.e. as an N-degron, does not require the adoption of a higher order structure over the primary sequence, but is in contrast promoted when a disordered region allows for flexibility and protrusion of the N-terminal degron [13]. This condition of substrate recognition by NERD components predestines these enzymes for the study by the SPOT method.

So far, SPOT assays have been applied to characterize the substrate specificity of the ClpAP-specific adapter protein ClpS from *E. coli* [5] and of the unique NERD ubiquitin ligase Ubr1p from *Saccharomyces cerevisiae* [6, 7]. Furthermore, the importance of the penultimate amino acid for recognition by NERD ubiquitin ligases in *S. cerevisiae* and mouse was revealed by SPOT assay application [8].

The peptide synthesis using the ResPep SL method, side chain de-protection, and general protein-protein interaction screening was recently protocolled [11]. Here, it will be described how to perform binding assays on SPOT membranes with NERD enzymes, particularly E3 ubiquitin ligases. The presented protocol applies to peptide arrays synthesized by the ResPep SL method on acid-hardened cellulose membranes, derivatized with polyethylene glycol (PEG) spacers.

2 Materials

If not indicated otherwise, all solutions should be prepared freshly before every experiment. The studied protein should be available in a microgram scale as a purified, active enzyme solution of known concentration. If specific antibodies are unavailable, an epitope tag for detection should be strongly considered (*see* **Notes 1** and **2**).

2.1 General Equipment and Infrastructure

1. Synthesized SPOT array on PEG-functionalized cellulose (*see* **Notes 3**, **4**, and **5**).
2. Rocking platform for incubation at room temperature and in a 4 °C environment.
3. Rolling device or a tube rotator for 3D shaking of tubes (e.g. SB3 (Stuart) equipped with appropriate tube holders).
4. Vapor-proof polypropylene (PP) box (e.g. a sealable lunch box).
5. Polyethylene boxes of the area of the SPOT membrane and about 2 cm in height (e.g. Hartenstein, #AD01) (*see* **Note 6**).
6. Flat tweezers for membrane transfer.
7. Pipettes accurately delivering 2.5 or up to 200 μL, depending on the concentration of the stored enzyme solution.
8. Flasks with screw caps comprising at least 250 mL.

2.2 Membrane Activation and Blocking

1. Methanol or ethanol.
2. Washing/binding buffer: any buffer system in which the protein of interest (POI) is known to be active; alternatively: TBST (20 mM Tris–HCl, pH 7.4, 135 mM NaCl, 0.1% (v/v) Tween 20) can be tried as the basis for a binding buffer and stored as a 2× stock solution. According to protein requirements, it should be supplemented before use with reducing agents (such as dithiothreitol) and compatible solutes (*see* **Notes 7**, **8**, and **9**).
3. Blocking buffer: Binding buffer (*see* **item 2**; **Note 10**) containing a blocking agent (e.g. 3% (w/v) PVP40, 3% (w/v) BSA or up to 10% (w/v) milk powder).

2.3 Enzyme Binding on SPOT Membranes

1. Binding buffer (*see* **item 2** in Subheading 2.2) (*see* **Notes 7**, **8**, and **11**).
2. Optional: protease inhibitors. If the purified protein samples used for binding assays might contain protease contaminations or the ambient air bears high titers of bacterial or fungal contaminants, use of protease inhibitors is indicated as follows: 1 mg/mL Pepstatin A (Fluka, store at −80 °C), 1 mg/mL 4-(2-aminoethyl)benzenesulfonyl fluoride hydrochloride (AEBSF, Santa Cruz, sc-202041B; store at −80 °C).

2.4 Western Blotting of SPOT-Array-Bound Proteins

1. Cathode buffer: 25 mM Tris-base, 40 mM 6-aminohexanoic acid (6-aminocaproic acid/ε-aminocaproic acid, Santa Cruz, sc-202146), 0.01% (w/v) SDS, 20% (v/v) methanol (pH of the solution should be 9.4).
2. Anode buffer I: 30 mM Tris-base, 20% (v/v) methanol, (pH should be 10.4).
3. Anode buffer II: 300 mM Tris-base, 20% (v/v) methanol.
4. Semi-dry blot apparatus (such as Bio-Rad Trans-Blot SD Semi-Dry Transfer Cell) with power supply.
5. Filter paper (Whatman).
6. PVDF membrane (Amersham/GE Healthcare).

3 Methods

The experiment takes in total about 2 days.

3.1 Membrane Activation and Blocking

Before performing each binding assay, the membrane must be activated in order to fully hydrate the coupled peptides. This can be done directly after side chain de-protection (for method refer to [18]) before an assay.

1. If the membrane was stored de-protected at refrigerated conditions, let it reach room temperature before exposing it to the procedure.
2. Place the membrane into a vapor-tight box and cover it completely with methanol or ethanol and incubate until the visible spots disappear (at least 1 min, up to half an hour) (*see* **Note 12**).
3. Wash and equilibrate the membrane at least three times for 10 min with binding buffer. Be quick upon transfer from alcohol to the buffer since methanol and ethanol are volatile and the membrane must be prevented from drying at this step (*see* **Note 13**).
4. Cover the membrane completely with appropriate blocking solution and incubate on a rocker for at least 1 h at room temperature or overnight at 4 °C (*see* **Notes 14** and **10**).

3.2 Enzyme Binding on SPOT Membranes

Ubiquitin E3 ligases are characterized by relatively high dissociation constants [19] and one may face problems when trying to approach their substrate binding capacities by pull-down experiments. This issue is circumvented in a SPOT binding assay through extremely high local substrate concentrations (*see* Fig. 1). In epitope mapping experiments, peptide–antibody interactions with dissociation constants as high as 1–0.1 mM remained detectable [2].

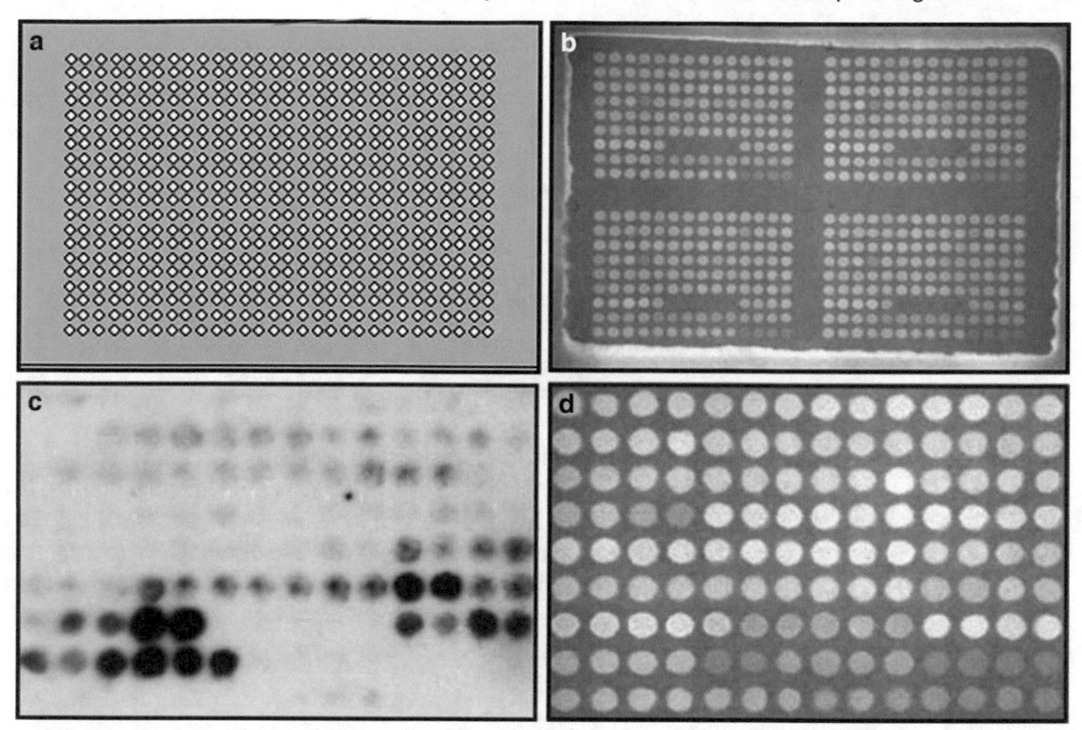

Fig. 1 SPOT array seen at different stages of synthesis and assay. (**a**) Shown is the standard grid of 600 positions available for peptide spots to be synthesized on one cellulose membrane predefined by the ResPepSL software. (**b**) Example of one membrane divided into four identical subsets of each 126 spots seen under UV light directly after SPOT synthesis. The peptides appear as light or gray spots according to the presence of UV absorbing groups of the amino acid residues and side chain protection groups. (**c**) Immunodetection of a recombinant enzyme from the plant NERD after binding assay with one of the four membranes shown in (**b**) and electrotransfer. (**d**) UV light view of the SPOT membrane used in (**c**) after side chain de-protection

1. Prepare at least 100 mL of the binding buffer.
2. Add your recombinant enzyme of interest at a starting concentration of 50 nM (*see* **Note 15**) to a volume of binding buffer that is sufficient to completely cover the membrane. Incubate the solution for 5 min under rotation.
3. Add the binding buffer containing enzyme to the blocked membrane and incubate for 1–3 h on a rocking platform at room temperature (*see* **Notes 16** and **17**).
4. Wash the membrane 3–6 times for 10 min with washing buffer.

3.3 Western Blotting of SPOT-Array-Bound Proteins

In order to immobilize the relatively weak binding enzyme for antibody incubations, the NERD SPOT assay usually involves electrotransfer of the bound protein to a PVDF membrane. Here, the three buffer blot system according to [21] is the method of choice for semi-dry electrotransfer (Fig. 2). Unlike when blotting an SDS-PAGE gel, the blotted protein from a SPOT membrane is only very briefly incubated with the transfer buffer, including only mild

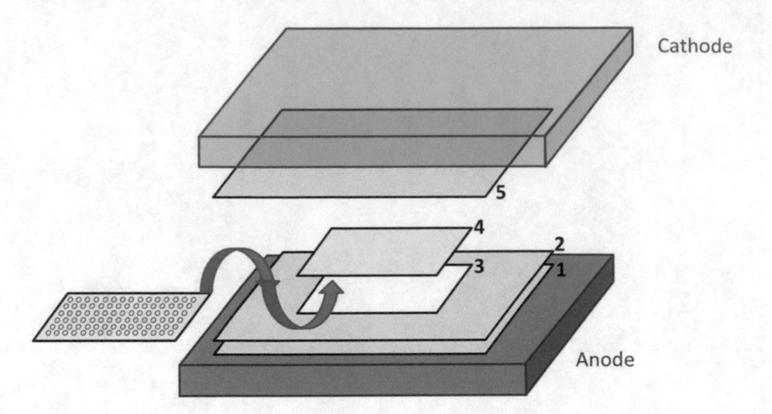

Fig. 2 Assembly of the semi-dry western blot using the three buffer blot system. Up to three layers of filter papers soaked in each Anode II buffer (*1*) and Anode I buffer (*2*) are placed on top of the anode (only one layer each is shown). The activated PVDF membrane (*3*) is equilibrated in Anode I buffer and faces the SPOT side of the SPOT membrane (*4*) bearing the POI. This is covered by again up to three filter papers soaked in cathode buffer (*5*) and mounted by the cathode. The system was described by Kyhse-Anderson, 1984

charging by SDS, and can be considered to be in a partly native form. For this reason, the migration properties of one enzyme might differ depending on its stability and isoelectric point.

1. Pre-wet up to three filter papers in each cathode, anode I, and anode II buffer.
2. Activate a PVDF membrane by incubation in methanol for at least 1 min.
3. Equilibrate the PVDF membrane in anode I buffer.
4. Equilibrate the POI-charged SPOT membrane briefly (max. 1 min) in cathode buffer.
5. Assemble the blot sandwich as follows and strictly avoid air bubbles between all layers:
    - Filter papers with anode II buffer facing the anode.
    - Up to three layers of filter papers soaked with anode I buffer.
    - PVDF membrane.
    - SPOT membrane with the protein-bound site facing the PVDF membrane.
    - Up to three layers of filter papers soaked in cathode buffer facing the cathode.
6. Perform the electrotransfer for 30 min at 0.8 mA/cm^2 of SPOT membrane.

7. Transfer the PVDF membrane to your favored western blot blocking solution (*see* **Note 18**).
8. Transfer the SPOT membrane to distilled water for subsequent stripping and reuse (for protocol, *see* [22]).

The PVDF membrane can be treated with antibodies specific to the target protein and detected like a normal western blot.

4 Notes

1. When using affinity tags, be aware of the fact that those may have binding capacity to parts of your membrane. A negative control with the affinity-tag protein itself is recommended. Particular care must be taken when using maltose binding protein which has strong affinity for cellulose units resulting in a massive background signal also on the PEG-ylated membranes. This can be blocked by addition of maltose (10 mM) to the binding buffer.
2. In any case, especially if there is no positive control present on the SPOT membrane, the activity of the used protein should be proven by additional methods like enzymatic assays.
3. The Synthesis of a SPOT membrane using the ResPep SL method from INTAVIS (Cologne, Germany) was recently described [11]. A customized SPOT membrane bearing the requested peptide sequences can be purchased from JPT Peptide Technologies (Berlin, Germany).
4. Concerning array size, it should be taken into account that the cost-effectiveness of the SPOT method will improve with increased number of different peptides on a membrane. This makes SPOT assays excellently suited for broad affinity screens, rather than the testing of a small set of predefined substrate candidates. Subsets of 120 peptides were described in the NERD field [6]. This array size also appears reasonable for its workable dimensions (about 6×4 cm^2; spot diameter of ~0.4 cm). The standard grid provided by the INTAVIS ResPep SL software is defined by 600 possible spot positions for one customary membrane (*see* Fig. 1). This implies that a set of 120 peptides can be synthesized four times in parallel on one membrane, giving rise to an appropriate set of membrane replicates with sufficient interspace left for cutting the membrane in four pieces.
5. The optimal range of peptide length with respect to synthesis success is stated between 6 and 15 amino acids, with higher amounts of amino acid couplings increasing the probability of possible side reactions [12]. For NERD E3 ubiquitin ligase screens, peptides of 5–13 residues in length were applied. In theory, substrate recognition by NERD components should depend

entirely on the very N-terminal amino acids and peptides of five amino acids are theoretically sufficient. However, one might miss the potential influence of more distal residues which might affect the total peptide properties by charge and hydrophobicity.

6. The size of the box should not exceed the membrane diameters, otherwise higher volumes of the protein solution will have to be incubated with the membrane.
7. The optimal binding condition for each tested enzyme will have to be defined and fine-tuning of the buffer conditions has the potential to alter binding properties. For instance, hydrophobic interactions by E3 ligases may tolerate salt concentrations up to 1 M [20], whereas this will obviously interfere with ionic binding situations. The opposite effect may be achieved by varying detergent strength of the binding buffer.
8. So far, binding buffer systems based on Tris [5, 7], MES [6], or HEPES buffers [8] have been reported to work successfully for SPOT assays investigating NERD components. If the enzyme of interest is known to be active in a certain buffer system, e.g. the storage buffer, this will be the first condition to use in the SPOT assay. For enzymes sensitive to pH shifts it should be considered that the assay will be performed at room temperature which will cause changes in the pH of buffers such as Tris when the enzyme is usually handled at low temperature conditions.
9. Frequently in the literature, a compatible solute like 5% (w/v) sucrose or 10% (v/v) glycerol is also present in the buffer systems.
10. When blocking the membrane overnight, avoid the use of sugars in the blocking solution, since this might promote bacterial growth.
11. It can be advisable to supplement the binding buffer (*see* **item 2** in Subheading 2.2) with low amounts of the blocking agent (*see* **item 3** in Subheading 2.2). This is especially the case if the protein tends to bind to plastic surfaces.
12. Incompletely hydrated peptides remain apparent on the membrane as white spots. Activation needs to be prolonged until the visible spots disappear. Depending on the overall hydrophobicity of each peptide sequence, the time required can differ between the spots.
13. From this step on, polyethylene boxes can be used.
14. The PEG-derivatized cellulose membrane is considered to give very little background binding. However, blocking of the membrane is recommended in order to inactivate unspecific binding sites.

15. Due to the high binding capacity of SPOT membranes, the protein concentration in the binding buffer can be as low as 20 nM depending on the enzyme binding properties. However, if the peptide set comprises lots of strong binders, enough protein has to be supplied to prevent depletion of the solution and overall loss of signal strength. On the other hand, it will be easier to discriminate between the interaction strength of different spots when using lower amounts of enzyme to be studied.
16. Usually, the binding assay is robust enough to tolerate also residuals of the blocking solution and the membrane does not necessarily need to be washed with the washing buffer before incubation with the enzyme mixture.
17. The duration of the binding depends mostly on the protein binding strength, the concentration, and the stability of the tested protein.
18. If high amounts of protein are bound to the SPOT membrane, it can be subjected to another turn of blotting giving rise to western blots with more stringent protein signals. Often, only this sequential blotting gives clear results due to unclear loading or binding status of the spots and unclear binding affinity of the POI to the highly concentrated peptides.

Acknowledgements

We thank Christian Behn from INTAVIS for personal advices on the ResPep SL SPOT peptide synthesis and we are grateful to Petra Majovsky and Wolfgang Hoehenwarter for constant support in mass spectrometry and proteome analytics. This work was supported by a grant for setting up the junior research group of the *ScienceCampus Halle—Plant-based Bioeconomy* to N.D., a Ph.D. fellowship of the *ScienceCampus Halle* to M.K. Financial support came from the Leibniz Association, the state of Saxony-Anhalt, the Deutsche Forschungsgemeinschaft (DFG) Graduate Training Center GRK1026 "*Conformational Transitions in Macromolecular Interactions*" at Halle, and the Leibniz Institute of Plant Biochemistry (IPB) at Halle, Germany.

References

1. Wenschuh H, Volkmer-Engert R, Schmidt M, Schulz M, Schneider-Mergener J, Reineke U (2000) Coherent membrane supports for parallel microsynthesis and screening of bioactive peptides. Biopolymers 55(3):188–206
2. Hilpert K, Winkler DF, Hancock RE (2007) Peptide arrays on cellulose support: SPOT synthesis, a time and cost efficient method for synthesis of large numbers of peptides in a parallel and addressable fashion. Nat Protoc 2(6):1333–1349

3. Thiele A, Stangl GI, Schutkowski M (2011) Deciphering enzyme function using peptide arrays. Mol Biotechnol 49(3): 283–305
4. Thiele A, Zerweck J, Schutkowski M (2009) Peptide arrays for enzyme profiling. Methods Mol Biol 570:19–65
5. Erbse A, Schmidt R, Bornemann T, Schneider-Mergener J, Mogk A, Zahn R, Dougan DA, Bukau B (2006) ClpS is an essential component of the N-end rule pathway in Escherichia coli. Nature 439(7077):753–756
6. Choi WS, Jeong BC, Joo YJ, Lee MR, Kim J, Eck MJ, Song HK (2010) Structural basis for the recognition of N-end rule substrates by the UBR box of ubiquitin ligases. Nat Struct Mol Biol 17(10):1175–1181
7. Hwang CS, Shemorry A, Varshavsky A (2010) N-terminal acetylation of cellular proteins creates specific degradation signals. Science 327(5968):973–977
8. Kim HK, Kim RR, Oh JH, Cho H, Varshavsky A, Hwang CS (2014) The N-terminal methionine of cellular proteins as a degradation signal. Cell 156(1-2):158–169
9. Bachmair A, Finley D, Varshavsky A (1986) In vivo half-life of a protein is a function of its amino-terminal residue. Science 234(4773):179–186
10. Varshavsky A (1996) The N-end rule: functions, mysteries, uses. Proc Natl Acad Sci U S A 93(22):12142–12149
11. Yim YY, Betke K, Hamm H (2015) Using peptide arrays created by the SPOT method for defining protein-protein interactions. Methods Mol Biol 1278:307–320
12. Winkler DF, Hilpert K, Brandt O, Hancock RE (2009) Synthesis of peptide arrays using SPOT-technology and the CelluSpots-method. Methods Mol Biol 570:157–174
13. Bachmair A, Varshavsky A (1989) The degradation signal in a short-lived protein. Cell 56(6):1019–1032
14. Suzuki T, Varshavsky A (1999) Degradation signals in the lysine-asparagine sequence space. EMBO J 18(21):6017–6026
15. Gibbs DJ, Lee SC, Isa NM, Gramuglia S, Fukao T, Bassel GW, Correia CS, Corbineau F, Theodoulou FL, Bailey-Serres J, Holdsworth MJ (2011) Homeostatic response to hypoxia is regulated by the N-end rule pathway in plants. Nature 479(7373):415–418
16. Licausi F, Kosmacz M, Weits DA, Giuntoli B, Giorgi FM, Voesenek LA, Perata P, van Dongen JT (2011) Oxygen sensing in plants is mediated by an N-end rule pathway for protein destabilization. Nature 479(7373):419–422
17. Mendiondo GM, Gibbs DJ, Szurman-Zubrzycka M, Korn A, Marquez J, Szarejko I, Maluszynski M, King J, Axcell B, Smart K, Corbineau F, Holdsworth MJ (2016) Enhanced waterlogging tolerance in barley by manipulation of expression of the N-end rule pathway E3 ligase PROTEOLYSIS6. Plant Biotechnol J 14:40
18. Gausepohl H, Behn C (2002) Automated synthesis of solid-phase bound peptides. In: Koch J, Mahler M (eds) Peptide arrays on membrane supports, vol 4, Springer lab manuals. Springer, New York, NY, pp 55–68
19. Ayad NG, Rankin S, Ooi D, Rape M, Kirschner MW (2005) Identification of ubiquitin ligase substrates by in vitro expression cloning. Methods Enzymol 399:404–414
20. Tasaki T, Zakrzewska A, Dudgeon DD, Jiang Y, Lazo JS, Kwon YT (2009) The substrate recognition domains of the N-end rule pathway. J Biol Chem 284(3):1884–1895
21. Kyhse-Andersen J (1984) Electroblotting of multiple gels: a simple apparatus without buffer tank for rapid transfer of proteins from polyacrylamide to nitrocellulose. J Biochem Biophys Methods 10(3-4):203–209
22. Cretich M, Longhi R, Corti A, Damin F, Di Carlo G, Sedini V, Chiari M (2009) Epitope mapping of human chromogranin A by peptide microarrays. Methods Mol Biol 570:221–232

Part II

Ubiquitin-like Protein Conjugation, Deconjugation, and Cell Imaging Studies

Chapter 8

SUMO Chain Formation by Plant Enzymes

Konstantin Tomanov, Ionida Ziba, and Andreas Bachmair

Abstract

SUMO conjugation is a conserved process of eukaryotes, and essential in metazoa. Different isoforms of SUMO are present in eukaryotic genomes. *Saccharomyces cerevisiae* has only one SUMO protein, humans have four and *Arabidopsis thaliana* has eight, the main isoforms being SUMO1 and SUMO2 with about 95 % identity. Functionally similar to human SUMO2 and SUMO3, Arabidopsis SUMO1 and 2 can form chains, even though they do not possess a consensus SUMOylation motif. The surprising finding that plants have dedicated enzymes for chain synthesis implies a specific role for SUMO chains in plants. By the cooperative action with SUMO chain recognizing ubiquitin ligases, chains might channel substrates into the ubiquitin-dependent degradation pathway.

A method is described to generate SUMO chains, using plant enzymes produced in *E. coli*. In vitro SUMO chain formation may serve for further analysis of SUMO chain functions. It can also provide an easy-to-synthesize substrate for SUMO-specific proteases.

Key words Small ubiquitin-related modifier SUMO, SUMO chains, SUMO-specific protease activity, Protein expression and purification, Maltose binding protein fusion

1 Introduction

In vitro activity of enzymes is of great help in understanding enzyme reactions, and for generation of reaction products for further studies. We describe a protocol for synthesis of chains formed by the small ubiquitin-related modifier SUMO. In vitro SUMO conjugation requires the heterodimeric SUMO activating enzyme (E1, SAE), and the small SUMO conjugating enzyme (E2, SCE). The formation of SUMO-SUMO linkages is greatly enhanced by addition of PIAL, which acts as a chain forming SUMO ligase. The enzymes are of plant (*Arabidopsis thaliana*) origin, and work well with the most abundant plant SUMO isoforms, SUMO1, 2 and 3, as substrates. The differences between SUMO1,2 versus SUMO3 are considerable, suggesting that SUMO isoforms from other organisms might also be substrates for in vitro chain formation by the plant enzymes. We also show that plant SUMO chains are suitable substrates for SUMO-specific proteases from bacteria, plants, and humans.

L. Maria Lois and Rune Matthiesen (eds.), *Plant Proteostasis: Methods and Protocols*, Methods in Molecular Biology, vol. 1450, DOI 10.1007/978-1-4939-3759-2_8, © Springer Science+Business Media New York 2016

2 Materials

2.1 Components of the SUMOylation System

1. SUMO1 (AT4G26840), used N-terminally tagged in a pET9d vector.
2. SUMO activating enzyme (SAE), consisting of the smaller His-tagged subunit SAE1b (AT5G50680) and the larger subunit SAE2 (AT2G21470), expressed from a bicistronic construct in the pET9d vector.
3. SUMO conjugating enzyme, SCE1 (AT3G57870), expressed untagged in the pET9d vector.
4. PIAL1 (AT1G08910) or PIAL2 (AT5G41580) E4 ligase, used N-terminally tagged with MBP and expressed from the pMAL-c2 vector (NEB).

2.2 Expression of Recombinant Plant SUMOylation Enzymes in E. coli

1. Lysogeny broth (LB): Weigh 10 g peptone, 5 g NaCl and 5 g granulated yeast extract. Add 800 mL deionized water. Adjust the pH to 7.2 using 1 M NaOH. Around 1 mL should be enough. Make up to 1 L with water and autoclave. Store at room temperature.
2. Phosphate-buffered saline (PBS): 137 mM NaCl, 2.7 mM KCl, 10 mM Na_2HPO_4, 2 mM KH_2PO_4, pH 7.4. Weigh 8 g NaCl, 0.2 g KCl, 1.44 g Na_2HPO_4, 0.24 g KH_2PO_4. Add 800 mL deionized water. Adjust pH to 7.4 with HCl. Make up to 1 L with water and autoclave. Store at room temperature.
3. Isopropyl β-d-1-thiogalactopyranoside (IPTG): 1 M. Weigh 2.38 g IPTG and dissolve in 10 mL deionized water. Aliquots are stored at −20 °C.
4. Phosphate buffer (for the expression of SCE1): mix 68.3 mL 50 mM NaH_2PO_4 with 31.5 mL 50 mM Na_2HPO_4.
5. Culture tubes.
6. Shaking platform.
7. Erlenmeyer flasks.
8. Ultracentrifugation tubes (for the expression of SCE1).
9. Centrifuge with temperature control, able to reach at least 4500 × *g* with 200 mL samples (Sorvall etc.).
10. Ultracentrifuge (for the expression of SCE1).
11. Sonicator (or other cell lysis apparatus, *see* **Note 1**).

2.3 Purification of His-Tagged Proteins (See Notes 2 and 3)

1. Binding buffer: 50 mM NaH_2PO_4, 300 mM NaCl, 10 mM imidazole, 10% glycerol, 0.5% Triton X-100, pH 8.0.
2. Wash buffer: 50 mM NaH_2PO_4, 300 mM NaCl, 20 mM imidazole, pH 8.0.
3. Elution buffer: 50 mM NaH_2PO_4, 300 mM NaCl, 250 mM imidazole, pH 8.0.

2.4 Purification of MBP-Tagged Proteins

1. Column buffer: 20 mM Tris–HCl, 200 mM NaCl, 1 mM EDTA, pH 7.4. Mix 1 mL 1 M Tris–HCl pH 7.4, 2 mL 5 M NaCl and 100 μL 0.5 M EDTA. Fill up to 50 mL with deionized water.
2. Elution buffer: 20 mM Tris–HCl, 200 mM NaCl, 1 mM EDTA, 10 mM maltose, pH 7.4. Add 34 mg maltose to 10 mL column buffer. Make aliquots and store them at −20 °C.
3. 75% glycerol: Mix 75 mL glycerol with 25 mL deionized water. Autoclave.

2.5 Purification of Untagged SCE1

1. Dithiothreitol (DTT): 1 M. Weigh 1.54 g DTT and dissolve in 10 mL deionized water. Make aliquots and store them at −20 °C.
2. SUMO buffer: 20 mM Tris–HCl, 5 mM $MgCl_2$, pH 7.4.

2.6 FPLC Purification

All buffers are filter sterilized, which also helps to degas them for the FPLC machine.

1. MonoQ buffer A: 1 mM DTT, 50 mM Tris–HCl, 10 mM NaCl, 20% v/v glycerol, pH 7.5.
2. MonoQ buffer B: 1 mM DTT, 50 mM Tris–HCl, 1 M NaCl, 20% v/v glycerol, pH 7.5.
3. MonoS buffer A: 1 mM DTT, 10 mM Na-PO_4, 10 mM NaCl, 20% v/v glycerol, pH 6.5.
4. MonoS buffer B: 1 mM DTT, 10 mM Na-PO_4, 1 M NaCl, 20% v/v glycerol, pH 6.5.

2.7 In Vitro SUMO Chain Formation

1. 10× SUMO buffer: 200 mM Tris–HCl, 50 mM $MgCl_2$, pH 7.4. Mix 2 mL 1 M Tris–HCl pH 7.4 and 500 μL 1 M $MgCl_2$. Fill up to 10 mL with deionized water, aliquot, and store at −20 °C.
2. ATP solution: 20 mM Hepes, 100 mM ATP, 100 mM $Mg(OAc)_2$, pH 7.4.

3 Methods

3.1 Expression of Recombinant Plant SUMOylation Enzymes in E. coli

1. Inoculate a single colony into 3 mL LB with appropriate antibiotics (*see* **Note 4**) and grow the culture overnight with shaking at 200 rpm at 37 °C (*see* **Note 5**).
2. Inoculate 200 mL fresh LB with 2 mL of the overnight culture and incubate at 37 °C on a shaking platform at 200 rpm. When the OD_{600} reaches 0.6–0.8, take a 500 μL sample for electrophoresis and induce protein expression by adding IPTG to a final concentration of 1 mM. Keep shaking at 37 °C for 3 h.
3. Withdraw another 500 μL sample and harvest the cells by centrifuging the culture for 20 min at 4500 × *g* and 4 °C. Discard the supernatant.

4. Resuspend the pellet in 5 mL PBS and centrifuge again for 20 min at 4500 × *g* and 4 °C (When expressing SCE1, resuspend the pellet in phosphate buffer pH 6.5, distribute into ultracentrifugation tubes, and freeze at −80 °C overnight).
5. Discard the supernatant and freeze the pellet at −20 °C overnight.

3.2 Purifying His-Tagged Proteins (SUMO and SAE)

1. Thaw the pellet on ice.
2. Resuspend in 5 mL Binding buffer and add 1 μg/mL aprotinin and 1 μg/mL leupeptin (*see* **Note 6**). Keep on ice for 15 min.
3. Lyse the cells. We use sonication, three times 30 s (*see* **Note 1**).
4. Centrifuge the suspension for 20 min at 4500 × *g* and 4 °C.
5. While the centrifugation is ongoing, add 200 μL 50% Ni^{2+}-NTA Sepharose slurry to a Bio-Rad PolyPrep Chromatography column with 2 mL bed volume. Wash with 3 column volumes (CV) deionized water and equilibrate with 6 CV Binding buffer. Cap the lower end.
6. When the centrifugation is finished, withdraw a 10 μL sample from the supernatant. Carefully pipette the rest of the liquid to the chromatography column without touching the pellet (*see* **Note 7**). Add 10 μg DNase I and cap the upper end of the column (*see* **Note 8**).
7. Incubate the column for 30–45 min on a rotating wheel at 4 °C.
8. Remove both the upper and the lower cap and collect the flow-through in a 15 mL Falcon tube. Withdraw a 10 μL sample.
9. Wash the slurry with 20 CV Wash buffer. Collect the wash. Withdraw a 10 μL sample.
10. Elute with 3 × 1 CV Elution buffer, collecting the fractions separately. Withdraw a 10 μL sample from each fraction.
11. Check the expression level and purity of the protein by SDS-PAGE. Centrifuge the 500 μL samples from the previous day (*see* Subheading 3.1, **steps 2** and **3**) in a tabletop centrifuge at full speed for 1 min. Discard the supernatant and resuspend the pellet in 50 μL Laemmli sample buffer. Add 10 μL Laemmli sample buffer to the 10 μL samples. Heat the samples at 95 °C for 5 min. Spin the tubes down and load 20 μL from each on a gel.
12. Add glycerol to a final concentration of 20% to the best eluted fractions. Freeze in liquid nitrogen and store at −80 °C.
13. Optional: For additional purity, perform anion exchange on SUMO using a MonoQ column. Dilute the SUMO fractions 1:10 with MonoQ buffer A and load the solution in an FPLC machine. SUMO, with its amino-terminal extension (*see* **Note 12**), elutes at around 20–30% of MonoQ buffer B.

3.3 Purification of MBP-Tagged Proteins (PIAL1 and PIAL2)

1. Thaw the pellet on ice.
2. Resuspend in 5 mL Column buffer and add 1 μg/mL aprotinin and 1 μg/mL leupeptin (*see* **Note 6**). Keep on ice for 15 min.
3. Lyse the cells. We use sonication, three times 30 s (*see* **Note 1**).
4. Centrifuge the suspension for 20 min at 4500 × *g* and 4 °C.
5. During the centrifugation, add 200 μL amylose resin to a Bio-Rad PolyPrep Chromatography column with 2 mL bed volume. Wash with 3 CV deionized water and equilibrate with 6 CV Column buffer. Cap the lower end.
6. When the centrifugation is finished, withdraw a 10 μL sample from the supernatant. Carefully pipette the rest of the liquid to the chromatography column without touching the pellet (*see* **Note 7**). Add 10 μg DNase I and cap the upper end of the column (*see* **Note 8**).
7. Incubate the column for 30–45 min on a rotating wheel at 4 °C.
8. Remove both the upper and the lower cap and collect the flow-through in a 15 mL Falcon tube. Withdraw a 10 μL sample.
9. Wash the slurry with 12 CV Column buffer. Collect the wash. Withdraw a 10 μL sample.
10. Elute with 3 × 1 CV Elution buffer, collecting the fractions separately. Withdraw a 10 μL sample from each fraction.
11. Check the expression level and purity of the protein by SDS-PAGE. Centrifuge the 500 μL samples from the previous day in a tabletop centrifuge at full speed for 1 min. Discard the supernatant and resuspend the pellet in 50 μL Laemmli sample buffer. Add 10 μL Laemmli sample buffer to the 10 μL samples. Heat the samples at 95 °C for 5 min. Spin the tubes down and load 20 μL from each on a gel.
12. Add glycerol to a final concentration of 20% to the best eluted fractions. Freeze in liquid nitrogen and store at −80 °C.

3.4 Purification of Untagged Proteins (SCE1)

1. Thaw the pellet on ice.
2. Add 1 μg/mL aprotinin, 1 μg/mL leupeptin and DTT to a final concentration of 10 mM (*see* **Note 9**).
3. Centrifuge the suspension for 1 h at 100,000 × *g* and 4 °C. Collect the supernatant.
4. Dilute the supernatant 1:10 with MonoS buffer A and load the solution in the FPLC machine. The SCE1 elutes around 15–20% of MonoS buffer B.
5. Load the fraction(s) containing the protein on a VivaSpin 500 column and centrifuge for 15 min at full speed in a tabletop cen-

trifuge. Discard the flow-through and fill the top compartment with SUMO buffer. Repeat four times.

6. Add glycerol to a final concentration of 20%. Freeze in liquid nitrogen and store at −80 °C.

3.5 In Vitro SUMO Chain Formation. Small Scale Reaction for Western Blot Detection (Fig. 1)

1. Calculate how much of each protein is needed for 2 μM SAE, 1.75 μM SCE1, 14 μM SUMO1, 1.5 μM PIAL1 or PIAL2 E4 ligase, and how much water should be added to 20 μL (*see* **Note 10**).
2. Pipette 2 μL 10× SUMO buffer into an Eppendorf tube. Add the water, and then the enzymes, as well as 5 mM ATP (*see* **Note 11**).
3. Incubate the enzyme mix for 2 h at 30 °C.
4. Add 20 μL Laemmli sample buffer, heat the samples for 5 min at 95 °C and load 20 μL on an SDS-PAGE gel.
5. Transfer the proteins to a membrane and visualize with appropriate antibodies (*see* **Note 12**).

3.6 In Vitro SUMO Chain Formation. Large Scale Reaction for Isolation of SUMO Chains

1. Calculate how much of each protein is needed for 2 μM SAE, 1.75 μM SCE1, 14 μM SUMO1, 1.5 μM PIAL1 or PIAL2 E4 ligase and how much water should be added to 250 μL (*see* **Note 10**).

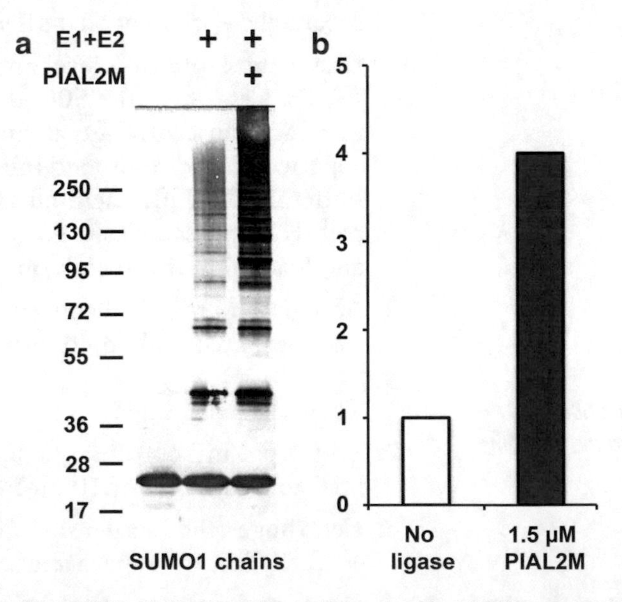

Fig. 1 In vitro SUMO chain formation with and without PIAL enzymes [1]. (**a**) Reaction with tagged *A. thaliana* SUMO1 in the presence of PIAL2M (*right lane*) results in longer and more chains than without PIAL2M. Protein blot of reaction components is visualized with antibody against the tag on SUMO1. The *left lane* shows input SUMO1 preparation. (**b**) Quantification of gel blot with ImageJ indicates fourfold enhancement of chain formation by PIAL2M

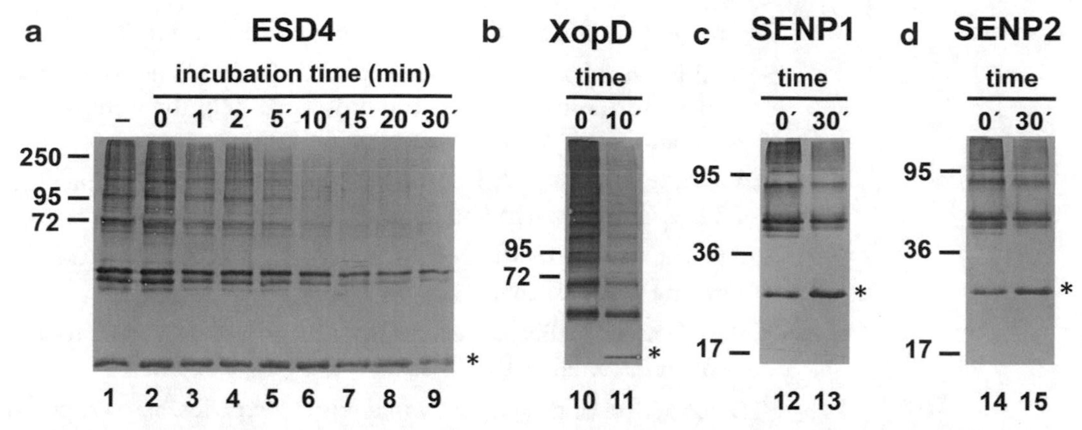

Fig. 2 Use of SUMO chains as substrates for SUMO-specific proteases. Protein gel blots after incubation of different proteases with purified SUMO chains made from tagged *A. thaliana* SUMO1. (**a**) Fragment of plant SUMO protease ESD4 was used for chain hydrolysis [2]. *Lane 1*, incubation without protease. *Lanes 2–9*, incubation with protease after 0 min, 1 min, 2 min, 5 min, 10 min, 15 min, 20 min, and 30 min, respectively, at an incubation temperature of 20 °C. (**b**) Fragment of bacterial protease XopD was used in the incubation [3, 4]. *Lanes 10* and *11* show reaction after 0 and 10 min at incubation temperature 30 °C. (**c**) Assay with commercial preparation of human SUMO protease SENP1. *Lanes 12* and *13* show reaction products after 0 min and 30 min, respectively, with incubation temperature 37 °C. (**d**) Assay with commercial preparation of human SUMO protease SENP2. *Lanes 14* and *15* show reaction products after 0 min and 30 min, respectively, incubation temperature was 37 °C. The *asterisk* indicates the position of monomeric SUMO

2. Pipette 25 μL 10× SUMO buffer into an Eppendorf tube. Add the water, and then the enzymes, as well as 5 mM ATP (*see* **Note 11**).
3. Incubate the enzyme mix for 2 h at 30 °C.
4. Load the reaction into an FPLC machine for anion exchange with a MonoQ column (*see* **Note 13**).
5. Check the purity of the SUMO chains, as well as the separation from monomeric SUMO, with a Western blot (*see* **Note 12**).
6. Store SUMO chains in liquid nitrogen, since they are not preserved at −80 °C.

3.7 SUMO Protease Activity Test Using SUMO Chains (Fig. 2)

1. Incubate 10 μM of the FPLC purified SUMO chains with 2 μM of the tested protease and 1 mM DTT in 1× SUMO buffer (*see* **Notes 14** and **15**).
2. Terminate the reaction by adding Laemmli sample buffer and heating to 95 °C. The samples can be analyzed on a Western blot using anti-SUMO antibodies (*see* **Note 12**).

4 Notes

1. We use a Bandelin Sonoplus HD70 sonicator with an MS73 tapered probe, set at 50% cycle and 60% intensity, for three times 30 s bursts on ice with 30 s rest on ice between the bursts.

2. The best results are always obtained with fresh buffers.
3. SAE is a heterodimer and we express it from a bicistronic construct. Add 5 mM ATP to the buffers to help the complex to form and purify in stoichiometric amounts.
4. We use working concentrations of 100 mg/L ampicillin, 25 mg/L kanamycin and 25 mg/L chloramphenicol.
5. Make sure to start the culture later than 16:00, otherwise the bacteria will start dying off.
6. The thawed pellet never resuspends completely due to the broken bacterial cell walls and long DNA filaments.
7. The suction of the pipette will disturb the pellet, so pay special attention towards the end.
8. We seal the caps with parafilm as an additional precaution against leakage.
9. DTT protects the active site cysteine during the purification but has to be removed prior to the enzymatic reactions because it interferes with formation of the SUMO-SAE and/or SUMO-SCE thioester bond.
10. The starting concentration of SCE1 determines the speed of the reaction, while PIAL1 and PIAL2 enhance the SUMO chain formation.
11. The reaction needs ATP to proceed, so add the SAE and the ATP last. Do not use a reaction without ATP as a negative control because the SAE was purified in the presence of ATP.
12. We use a PVDF membrane. In addition to the His tag, the amino-terminal extension of the SUMO we use has a Strep tag and can be detected with Strep-tactin (IBA). PIAL1 and PIAL2 can be detected with an anti-MBP antibody (NEB). For the detection of SCE1, we have a home-made antibody.
13. In some cases, the addition of 0.1 % Triton X-100 can improve the yield.
14. In our hands, a 20 μL reaction worked best.
15. To test XopD, the reactions were incubated for 10 min at 30 °C. For ESD4, the incubation time was 15 min at 20 °C, and for SENP1 and SENP2 the samples were incubated for 30 min at 37 °C.

Acknowledgment

Work on SUMO conjugation in the authors' laboratory was supported by the Austrian Science Fund FWF (grant P25488).

References

1. Tomanov K, Zeschmann A, Hermkes R, Eifler K, Ziba I, Grieco M et al (2014) Arabidopsis PIAL1 and 2 promote SUMO chain formation as E4 type SUMO ligases, and are involved in stress responses and sulfur metabolism. Plant Cell 26:4547–4560
2. Murtas G, Reeves PH, Fu YF, Bancroft I, Dean C, Coupland G (2003) A nuclear protease required for flowering-time regulation in Arabidopsis reduces the abundance of SMALL UBIQUITIN-RELATED MODIFIER conjugates. Plant Cell 15:2308–2319
3. Hotson A, Chosed R, Shu H, Orth K, Mudgett MB (2003) Xanthomonas type III effector XopD targets SUMO-conjugated proteins in planta. Mol Microbiol 50:377–389
4. Canonne J, Marino D, Noel LD, Arechaga I, Pichereaux C, Rossignol M et al (2010) Detection and functional characterization of a 215 amino acid N-terminal extension in the Xanthomonas type III effector XopD. PLoS One 5:e15773

Chapter 9

Kinetic Analysis of Plant SUMO Conjugation Machinery

Laura Castaño-Miquel and L. Maria Lois

Abstract

Plants display a high diversification degree of the SUMO conjugation machinery, which could confer a biological specialization of the different isoforms. For instance, the two essential *Arabidopsis* SUMO isoforms, SUMO1/2, display the highest conjugation rate when compared to SUMO3 and 5, suggesting that their specific biochemical properties may be linked to their biological specialization. In order to study the biochemical properties of plant SUMO conjugation systems, quantitative biochemical assays must be performed. We will present a detailed protocol for reconstituting an in vitro SUMO conjugation assay covering all steps from protein preparation to assay development.

Key words SUMO, E1 SUMO-activating enzyme (SAE2/SAE1), E2 SUMO-conjugating enzyme, Catalase 3 C-terminal domain, In vitro SUMOylation assay, Thioester

1 Introduction

In eukaryotic cells, posttranslational modifications by SUMO (Small Ubiquitin-like MOdifier) modulates protein activity through regulation of subcellular localization, protein activity and stability, and protein–protein interactions [1]. SUMO conjugation initiates with SUMO activation, which is a two-step ATP-dependent reaction catalyzed by the heterodimeric E1-activating enzyme, SAE1/SAE2. SUMO activation is the first control point in the selection of the SUMO/Ubl (Ubiquitin-like) modifier to enter the conjugation pathway [2, 3]. SAE2 displays four functional domains: adenylation, catalytic cysteine, ubiquitin fold (UFD), and C-terminal domains and SAE1 contributes the essential Arg21 to the adenylation domain [4]. The adenylation domain is responsible for SUMO recognition and SUMO C-terminus adenylation. In a second step, the SUMO C-terminal adenylate establishes a thioester bond with the E1 catalytic cysteine. After the thioester bond is formed, SUMO can be transferred to the E2-conjugating enzyme in a reaction that requires E2 recruitment through the $SAE2^{UFD}$ domain [5] in collaboration with the $SAE2^{Cys}$ domain [6]. The E2-conjugating

L. Maria Lois and Rune Matthiesen (eds.), *Plant Proteostasis: Methods and Protocols*, Methods in Molecular Biology, vol. 1450, DOI 10.1007/978-1-4939-3759-2_9, © Springer Science+Business Media New York 2016

enzyme is competent for transferring SUMO to the substrate, although this reaction is facilitated by E3 ligases. SUMO conjugation is a reversible modification and the same proteases responsible for the processing of the SUMO immature form (peptidase activity) are also involved in its removal from the substrate (isopeptidase activity) [7].

In plants, SUMO conjugation controls plant development [8, 9] and it has a major role in the modulation of plant responses to hormones [10], development [11, 12], abiotic stress [13–16], and defense responses to pathogens [17, 18]. These plant biological processes regulated by SUMOylation have been uncovered by the analysis of proteases, ULP, and SUMO E3 ligase mutant plants [19]. Among them, the most studied mutants are the *siz1 and mms21* E3 ligases and the *esd4* ULP protease, which display pleiotropic growth defects and reduced viability. *siz1* and *mms21* null mutants display a reduction in endogenous SUMO conjugate accumulation [20–22], while an overaccumulation of SUMO conjugates is found in *esd4* mutant [8]. Even though they have opposite molecular effects, the physiological outcome of these mutations is very similar and, surprisingly, this pleiotropic phenotype is the result of an overaccumulation of salicylic acid in both *siz1* and *esd4* mutant plants [23]. These results indicate that SUMO conjugation homeostasis is under a tight control and over- or under-accumulation of SUMO conjugates results in a misregulation of essential processes.

The critical SUMO homeostasis in vivo can be achieved through regulation of SUMOylation machinery activity. Accordingly, biochemical studies have shown that SUMO conjugation machinery is a complex system in plants. In *Arabidopsis*, functional diversity has been found in SUMO proteases, SUMO isoforms, SUMO-activating enzyme E1, and SUMO E3 ligases. *Arabidopsis* SUMO proteases display distinct specific activities toward the existing SUMO isoforms, SUMO1, 2, 3, and 5 [24], which also display distinct biochemical properties that might influence their conjugation in vivo and biological function. The conjugation system seems to have evolved for assuring the conjugation of the essential SUMO1/2 isoforms. In this mechanism, the E1-activating enzyme would have a crucial role by conferring SUMO paralog specificity [3], in addition to a rate limiting role of SUMO activation during the conjugation cascade [25].

In order to understand the functional relevance of SUMO conjugation machinery diversification among plantae kingdom, performing accurate biochemical assays is a crucial approach. We provide a detailed protocol for performing SUMO conjugation assays, from protein preparations to quantification of kinetic data.

2 Materials

2.1 Expression and Purification of Enzymes

1. Plasmids: pET-15b (Novagen), pET-28a (Novagen), pGEX-6p (GE healthcare) (or similar).
2. LB agar plates: 10 g bacto-tryptone, 5 g bacto-yeast extract, and 10 g NaCl in 1 L water, adjust to pH 7.5 with NaOH and add 15 g LB agar powder. After autoclave add the appropriate antibiotic and pour the LB agar into sterile Petri dishes.
3. Antibiotics stock solutions: 100 mg/mL ampicillin in water, 50 mg/mL kanamycin in water, and 34 mg/mL chloramphenicol in ethanol. Sterilize all antibiotics by filtration and store in 1 mL aliquots at −20 °C. Make a 1/1000 dilution for reaching the working concentration.
4. 2× TY medium: 16 g bacto-tryptone, 10 g bacto-yeast extract, and 5 g NaCl in 1 L water and autoclave.
5. Isopropyl-β-D-thiogalactopyranoside (IPTG): 0.1 M IPTG in water, sterilize by filtration and store in aliquots at −20 °C.
6. Protease inhibitors stock: 0.1 M PMSF (phenylmethanesulphonylfluoride) in ethanol, 1 mg/mL pepstatin in ethanol, and 1 mg/mL leupeptin in ethanol. All solutions are stored at −20 °C in small aliquots.
7. Lysis buffer: 20% sucrose, 50 mM Tris–HCl pH 8.0, 350 mM NaCl, 10 mM $MgCl_2$, 1 mM β-mercaptoethanol, 0.1% NP40 (v/v), 50 μg/μL DNAsa, 1 mg/mL lysozyme, 1 mM PMSF, 1 μg/mL leupeptin, and 1 μg/mL pepstatin.
8. IMAC Sepharose 6 Fast Flow (17-0921-07, GE healthcare) or similar.
9. Equilibration buffer I: 50 mM Tris–HCl pH 8.0, 350 mM NaCl, 1 mM β-mercaptoetanol, and 20 mM imidazol.
10. Elution buffer I: 50 mM Tris–HCl pH 8.0, 350 mM NaCl, 1 mM β-mercaptoetanol, and 300 mM imidazol.
11. Dialysis membrane with a nominal 5 kDa molecular weight cut-off (MWCO).
12. Thrombin protease (27-0846-01, GE healthcare) prepare at 1 U/μL in PBS (140 mM NaCl, 2.7 mM KCl, 10 mM Na_2HPO_4, 1.8 mM KH_2PO_4, pH 7.3).
13. Size exclusion buffer: 50 mM Tris–HCl pH 8.0, 100 mM NaCl, and 1 mM β-mercaptoetanol.
14. Glutathione Sepharose 4B (GE healthcare, 17-0756-01).
15. Equilibration buffer II: 50 mM Tris–HCl pH 8.0, 350 mM NaCl.

16. Elution Buffer II: 50 mM Tris–HCl pH 8.0, 350 mM NaCl, and 10 mM reduced L-glutathione (BioXtra, ≥98.0 %, SIGMA-ALDRICH G6529).
17. Protein chromatography: AKTA-FPLC system with preparative gel filtration columns (HiLoad 16/60 Superdex 75 prep grade (120 mL) 17-1068-01 and HiLoad 16/60 Superdex 200 prep grade (120 mL) 17-1069-01; GE healthcare).
18. Centrifugal device concentrators: 10 and 50 kDa cut-off filters (Amicon).
19. Nylon or cellulose acetate membrane syringe filters 0.2 μm pore size.
20. PD-10 Desalting columns (GE Healthcare).
21. Bradford protein assay (Bio-Rad).

2.2 SUMO Conjugation Assays

1. Reaction Buffer 5×: 250 mM NaCl, 100 mM HEPES-NaOH pH 7.5, 0.5 % Tween-20, and 25 mM $MgCl_2$.
2. ATP solution: 100 mM ATP dissolved in 1 M Tris–HCl pH 7.5.
3. SDS loading buffer 6×: 0.5 M Tris–HCl pH 6.8, 10% SDS, 30 % glycerol, and 0.012 % bromophenol blue. Store in 0.5 mL aliquots at −20 °C.
4. 4–12 % gradient polyacrylamide Bis-Tris gels (Invitrogen).
5. Transfer buffer: 48 mM Tris–HCl, 39 mM glycine and 10 % (v/v) methanol for semi-dry unit.
6. Blocking buffer: 3 % (w/v) nonfat dry milk in TBST buffer.
7. TBST buffer: 20 mM Tris–HCl pH 7.5, 137 mM NaCl, and 0.1 % (v/v) Tween 20.
8. Primary antibody: antibody anti-GST (Sigma, G7781) used at 1:2500 dilution in blocking buffer.
9. Secondary antibody: anti-rabbit IgG horseradish peroxidase linked whole antibody (GE healthcare, NA934) used at 1:5000 dilution in blocking buffer.
10. ECL Prime Western Blotting Detection Reagent (GE healthcare, RPN2232) or similar.
11. Chemiluminescence imaging system such as LAS4000 (Fujifilm).

3 Methods

Reconstituted SUMO in vitro reaction allows the study of the biochemical properties of SUMO machinery components. We use GST-*At*CAT3Ct as a substrate (Fig. 1), which is modified by SUMO at the Lys-423 leading to a detectable shift of 15 kDa. This posttranslational modification is visualized by SDS-PAGE

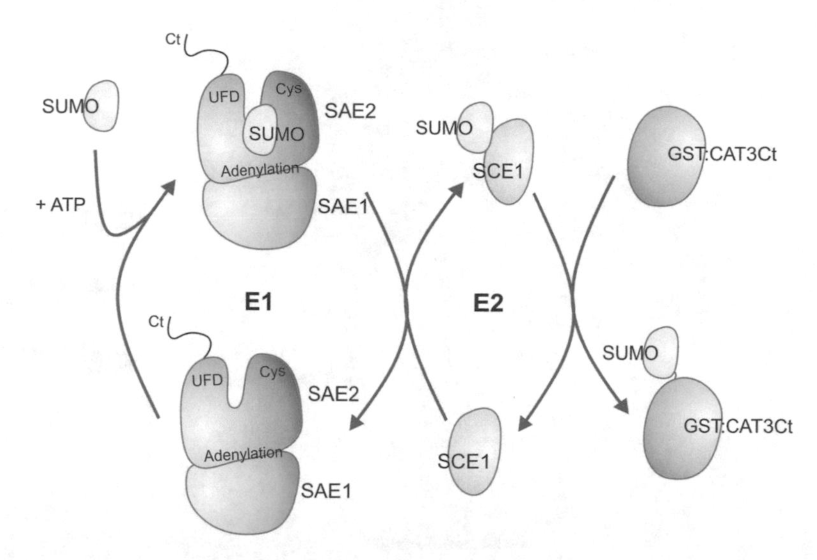

Fig. 1 SUMO conjugation assay. Kinetics analysis of SUMO isoforms and E1 isoforms was performed by monitoring SUMO conjugation to the C-terminal domain of catalase 3 (comprising amino acids 419–492), fused to GST, in the absence of SUMO E3 ligases

followed by gel Coomassie Blue staining or immunoblotting. The diversification of the SUMOylation system in plants is higher than in mammals, suggesting the existence of different molecular properties for each isoform that might affect their in vivo conjugation and biological function. To address this issue, we have developed an efficient time-course assay to facilitate analysis of SUMO conjugation in vitro. The assay is done with all the purified SUMOylation system components, as described through Subheadings 3.1–3.4, except for the E3 SUMO ligase enzyme.

3.1 Preparation of Recombinant SUMO Machinery Components: Expression and Purification of SUMO Isoforms and the SUMO-Conjugating Enzyme SCE1

cDNA-encoding SUMO proteins in their mature form *At*SUMO1, *At*SUMO2, *At*SUMO3, and *At*SUMO5 was obtained from 2-week-old plants and cloned into pET28a (Novagen) [3]. DNA encoding full-length *At*SCE1 was acquired from the ABRC (Ohio State University, Columbus) and cloned into pET28a [10]. All genes were cloned into pET28a to generate N-terminal thrombin-cleavage His_6-fusion proteins (Figs. 2 and 3).

1. Transform the plasmids encoding the HIS-tagged proteins into *E. coli* BL21 Codon Plus RIL (Stratagene) competent cells (*see* **Note 1**).
2. Transformed cells are selected on LB agar plates supplemented with 50 μg/mL kanamycin and 34 μg/mL cloramphenicol.
3. Pick a fresh single colony and inoculate 3 mL of 2× TY medium containing kan/chlr during 6 h at 37 °C with vigorous shaking (≈250 rpm).

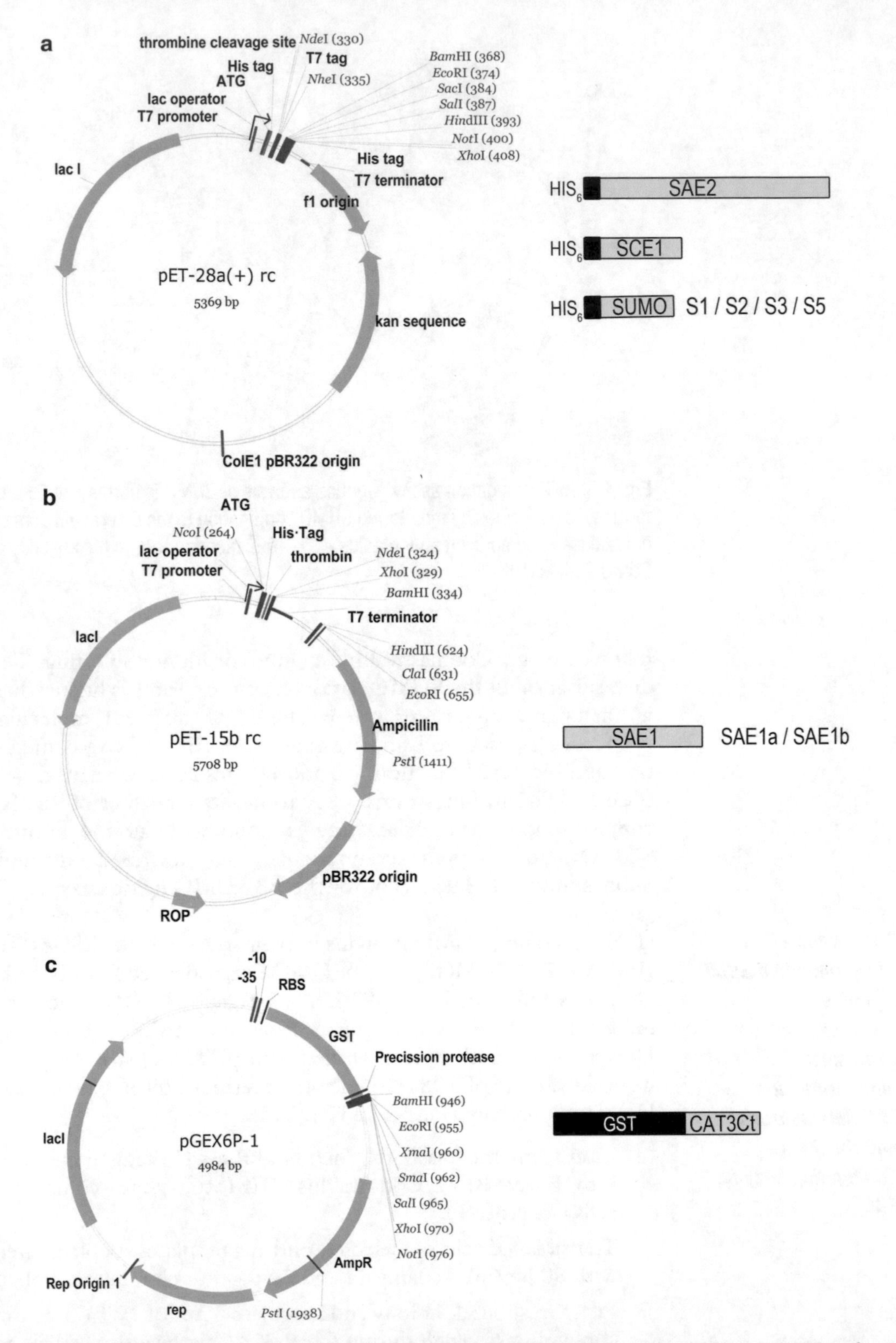

Fig. 2 DNA plasmids used for production of recombinant protein in *E. coli*. (**a**) SAE2, SCE1, and SUMO isoforms were cloned in the pET28a vector for generating HIS-tagged protein fusions. (**b**) SAE1a and b isoforms were cloned into the pET15b vector in the NcoI cloning site in order to produced untagged versions. (**c**) CAT3 C-terminal domain was cloned into the pGEX-6P vector for generating GST-protein fusions

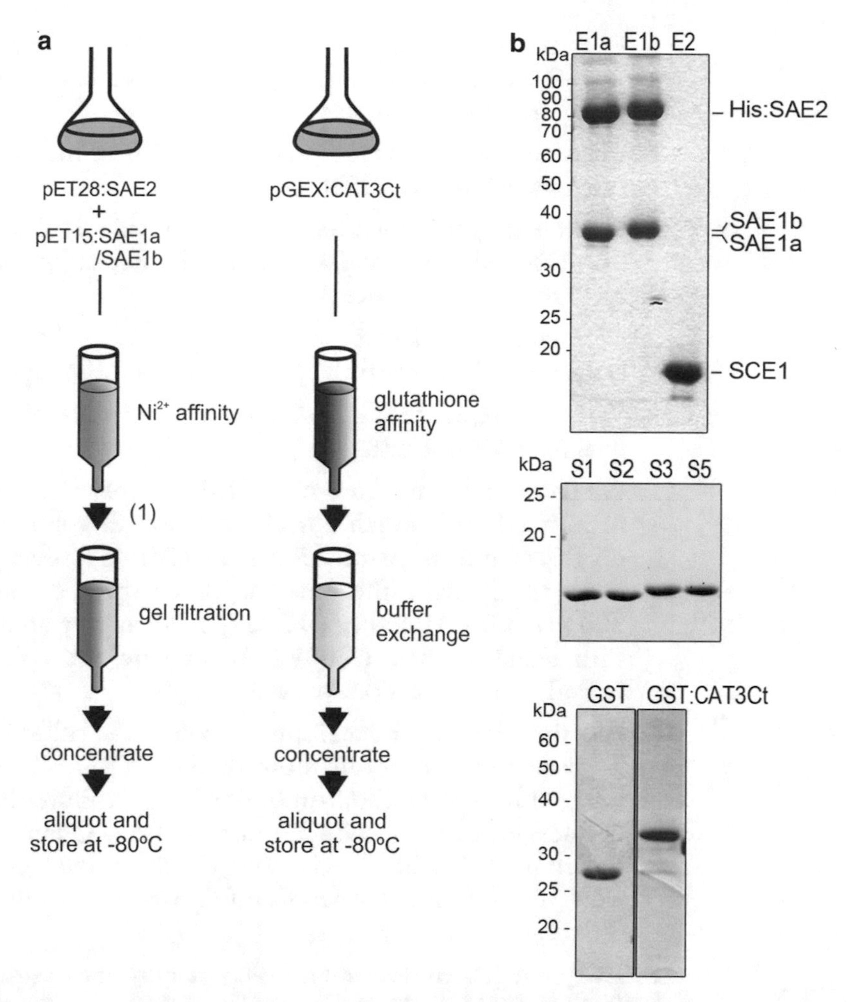

Fig. 3 Purification of SUMO conjugation assay components. (**a**) All proteins were expressed in independent *E. coli* BL21 cultures except for SAE2/SAE1a and SAE2/SAE1b that were coexpressed in order to purify the corresponding E1 heterodimer isoform. HIS-tagged fusion proteins were Ni^{2+}-affinity purified and eluted fractions were further purified through gel filtration chromatography. (1) After Ni^{2+}-affinity purification, HIS-tag was removed by thrombin digestion except for the E1 heterodimer sample. GST and GST-CAT3Ct were purified by glutathione-affinity chromatography followed by buffer exchange chromatography. For each purification experiment, fractions showing the highest purity degree were pooled together, concentrated, and aliquots stored at −80 °C. (**b**) Before storage, all samples were quantified by Bradford and purity analyzed by SDS-PAGE followed by Coomasie-Blue staining. Samples from representative purification experiments are shown

4. Dilute 1:50 the preculture into 60 mL of 2×TY with kan/chlr and grow overnight at 37 °C and 250 rpm.
5. Next morning, inoculate 0.5 L culture of 2×TY making a 1:50 dilution of the saturated overnight culture, and grow the bacterial culture at 37 °C and 250 rpm until an OD_{600nm} of 0.6–0.8 is reached. Induce protein expression by adding IPTG to final concentration of 0.5 mM. Cultures are grown for another 4 h at 28 °C and 250 rpm (*see* **Note 2**).

6. Harvest cells by centrifugation (6000 × *g* for 15 min at RT) and discard supernatant. At this point, cells can be extracted or kept it at −80 °C until use.
7. Thaw cell pellet and resuspend with 1/20 of the original culture volume in lysis buffer.
8. Sonicate the cell suspension in ice-water bath. Sonication cycle: 30 s ON, 30 s OFF at 10% amplitude. Repeat the sonication cycle six times (*see* **Note 3**).
9. Centrifuge the cell lysate at 39,000 × *g* for 1 h at 4 °C to remove cellular debris. Discard the pellet and retain the supernatant.
10. Pass the sample through 0.2 μm filter and add imidazole to a final 20 mM concentration (*see* **Note 4**).
11. Pre-pack a column with 3 mL of IMAC sepharose (50% slurry in 20% ethanol, which corresponds to 1.5 column volume, CV) (*see* **Note 5**). Add 7.5 mL (5 CV) of distilled water in order to eliminate the ethanol. To charge the column add 300 μL of 0.1 M $NiSO_4$ (0.2 CV) following by another wash with distilled water (5 CV). Equilibrate the column with 7.5 mL of equilibration buffer I (5 CV).
12. Pass the protein extract (input) through the column by gravity flow and collect the flow through (FT). Perform one wash with the equilibration buffer (5 CV). Elute the protein in fractions of 1 mL with Elution buffer I. Quantify elution fractions by Bradford assay and check purified proteins by 12% SDS-PAGE gel separation followed by Coomassie Blue staining.
13. Pool together elution fractions containing the desired protein (His_6-*At*SUMO1/2/3/5 or His_6-*At*SCE1), add thrombin, transfer to a dialysis membrane, and dialyze overnight at 4 °C against size exclusion buffer. Add 10 units of thrombin per 1 mg of the protein (*see* **Note 6**).
14. Concentrate the sample using 10 kDa cut-off filters (SUMO and SCE1 predicted MW are approximately 11 kDa and 18 kDa, respectively) to a final volume close to 1 mL. Filtrate the sample through 0.2 μm syringe membrane filter and apply to a gel filtration column, Superdex 75, equilibrated with size exclusion buffer. Fractions of 1 mL are collected and 10 μL aliquots corresponding to the eluted protein peak are analyzed by 12% SDS-PAGE. Those fractions containing the pure protein (SUMOs or SCE1) are pooled together and concentrated until 5–10 mg/mL using a centrifugal device (10 kDa cut-off). Freeze small aliquots in liquid nitrogen and store at −80 °C until use (*see* **Notes 7** and **8**).

3.2 Preparation of Recombinant SUMO Machinery Components: Expression and Purification of E1-Activating Enzyme Isoforms (SAE2/SAE1a and SAE2/ SAE1b)

The SUMO E1-activating enzyme is a heterodimer consisting of a large subunit, SAE2, and a small subunit, SAE1. *Arabidopsis* presents two isoforms of the SUMO E1 (E1a and E1b), which differ in the small subunit composition, SAE1a or SAE1b. In this case, purification of the dimeric E1 complex is performed by coexpression of His_6-tagged SAE2 and untagged SAE1a or SAE1b in *E. coli*, which were previously cloned in pET28a (SAE2) and pET15b (SAE1a/b). cDNA encoding SAE2 and SAE1a/b was obtained from 2-week-old plants and cloned into the expression plasmids [3] (Figs. 2 and 3).

1. For protein expression, cell lysis and protein purification through the IMAC, follow the steps described in **steps 1–12** in Subheading 3.1. Add the appropriate antibiotics in all steps of growing bacterial cell cultures: kanamycin (pET28a), ampicillin (pET15a), and chloramphenicol for the bacterial strain.
2. Analyze the elution fractions on Coomassie Blue stained 10% SDS gel. SAE2 migrates at 80 kDa while SAE1a/b migrates at 37 kDa.
3. Pool together the fractions containing the E1a/b heterodimer and dialyze the samples overnight against the size exclusion buffer (*see* **Note 9**).
4. Concentrate the mixture using a 50 kDa cut-off filter to a final volume of 1 mL and filtrate through 0.2 μm filter.
5. Load the sample onto a preparative Superdex 200 gel filtration column, equilibrated with size exclusion buffer. Fractions of 1 mL are collected during the chromatography.
6. Analyze by 10% SDS-PAGE gel 10 μL of the fractions corresponding to the E1a/b elution peak. Pool the fractions were SAE2/SAE1a or SAE2/SAE1b display a 1:1 stoichiometry and concentrate using 50 kDa cut-off filters to 20–50 mg/mL final concentration. Freeze small aliquots in liquid nitrogen and store at −80 °C until use.

3.3 Preparation of Recombinant SUMO Conjugation Substrate: Expression and Purification of GST-CAT3Ct

As an efficient substrate we use the catalase 3, which has a SUMOylation consensus site, Lys 423, fully exposed on the protein surface and located at the C-terminal domain (comprising amino acids 419–492). The cDNA encoding the C-terminal tail of CAT3 (419–472) was obtained from 2-week-old plants and cloned into pGEX-6p-1 to obtain an N-terminal GST (glutathione transferase)-fusion protein (Figs. 2 and 3).

1. For the protein expression and cell lysis, follow the procedure described in **steps 1–10** in Subheading 3.1. Add the appropriate antibiotics: ampicillin (pGEX-6p1) and chloramphenicol (for the bacterial strain).

2. Pass the sample through 0.2 μm filter before loading it to the affinity column (*see* **Note 4**).
3. Pre-pack a column with 3 mL of gluthatione-Sepharose (50% slurry in 20% ethanol, which corresponds to 1.5 CV). Add 7.5 mL (5 CV) of distilled water in order to remove the ethanol. Equilibrate the column with 5 CV of the equilibration buffer II.
4. Pass the protein extract (input) through the column by gravity flow and collect the FT. Perform one wash with the equilibration buffer 2 (5 CV). Elute the protein in fractions of 1 mL with elution buffer II. Quantify protein content by Bradford assay and analyze eluted fractions by Coomassie Blue staining in 12% SDS-PAGE gel.
5. Fractions containing GST-AtCAT3Ct are pooled together and desalting by passing the mixture into prepacked disposable PD-10 desalting column (GE healthcare) equilibrated with size exclusion buffer.
6. The recovered sample is concentrated using 10 kDa cut-off filters (GST-*At*CAT3Ct has a predicted MW of 34 kDa) to 5 mg/mL, flash-frozen into liquid nitrogen and stored at −80 °C in small aliquots.

3.4 In Vitro SUMOylation Assays for Analyzing Distinct SUMO Isoforms

In order to test the efficiency of each SUMO isoform for conjugation to the substrate, SUMO isoforms are used in independent in vitro assays. For simplicity, major qualitative differences in SUMO conjugation rate can be identify in single time point assays through a temperature range [3]. Once identified the SUMO isoforms displaying the highest differences in conjugation efficiency, a time-course assay is performed for quantifying conjugation kinetics. In *Arabidopsis*, *At*SUMO1, *At*SUMO3, and *At*SUMO5 are the isoforms that differ dramatically in their conjugation capacity. For performing kinetics studies, SUMO conjugation is assayed at two temperatures, 37 and 42 °C, and reaction products are analyzed at several time points: 0, 10, 20, 30, and 60 min. In order to minimize technical differences between independent reactions, a master mix is prepared by adding all the common components in the reaction mixture. The preparation of a reaction master mix for one incubation temperature is as follows (the volumes should be scaled up according to the number of temperatures to be assayed):

1. Prepare a master mix reaction by mixing the following components in the indicated order. Add H_2O, 5× Reaction Buffer, 0.5 μM *At*SAE2/*At*SAE1a, 0.5 μM *At*SCE1 and 5 μM GST-*At*CAT3Ct calculated to a final 360 μL reaction volume, although the master mix final volume is adjusted to 330 μL at this point (*see* **Note 10**).

2. Divide the preparative master mix in three PCR tubes (110 μL per tube) and add 2 μM of AtSUMO1, AtSUMO3, or AtSUMO5, calculated to a final reaction volume of 120 μL (110 μL master mix + 10 μL SUMO isoform to be tested).
3. Always include a control reaction without ATP. Transfer 20 μL of each of the three performed reactions into a new PCR tube as a negative control. A total of 6 tubes corresponding to *At*SUMO1, *At*SUMO2, *At*SUMO3, and the respective negative controls are obtained for each temperature (*see* **Note 11**).
4. Start the reaction by adding 1 μL of 100 mM ATP (1 mM final concentration) into the reaction tubes, except for the control reactions, mix gently, and collect the reaction mixture on the bottom of the tube by a short spin (*see* **Note 12**).
5. Immediately take the first time-course point (0 min), and stop the reaction. Stop reactions by removing 20 μL of the reaction mixture at a given time point, transfer them to a new centrifuge tube containing 4 μL of SDS 6× loading buffer and heat for 10 min at 70 °C.
6. Transfer the reaction tubes to a PCR machine with a gradient temperature program and incubate at 37 °C or/and 42 °C (depending on the experimental conditions being assayed).
7. Remove 20 μL of the reactions at the specified times and stop the reactions.
8. Stop negative control reactions at the last time-course point, 60 min.
9. For SUMO conjugation efficiency quantification, resolve reaction products by SDS-PAGE. For facilitating comparative quantification among SUMO isoform conjugation rate, analyze, in the same protein gel, time-course reaction aliquots incubated at the same temperature and containing either AtSUMO1 or AtSUMO3 or AtSUMO5.
10. Load 12 μL of each time point denatured sample on a Novex 4–12% Bis-Tris gradient gels and perform electrophoresis in MOPS running buffer.
11. Blot proteins into PDVF membranes using a semi-dry transfer for 30 min at 20 V at room temperature in transfer buffer.
12. Block the membrane for 1 h in blocking buffer solution at room temperature.
13. Incubate the PVDF membranes with a primary antibody solution against GST (anti-GST polyclonal antibody) in blocking buffer overnight at 4 °C (*see* **Note 13**).
14. Rinse the blots three times for 10 min with the TBST solution to remove the excess of the primary antibody unbound at the membrane.

15. Incubate the blots in the secondary antibody for 45 min at room temperature.
16. Wash the PVDF membranes three times for 10 min with TBST solution.
17. Apply the chemiluminescent substrate, ECL-prime reagent, to the blot according to the manufacturer's recommendation.
18. Capture the chemiluminescent signal with the LAS4000 imaging system.

The same procedure can be applied for performing kinetics studies of other SUMO conjugation machinery components. For each specific case, the reaction master mix will be modified accordingly. In case of analyzing SUMO E1-activating enzyme kinetics, the reaction mixture will contain AtSUMO2, AtSCE1 and GST-AtCAT3Ct, and AtE1a (AtSAE2/AtSAE1a) or AtE1b (AtSAE2/AtSAE1b) will be added after distributing the reaction master mix into independent tubes.

3.5 Quantification of SUMO Conjugation Kinetics

In order to calculate the SUMOylation efficiency specific to SUMO isoform or E1 isoform present in the assay, chemiluminescent signal is quantified using the quantitative image analysis software Multigauge. In this protocol, as an example, quantification of the E1a or E1b in vitro assays is explained (Fig. 4).

1. Gel images are processed and quantified with Multi Gauge software (*see* **Note 14**). For simplification, only the GST:CAT3-monoSUMO adduct is quantified.
2. Draw a region of interest (ROI) using the drawing tools. We recommend using the rectangle shape. Draw a rectangle that encloses the largest band of interest and use the same box area for enclosing the rest, including an area of the membrane without signal, although right above or below the bands being quantified, to be used as background (Fig. 4a).
3. Export original quantification raw data to an Excel sheet (or similar).
4. Subtract background signal from all data points (AU-B) (Fig. 4b).
5. Calculate the average signal obtained from each membrane (average of all data calculated in the previous step) and use it for normalizing the obtained values ((AU-B)/signal average). This procedure facilitates the reduction of technical variability resulting from differences in western blotting and chemiluminescent capture time among experiments (Fig. 4c).
6. Plot values onto a scattered graph and determine the time-course points that fit to the linear range (Fig. 4d).

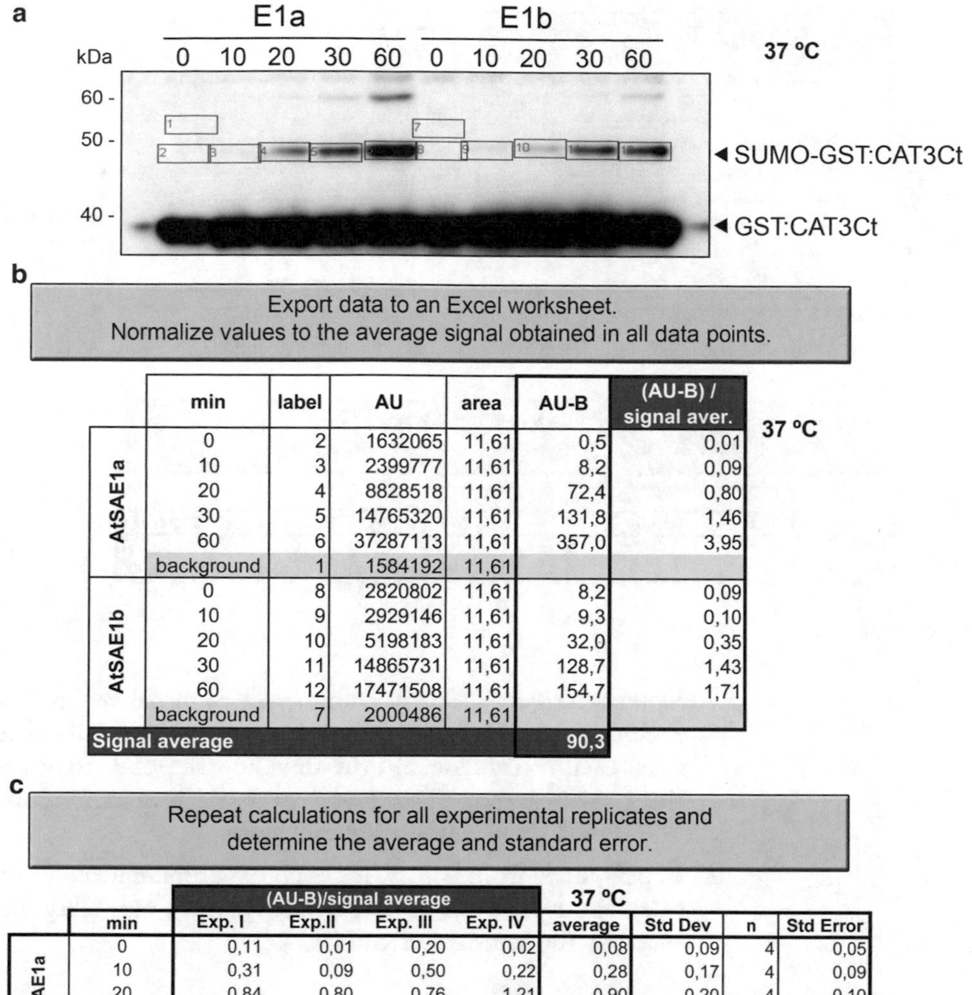

	min	label	AU	area	AU-B	(AU-B) / signal aver.
AtSAE1a	0	2	1632065	11,61	0,5	0,01
	10	3	2399777	11,61	8,2	0,09
	20	4	8828518	11,61	72,4	0,80
	30	5	14765320	11,61	131,8	1,46
	60	6	37287113	11,61	357,0	3,95
	background	1	1584192	11,61		
AtSAE1b	0	8	2820802	11,61	8,2	0,09
	10	9	2929146	11,61	9,3	0,10
	20	10	5198183	11,61	32,0	0,35
	30	11	14865731	11,61	128,7	1,43
	60	12	17471508	11,61	154,7	1,71
	background	7	2000486	11,61		
Signal average					90,3	

		(AU-B)/signal average							
	min	Exp. I	Exp.II	Exp. III	Exp. IV	average	Std Dev	n	Std Error
AtSAE1a	0	0,11	0,01	0,20	0,02	0,08	0,09	4	0,05
	10	0,31	0,09	0,50	0,22	0,28	0,17	4	0,09
	20	0,84	0,80	0,76	1,21	0,90	0,20	4	0,10
	30	1,60	1,46	1,38	2,43	1,72	0,48	4	0,24
	60	2,86	3,95	2,39	3,32	3,13	0,67	4	0,33
AtSAE1b	0	0,02	0,09	0,16	0,09	0,09	0,06	4	0,03
	10	0,26	0,10	0,34	0,25	0,24	0,10	4	0,05
	20	0,65	0,35	0,78	0,55	0,58	0,18	4	0,09
	30	1,21	1,43	1,24	0,86	1,18	0,24	4	0,12
	60	2,15	1,71	2,24	1,05	1,79	0,54	4	0,27

Fig. 4 Workflow of SUMO conjugation quantification. (**a**) Reaction products were resolved by SDS/PAGE and examined by immunoblot analysis with anti-GST antibodies. Luminescence signal was quantified using Multi Gauge software by using the rectangle selection tool for determining the ROIs (which all have the same area). (**b**) Data (AU) is exported to Excel, background signal is subtracted (AU-B column), and values are normalized to the average signal obtained in the particular dataset considering all data points ((AU-B)/signal aver. column). This data processing allows comparison between experimental replicates. (**c**) Average of results obtained from independent replicates is calculated. (**d**) Data obtained in (**c**) are plotted on scattered graphs and the reaction time window fitting on linear regression lines is selected for calculating the reactions slopes (e.g., for 37 °C incubation reaction linearity is maintained through 60 min, while for 42 °C incubation linearity is only maintained up to 30 min.). (**e**) Average of slopes obtained in independent replicates is calculated in order to compare multiple samples/assay conditions

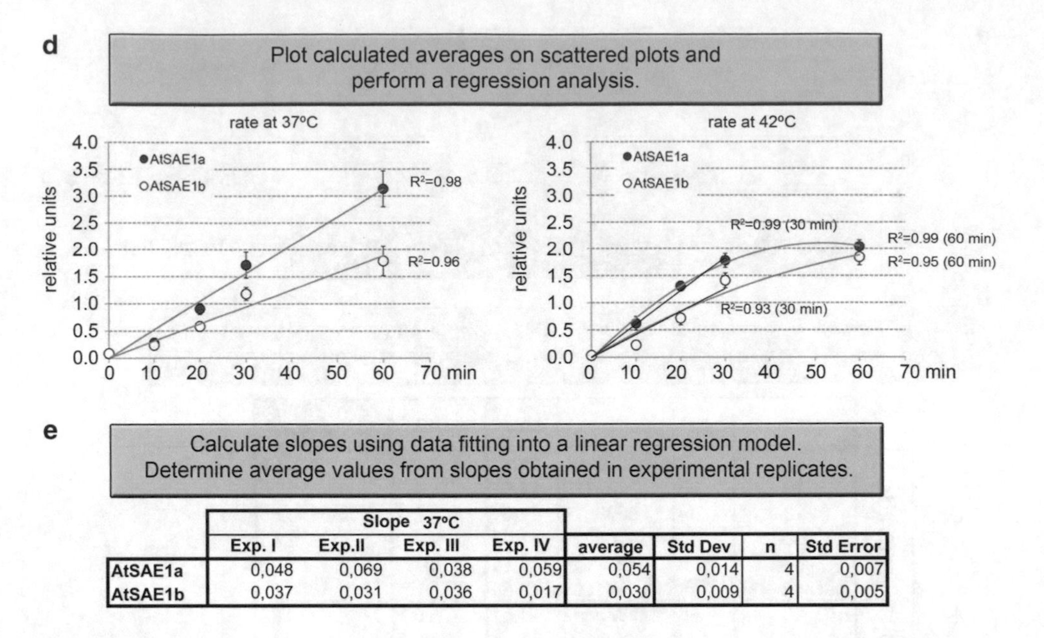

	Slope 37°C							
	Exp. I	Exp.II	Exp. III	Exp. IV	average	Std Dev	n	Std Error
AtSAE1a	0,048	0,069	0,038	0,059	0,054	0,014	4	0,007
AtSAE1b	0,037	0,031	0,036	0,017	0,030	0,009	4	0,005

Fig. 4 (continued)

7. Determine slopes, relative luminescence signal versus time, at each temperature for E1a or E1b using the normalized values from each membrane and the time-course points identified in the previous step. The slope of each line represents the GST:CAT3Ct SUMOylation efficiency.
8. Repeat steps from **2** to **7** for each experimental replicate and calculate the average of obtained slopes and variability as measured by the standard error (Fig. 4e).

4 Notes

1. *E. coli* BL21 Codon Plus RIL cells carry a plasmid pLysS (chloramphenicol resistance) that codifies for T7 lysozyme that represses the expression of the other genes under the T7 promoter but does not interfere with the protein expression induced by IPTG; allowing high efficiency of the protein of interest. This strain also contains extra copies of the argU, ileY, and leuW tRNA genes in order to avoid potential translation restrictions of heterologous proteins from organisms that have AT-rich genomes.
2. Take 1 mL sample before and after induction as a noninduced and induced control. Pellet cells by centrifugation and suspend it in 100 μL of cracking buffer. Store at −20 °C until SDS-PAGE analysis.
3. Cell disruption can be followed by Bradford quantification to ensure total cell lysis by sonication.

4. Filtration was performed using a reusable vacuum filtration system (e.g., Nalgene).
5. The amount of sepharose used has to be adapted to the scale of the experiment, the capacity of the sepharose being used (refer to manufacturer instructions), and the recombinant expression levels.
6. To ensure that digestion is complete analyze the reaction by SDS-PAGE before performing the next step. If partial digestion is detected, extend digestion by adding thrombin (add units according to the efficiency of the ON digestion).
7. Filtered and degas buffer solutions and samples are used in all chromatographic steps.
8. Each aliquot should be used only 3–4 times; extensive freeze-thawing cycles may lead to lose enzymatic activity of the native protein.
9. For this sample, we skip the thrombin digestion treatment.
10. When assembled in vitro reaction thaw samples on ice, centrifuge, and quantify aliquots by Bradford. If required, dilute enzymes using the 1× reaction buffer. Aliquots of 5× reaction buffer can be stored at −20 °C for 6 months.
11. Another control reaction can be done without the substrate, GST-AtCAT3Ct. In this case, using only a negative control at the highest temperature might be enough.
12. ATP aliquots are sensitive to the freeze-thawing cycles. Avoid more than 3 cycles.
13. Primary antibody might be applied to the PVDF membrane for 90 min at room temperature.
14. Image quantification accuracy will depend on the image acquisition system used. LAS4000 reader delivers a range cope from 0 to 65,535 before the image reaches saturation, while a scanned TIFF image delivers a range scope from 0 to 255 resulting in a sensitivity reduction. One interesting feature of Multi Gauge software is that allows contrast adjustment in order to visualize better the signals without varying the raw data that will be used for quantification.

Acknowledgments

This work was supported by the European Research Council (grant ERC-2007-StG-205927) and Departament d'Innovació, Universitats i Empresa from the Generalitat de Catalunya (Xarxa de Referència en Biotecnologia and 2014 SGR 447). L.C.M was supported by research contract through the CRAG. This article is based upon the work from COST Action (PROTEOSTASIS

BM1307), supported by COST (European Cooperation in Science and Technology). We thank Reyes Benlloch and Arnaldo L. Schapire for critical reading.

References

1. Wilkinson KA, Henley JM (2010) Mechanisms, regulation and consequences of protein SUMOylation. Biochem J 428(2):133–145
2. Walden H, Podgorski MS, Huang DT, Miller DW, Howard RJ, Minor DL Jr, Holton JM, Schulman BA (2003) The structure of the APPBP1-UBA3-NEDD8-ATP complex reveals the basis for selective ubiquitin-like protein activation by an E1. Mol Cell 12(6):1427–1437
3. Castaño-Miquel L, Seguí J, Lois LM (2011) Distinctive properties of *Arabidopsis* SUMO paralogues support the in vivo predominant role of AtSUMO1/2 isoforms. Biochem J 436(3):581–590. doi:10.1042/bj20101446
4. Lee I, Schindelin H (2008) Structural insights into E1-catalyzed ubiquitin activation and transfer to conjugating enzymes. Cell 134(2): 268–278
5. Lois LM, Lima CD (2005) Structures of the SUMO E1 provide mechanistic insights into SUMO activation and E2 recruitment to E1. EMBO J 24(3):439–451
6. Wang J, Hu W, Cai S, Lee B, Song J, Chen Y (2007) The intrinsic affinity between E2 and the Cys domain of E1 in ubiquitin-like modifications. Mol Cell 27(2):228–237
7. Gareau JR, Lima CD (2010) The SUMO pathway: emerging mechanisms that shape specificity, conjugation and recognition. Nat Rev Mol Cell Biol 11(12):861–871
8. Murtas G, Reeves PH, Fu YF, Bancroft I, Dean C, Coupland G (2003) A nuclear protease required for flowering-time regulation in *Arabidopsis* reduces the abundance of SMALL UBIQUITIN-RELATED MODIFIER conjugates. Plant Cell 15(10):2308–2319
9. Miura K, Lee J, Miura T, Hasegawa PM (2010) SIZ1 controls cell growth and plant development in *Arabidopsis* through salicylic acid. Plant Cell Physiol 51(1):103–113
10. Lois LM, Lima CD, Chua NH (2003) Small ubiquitin-like modifier modulates abscisic acid signaling in *Arabidopsis.* Plant Cell 15(6): 1347–1359
11. Sadanandom A, Adam E, Orosa B, Viczian A, Klose C, Zhang C, Josse EM, Kozma-Bognar L, Nagy F (2015) SUMOylation of phytochrome-B negatively regulates light-induced signaling in *Arabidopsis* thaliana. Proc Natl Acad Sci U S A 112:11108. doi:10.1073/pnas.1415260112
12. Nelis S, Conti L, Zhang C, Sadanandom A (2015) A functional Small Ubiquitin-like Modifier (SUMO) interacting motif (SIM) in the gibberellin hormone receptor GID1 is conserved in cereal crops and disrupting this motif does not abolish hormone dependency of the DELLA-GID1 interaction. Plant Signal Behav 10(2):e987528. doi:10.4161/15592324.2014.987528
13. Zhang S, Qi Y, Liu M, Yang C (2013) SUMO E3 Ligase AtMMS21 Regulates Drought Tolerance in *Arabidopsis* thaliana. J Integr Plant Biol 55:83
14. Lyzenga WJ, Stone SL (2012) Abiotic stress tolerance mediated by protein ubiquitination. J Exp Bot 63(2):599–616
15. Karan R, Subudhi PK (2012) A stress inducible SUMO conjugating enzyme gene (SaSce9) from a grass halophyte Spartina alterniflora enhances salinity and drought stress tolerance in *Arabidopsis.* BMC Plant Biol 12:187
16. Catala R, Ouyang J, Abreu IA, Hu Y, Seo H, Zhang X, Chua NH (2007) The *Arabidopsis* E3 SUMO ligase SIZ1 regulates plant growth and drought responses. Plant Cell 19(9): 2952–2966
17. van den Burg HA, Kini RK, Schuurink RC, Takken FL (2010) *Arabidopsis* small ubiquitin-like modifier paralogs have distinct functions in development and defense. Plant Cell 22(6): 1998–2016
18. Lee J, Nam J, Park HC, Na G, Miura K, Jin JB, Yoo CY, Baek D, Kim DH, Jeong JC, Kim D, Lee SY, Salt DE, Mengiste T, Gong Q, Ma S, Bohnert HJ, Kwak SS, Bressan RA, Hasegawa PM, Yun DJ (2007) Salicylic acid-mediated innate immunity in *Arabidopsis* is regulated by SIZ1 SUMO E3 ligase. Plant J 49(1):79–90
19. Lois LM (2010) Diversity of the SUMOylation machinery in plants. Biochem Soc Trans 38(Pt 1):60–64. doi:10.1042/BST0380060
20. Miura K, Rus A, Sharkhuu A, Yokoi S, Karthikeyan AS, Raghothama KG, Baek D, Koo YD, Jin JB, Bressan RA, Yun DJ, Hasegawa PM (2005) The *Arabidopsis* SUMO E3 ligase SIZ1 controls phosphate deficiency responses. Proc Natl Acad Sci U S A 102(21): 7760–7765

21. Ishida T, Fujiwara S, Miura K, Stacey N, Yoshimura M, Schneider K, Adachi S, Minamisawa K, Umeda M, Sugimoto K (2009) SUMO E3 Ligase HIGH PLOIDY2 Regulates Endocycle Onset and Meristem Maintenance in *Arabidopsis*. Plant Cell 21:2284–2297
22. Huang L, Yang S, Zhang S, Liu M, Lai J, Qi Y, Shi S, Wang J, Wang Y, Xie Q, Yang C (2009) The *Arabidopsis* SUMO E3 ligase AtMMS21, a homologue of NSE2/MMS21, regulates cell proliferation in the root. Plant J 60(999A):666–678
23. Villajuana-Bonequi M, Elrouby N, Nordstrom K, Griebel T, Bachmair A, Coupland G (2014) Elevated salicylic acid levels conferred by increased expression of ISOCHORISMATE SYNTHASE 1 contribute to hyperaccumulation of SUMO1 conjugates in the *Arabidopsis* mutant early in short days 4. Plant J 79(2):206–219. doi:10.1111/tpj.12549
24. Chosed R, Mukherjee S, Lois LM, Orth K (2006) Evolution of a signalling system that incorporates both redundancy and diversity: *Arabidopsis* SUMOylation. Biochem J 398(3):521–529
25. Castaño-Miquel L, Seguí J, Manrique S, Teixeira I, Carretero-Paulet L, Atencio F, Lois LM (2013) Diversification of SUMO-activating enzyme in *Arabidopsis*: implications in SUMO conjugation. Mol Plant 6(5):1646–1660. doi:10.1093/mp/sst049

Chapter 10

Expression, Purification, and Enzymatic Analysis of Plant SUMO Proteases

Gary Yates, Anjil Srivastava, Beatriz Orosa, and Ari Sadanandom

Abstract

The conjugation of SUMO can profoundly change the behavior of substrate proteins, impacting a wide variety of cellular responses. SUMO proteases are emerging as key regulators of plant adaptation to its environment because of their instrumental role in the SUMO deconjugation process. Here, we describe how to express, purify, and determine SUMO deconjugation activity of a plant SUMO protease.

Key words Protein purification, Protein expression, SUMO, Protease, Deconjugation, AtOTS1

1 Introduction

SUMOs (Small Ubiquitin-like MOdifiers) are proteins that modify the processes of proteins through covalent-linkage [1]. SUMO has been shown to have roles in many stress response pathways as they can be rapidly attached and detached from substrate proteins [2]. Much work has focus on the attachment sites, chain elongation, and substrate analysis, leaving the role of de-SUMOylation under studied. There are commonly two types of SUMO proteases described, ubiquitin-like-protein-specific proteases (Ulps) and Sentrin-specific proteases (SENPs), both types can act on "immature" SUMO by cleaving off the c-terminal to make it active and also by removing SUMO from its substrate [3].

Here, we describe methods to express, purify, and test putative plant SUMO proteases to confirm the enzymatic cleavage of a SUMO linked substrate. Expression of recombinant SUMO protease protein has been optimized and conditions for purification are described below. In addition to this, we show how to test the enzymatic activity of the SUMO protease in a simple assay that will result in cleavage of SUMO from its substrate or, as with this example, from a recombinant substrate. SUMO protease activity has been shown in vitro using the OTS1 protein from *Arabidopsis*

L. Maria Lois and Rune Matthiesen (eds.), *Plant Proteostasis: Methods and Protocols*, Methods in Molecular Biology, vol. 1450, DOI 10.1007/978-1-4939-3759-2_10, © Springer Science+Business Media New York 2016

(AtOTS1) and a His:SUMO:FLC fusion substrate [4, 5]. We use these components to describe ways in which to optimize each step to best suit the SUMO protease of choice.

2 Materials

2.1 Bacterial Culture and Purification of Arabidopsis OTS1 SUMO Protease

1. *AtOTS1* SUMO protease gene cloned into a vector with inducible promoter (pDEST17) and transformed into competent *E. coli* strain BL21.
2. Sterile 10 milliliters (ml) universal tubes, sterile 2 liter (l) flasks with indentations.
3. Liquid Broth (LB) media.
4. Antibiotics (for pDEST17 vector, Ampicillin).
5. Access to temperature controlled incubator.
6. Spectrophotometer (for measuring Optical Density).
7. Isopropyl-beta-D-thiogalactopyranoside (IPTG) at 1 Molar (M).
8. 1.5 ml centrifuge tubes.
9. Access to table top and ultra centrifuge machines.
10. Biohazard waste receptacle.
11. –20 °C freezer.
12. 4× Sodium dodecyl sulfate (SDS) loading buffer: (200 mM) Tris–HCl (pH 6.8), 400 mM Dithiothreitol (DTT), 8% SDS, 0.4% Bromophenol blue, 40% glycerol.
13. Access to heat block or water bath capable of 98 °C.
14. Bugbuster™ protein extraction reagent (Novogen).
15. Proteinase K (1 tablet/10 ml) (Roche/SigmaAldrich).
16. Bugbuster mix; 10 ml Bugbuster™ plus 1 Proteinase K tablet.
17. Coomassie Brilliant blue (Biorad).
18. 2 ml syringe.
19. Hypodermic needle.
20. 0.2–0.4 micrometer (μM) filter.
21. Sterile ultrapure water.
22. His-bind Resin and buffer kit: 8× Binding Buffer (8× = 4 M sodium chloride (NaCl), 160 mM Tris-hydrochloric acid (HCl), 40 mM imidazole, pH 7.9), 8× Wash Buffer (8× = 4 M NaCl, 480 mM imidazole, 160 mM Tris–HCl, pH 7.9), 8× Charge Buffer (8× = 400 mM nickel sulfate ($NiSO_4$)), 4× Elute Buffer (4× = 4 M imidazole, 2 M NaCl, 80 mM Tris–HCl, pH 7.9) (Novagen). All buffers are diluted in ultra pure water to make a working concentration of 1×.

2.2 SUMO Protease Assay

1. Access to a protein quantifier.
2. 20 microliter (μl) purified SUMO protease (recombinant protein).
3. Ice bucket.
4. 1× SUMO protease buffer (50 mM Tris–HCl, pH 8.0, 0.2% Igepal, 1 mM DTT).
5. Amicon Ultra 0.5 ml centrifugal filters (50 K concentrator columns) (MerckMillipore).
6. Purified SUMO substrate (for this example a recombinant HIS:SUMO:FLC protein was used, *see* refs. 4 and 5).

3 Methods

3.1 Expression of Recombinant SUMO Protease Protein

1. Using a single bacterial colony containing a vector with the cDNA for the protease to be expressed (in this example; pDEST17 with cDNA of AtOTS1), inoculate a 10 ml LB container for over night growth (16 h) at 37 °C supplemented with the appropriate antibiotic(s) (for AtOTS1 in pDEST17 Ampicillin was used—50 micro grams (μg) per ml of LB) (*see* **Notes 1 and 2**).
2. Prepare a sterile 2 l flask with 500 ml of LB with the appropriate antibiotics. Take 3 ml of the 10 ml over night culture and inoculate into the flask (*see* **Note 3**).
3. Grow the bacterial culture at 28 °C until optical density (OD_{600}) reaches 0.65 (*see* **Note 4**).
4. Once the OD_{600} of 0.65 is reached, take two 1 ml samples (labeled Total and Insoluble) in 1.5 ml micro centrifuge tubes (for further processing of these samples *see* **steps 6** and **8–11** in Subheading 3.1) and add 1 mM of IPTG to the remaining culture (*see* **Notes 5 and 6**).
5. At time points of 1, 2, and 3 h after adding IPTG, measure OD_{600} and adjust the volume so that samples contain the same number of cells (i.e., if OD_{600} is double take half the volume) take two samples in 1.5 ml micro centrifuge tubes and one 100 ml sample collected in a large centrifuge tube(s) (*see* **Note** 7).
6. Spin the samples in a tabletop centrifuge for 2 minutes (min) at 13,000 rpm (16,200 × *g*) decant supernatant and allow pellets to dry for 5–10 min (*see* **Note 8**).
7. Spin 100 ml samples at 5600 × *g* for 10 min then discard supernatant and let pellets air dry before freezing at −20 °C.
8. For Total extract, resuspended the pellet (labeled Total) in 60 μl of water and add 20 μl of 4× SDS loading buffer. Then heat samples at 98 °C for 3 min before placing on ice or freezing at −20 °C (*see* **Notes 9** and **10**).

9. Resuspend the pellet marked insoluble in 60 μl of Bugbuster mix and centrifuge for 2 min at 13,000 rpm (16,200 × *g*). Take supernatant to a new tube (labeled Soluble) and add 20 μl of 4× SDS loading buffer. To the left over pellet, add 60 μl of water and resuspend before adding 20 μl of 4× SDS loading buffer. Heat at 98 °C for 3 min (*see* **Notes 11** and **12**).
10. Samples are now ready for running on sodium dodecyl sulfate polyacrylamide gel electrophoresis (SDS PAGE).
11. Run two electrophoresis gels using one gel for coomassie staining and one for western blot analysis (*see* Fig. 1 and **Note 13**).

3.2 Purification of Expressed Protein Using Small Scale Batch Method

1. Using information from the SDS PAGE gels to find the best conditions, select the 100 ml pellet showing the best expression of the recombinant protein.
2. Weigh pellet and add 5 ml of Bugbuster mix per gram (100 ml should yield approx. 0.5 g of pellet).
3. Shake gently at room temperature for 15–20 min.
4. Centrifuge at 17,000 × *g* for 15 min at 4 °C.
5. Using a syringe and needle take up supernatant and pass through a 0.4 μm filter. Store on ice until **step 10** in Subheading 3.2 (*see* **Note 14**).

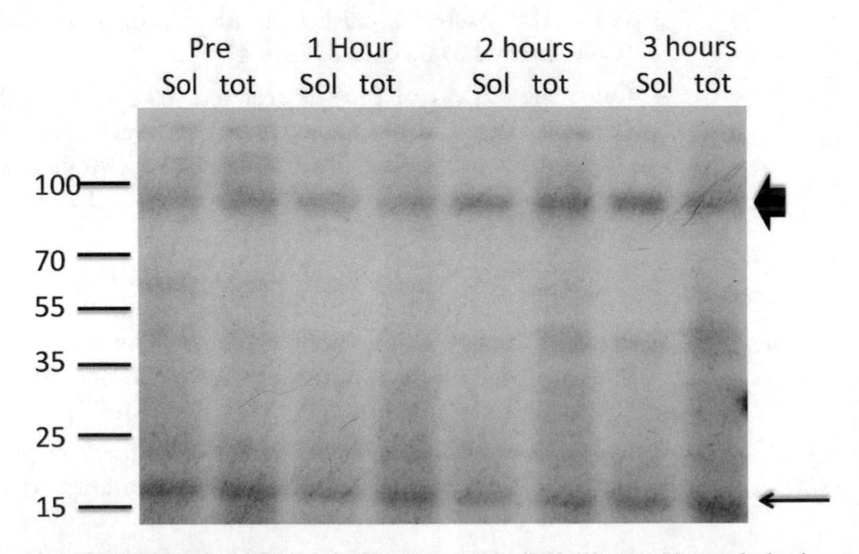

Fig. 1 Expression of AtOTS1 in ***E. coli*** BL21 strain is induced by IPTG. Western blot analysis shows AtOTS1:HIS protein can be seen just below the 100 kDa marker (indicated by the ***block arrow***) on an 8 % acrylamide gel. The IPTG promoter system in the pDEST17 Gateway destination vector system shows leaky expression in the BL21 strain as can be seen in the Preinduction lanes (Pre), at each 1 h time point after induction the AtOTS1:HIS band gets more intense. This indicates that expression increases over time as tested up to 3 h after addition of IPTG. Sol = soluble fraction, Tot = total extract. ***Thin arrow*** shows nonspecific binding by HIS antibody. ***Numbers*** on the ***right*** of panels indicate protein molecular weights in KiloDaltons

6. Add 400 μl of His-bind resin to a 1.5 ml micro centrifuge tube and spin in a table top centrifuge at 1000 × *g* for 1 min. Carefully remove and discard supernatant (*see* **Note 15**).
7. Wash with 800 μl of sterile water, invert several times, and spin at 1000 × *g* for 1 min, carefully removing and discarding supernatant after spinning. Repeat this step one more time (*see* **Note 16**).
8. Wash with 800 μl 1× charge buffer, invert several times and spin at 1000 × *g* for 1 min, carefully remove and discard supernatant. Repeat this step two more times.
9. Add 800 μl of 1× binding buffer, invert several times, and spin at 1000 × *g* for 1 min, remove and discard supernatant. Repeat this step one more time.
10. Add extract from **step 5** in Subheading 3.2 to the resin, invert several times, and incubate on ice for 5 min inverting several times every minute. Spin for 1 min at 600 × *g*. Discard supernatant.
11. Wash with 1.2 ml for 1× binding buffer, invert several times, and spin at 600 × *g* for 1 min, carefully remove and discard supernatant. Repeat this step two more times.
12. Wash with 1.2 ml for 1× wash buffer, invert several times, and spin at 600 × *g* for 1 min, carefully remove and discard supernatant. Repeat this step one more time.
13. Elute protein with 1.2 ml of 1× elution buffer, invert several times, and spin at 600 × *g* for 1 min, carefully remove and save supernatant. Repeat this step one more time.
14. Take 40 μl of eluted protein add 13.3 μl of 4× SDS loading buffer, and run on a SDS PAGE gel. *See* Fig. 2.

3.3 SUMO Protease Activity Assay Using a SUMO Linked Substrate

(The following four steps should be performed with the purified HIS:SUMO:FLC SUMO substrate in addition to the purified putative SUMO protease).

1. Load 400 μl of purified SUMO protease protein onto a buffer exchange/concentrator column. Centrifuge at 1000 × *g* for 4 min at 0 °C. Discard flow through. Repeat this step one more time using the same column (*see* **Notes 17** and **18**).
2. Wash column with 400 μl, of SUMO protease buffer, spin again at 1000 × *g* for 4 min at 0 °C and discard flow through.
3. Add 400 μl of SUMO protease buffer, place column upside down in a clean collection tube, and spin at 1000 × *g* for 5 min at 0 °C. Collect the elution and place on ice (*see* **Note 19**).
4. Measure concentration of purified protein and substrate protein.
5. Add 40 μg of both protease and substrate to a 1.5 ml centrifuge tube containing 20 μl of 10× SUMO protease buffer (*see* **Notes 20, 21, 22** and **23**).

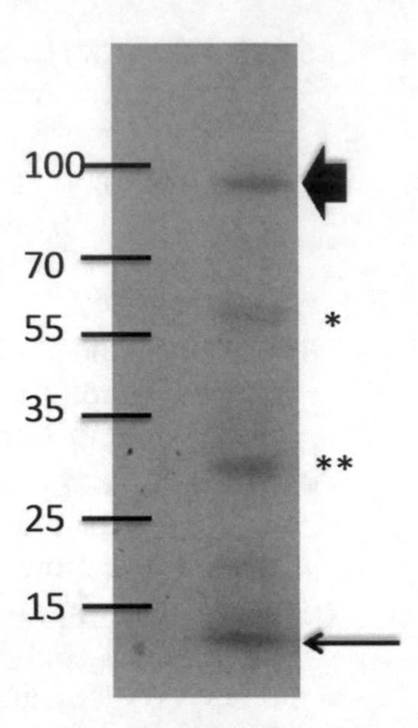

Fig. 2 Small scale batch purification of AtOTS1 using Ni-NTA Histidine bind resin. Histidine bind resin (Ni-NTA resin, Qiagen) was used to affinity purify the AtOTS1:His tagged protein, and results analysis by Western blotting with anti-HIS antibodies. ***Block arrow*** shows the purified AtOTS1:His protein. ***Thin arrow*** indicates free (or cleaved) His, and both * and ** show degradation products. ***Numbers*** on the ***right*** of panels indicate protein molecular weights in KiloDaltons

6. Make up to a final volume of 200 μl using ultrapure water and mix.
7. Take the same volume of protease and substrate as **step 5** in Subheading 3.3 and add each to 1.5 centrifuge tubes. Make up to 200 μl using ultrapure water.
8. Incubate at all at 30 °C, taking 20 μl samples at 3, 6, and 12 h.
9. To these samples add 6.6 μl of 4× SDS loading buffer and heat at 98 °C for 3 min. Store on ice or freeze until all samples are collected.
10. Run samples on a SDS PAGE gel and perform a western blot for analysis (*see* Fig. 3).

4 Notes

1. Optimal time for bacterial growth may vary depending on strains etc. Altering the conditions, such as incubation time and temperature, may yield better results. If expression is low try growing samples at 18, 23, 28, and 37 °C after adding IPTG.
2. Use of IPTG inducible promoter is recommended as the recombinant protein may interfere with normal bacterial growth. Inducible promoter constructs are not always silenced

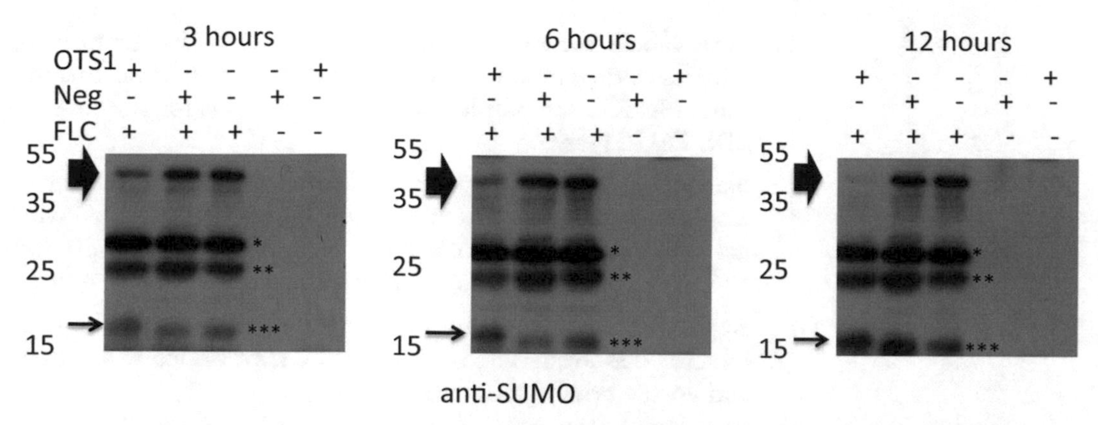

Fig. 3 AtOTS1 cleaves SUMO from HIS:SUMO:FLC substrate. Results of protease assay visualized by western blot analysis using anti-SUMO1 antibody. ***Block arrow*** indicates the substrate and the ***thin arrow*** shows cleaved HIS:SUMO. Over time the HIS:SUMO:FLC substrate is reduced in the presence of AtOTS1 but not in the presence of the negative control (Neg). * and ** show nonspecific bands appearing in the lanes with the substrate due to degradation of HIS:SUMO:FLC. *** shows untagged SUMO degradation. antibody ***Numbers*** on the ***right*** of panels indicate protein molecular weights in KiloDaltons

prior to induction, therefore some expression may be present in the preinduced samples.

3. Autoclave sterilization of media is recommend and antibiotics for selection should be made fresh to ensure the likely hood of only recombinant strain growth.
4. OD_{600} of bacteria prior to IPTG addition can be between 0.6 and 0.8, although we found 0.65 to be optimal here.
5. For ease later on, it is suggested that all tubes are labeled prior to taking samples, and all components made and ready in advance.
6. IPTG concentrations can be altered to find optimal expression, 1 mM is nearing the higher end for inducing expression. For initial testing of conditions, it is recommended to use three concentrations, e.g., 1, 0.5, and 0.1 mM. It is not uncommon for expression to be induced by concentrations as low as 0.05 mM, and less commonly as high as 4 mM.
7. Air drying of pellets is done by placing the inverted open tube on paper towel, and leaving until no liquid is left inside the tube, usually 5–10 min.
8. Bugbuster mix should be kept on ice at all times.
9. 100 ml samples can be taken as two 50 ml samples and spun in two 50 ml falcon tubes (later combined in the purification **steps 2** and **3**), or in one large 100–500 ml centrifuge container. Always make sure samples are evenly balanced before centrifugation.
10. Keep heated samples on ice if running SDS PAGE on the same day, or freeze at −20 °C for running another day.

11. Heat blocks are recommended over water baths for heating samples as they tend to ensure even distribution of heat to the tube. Heating samples helps denature proteins for running on SDS PAGE gels.
12. Standard western protocols are followed after SDS PAGE electrophoresis, and Coomassie staining is usually performed with gel in Coomassie blue solution shaking at 150 rpm for 30 min and then over night in destaining solution.
13. Bugbuster mix should be well vortex to ensure proteinase K tablet is dissolved. The mix should be kept on ice at all times and vortex briefly prior to each use.
14. Care has to be taken to ensure the pellet is not disturbed when up taking supernatant with needle and syringe.
15. Components of the His-Bind resin and buffer kit should be kept in fridge unless otherwise stated by manufactures protocol.
16. Removal of supernatant from His-bind resin is tricky, and extra care should be taken not to uptake the resin itself. Holding the tube in front of bright light makes seeing the layers easier.
17. Use of buffer exchange/concentrator columns follows manufacturer guidelines and adjustments may be required depending on the size of the purified SUMO protease. It is advised to follow the specification in the manufacturers protocol to best suit the protein purified.
18. Each time the concentrator column is spun it leaves a small volume of liquid in the bottom of the column, prior to the last spin pipette up and down when adding the final SUMO protease buffer.
19. Turning the tube upside down may cause liquid to pour out of the column, so its recommended that the collection tube goes on top of the column upside down, before inverting the two together.
20. It is better not to freeze the purified protease before using in the assay and although the protein may be ok after one flash freezing, it is recommend that the purification and protease assay are performed in the same day.
21. 150 mM of NaCl can be added to the SUMO protease buffer as some protease activity is enhanced by the presence of salt. This buffer can be added to the SUMO protease buffer and all steps preformed as described, this can be done at the same time as a buffer with no salt to see which buffer produces the best results.
22. Temperature of the assay can be reduced and incubation time increased if the purified protease is sensitive to heat (i.e., 4 °C for 16–24 h).

23. Concentration of proteins used in the assay can be doubled if no catalytic activity is seen, having made the other adjustments aforementioned.

Acknowledgment

European Research council provided grant-aided support to AS in the form of an ERC consolidator award.

References

1. Mahajan R, Delphin C, Guan T, Gerace L, Melchior F (1997) A small ubiquitin-related polypeptide involved in targeting RanGAP1 to nuclear pore complex protein RanBP2. Cell 88:97–107
2. Tempé D, Piechaczyk M, Bossis G (2008) SUMO under stress. Biochem Soc Trans 36: 874–878
3. Mukhopadhyay D, Dasso M (2007) Modification in reverse: the SUMO proteases. Trends Biochem Sci 32:286–295
4. Conti L, Price G, O'Donnell E, Schwessinger B, Dominy P, Sadanandom A (2008) Small ubiquitin-like modifier proteases OVERLY TOLERANT TO SALT1 and -2 regulate salt stress responses in *Arabidopsis*. Plant Cell 20:2894–2908
5. Murtas G, Reeves PH, Fu Y-F, Bancroft I, Dean C, Coupland G (2003) A nuclear protease required for flowering-time regulation in *Arabidopsis* reduces the abundance of SMALL UBIQUITIN-RELATED MODIFIER conjugates. Plant Cell 15:2308–2319

Chapter 11

Quantitative Analysis of Subcellular Distribution of the SUMO Conjugation System by Confocal Microscopy Imaging

Abraham Mas, Montse Amenós, and L. Maria Lois

Abstract

Different studies point to an enrichment in SUMO conjugation in the cell nucleus, although non-nuclear SUMO targets also exist. In general, the study of subcellular localization of proteins is essential for understanding their function within a cell. Fluorescence microscopy is a powerful tool for studying subcellular protein partitioning in living cells, since fluorescent proteins can be fused to proteins of interest to determine their localization. Subcellular distribution of proteins can be influenced by binding to other biomolecules and by posttranslational modifications. Sometimes these changes affect only a portion of the protein pool or have a partial effect, and a quantitative evaluation of fluorescence images is required to identify protein redistribution among subcellular compartments. In order to obtain accurate data about the relative subcellular distribution of SUMO conjugation machinery members, and to identify the molecular determinants involved in their localization, we have applied quantitative confocal microscopy imaging. In this chapter, we will describe the fluorescent protein fusions used in these experiments, and how to measure, evaluate, and compare average fluorescence intensities in cellular compartments by image-based analysis. We show the distribution of some components of the *Arabidopsis* SUMOylation machinery in epidermal onion cells and how they change their distribution in the presence of interacting partners or even when its activity is affected.

Key words Subcellular localization, Confocal microscopy, Fluorescence, Intensity, Quantification, SUMOylation

1 Introduction

Subcellular localization is essential to protein function since it determines the access of proteins to interacting partners and posttranslational modification machineries and enables the integration of proteins into functional biological networks [1].

Fluorescence microscopy is a powerful tool to study subcellular localization, protein–protein interactions, and intracellular dynamics of fluorophore tagged proteins [2]. The use of the green fluorescent protein (GFP) and its variants for generation of fluorescent fusion proteins facilitates the in vivo analysis of protein

L. Maria Lois and Rune Matthiesen (eds.), *Plant Proteostasis: Methods and Protocols*, Methods in Molecular Biology, vol. 1450,
DOI 10.1007/978-1-4939-3759-2_11,

dynamics relevant to cell biological processes [3]. Usually, for analysis of subcellular localization, the translational fusion of the protein of interest with a fluorescent protein is transiently expressed in plants cells and examined with confocal microscopy.

The subcellular distribution of many proteins can be influenced by binding to other biomolecules and by posttranslational modification, including SUMOylation, phosphorylation, acetylation, ubiquitylation, farnesylation, and proteolytic processing [1]. When subcellular redistribution is only partial, changes in fluorescence intensity in specific cellular components could be difficult to visually distinguish and, to circumvent this limitation, it is highly recommended to include quantitative evaluation of fluorescence images [4]. However, few works have addressed how to obtain accurate data of protein subcellular localization by quantitative confocal microscopy analysis, since the majority of subcellular localization studies have been qualitative in nature and nonrelated to plant cell biology research.

SUMO (Small Ubiquitin-like MOdifier) is a small protein that is covalently attached to lysine residues of target proteins via a reversible posttranslational modification. SUMO attachment is regulated by the sequential action of the heterodimer SUMO-activating E1-enzyme (SAE2/SAE1), the SUMO-conjugating E2-enzyme (SCE1), and E3-ligase enzymes [5]. As protein modifier, SUMO modulates protein activity through regulation of subcellular localization, protein activity and stability, and protein–protein interactions [6]. SUMOylation occurs predominantly in the nucleus, but nonnuclear proteins have also been identified as SUMO conjugation targets [7]. However, it is unclear whether SUMOylation enzymes are translocated out of the nucleus to catalyze SUMOylation in other cellular compartments. Interestingly, in mammals, both SUMO-E1 activating enzyme subunits have distinct functional nuclear localization signals, NLSs, although the NLS present at the E1 large subunit Uba2 is the only one required for the efficient import of the E1 complex into the nucleus [8]. Moreover, regulation of *Hs*E1 localization has been proposed to be also dependent on posttranslational modification by SUMO at the C-terminal domain, which would be required for its nuclear retention [9]. In addition to the SUMO machinery components, SUMO can modulate substrate subcellular localization through covalent modification of the substrate, or through noncovalent interactions mediated by SUMO interacting motifs, SIM, in the protein target, or both. A well reported example of subcellular distribution regulation by SUMO is the tumor suppressor PML. PML localizes in nuclear bodies and, in addition to be modified by SUMO, it contains a SUMO binding motif that is independent of its SUMOylation sites and necessary for nuclear bodies localization [10].

In plants, SUMO conjugation has been involved in the regulation of abiotic stress and defense responses, plant development, and flowering [11]. The Arabidopsis SUMO E1-activating enzyme displays nuclear localization like their human and yeast orthologues, consistent with the nuclear enrichment of SUMO targets identified in different studies [12, 13]. The E1 nuclear localization is determined by a conserved NLS located at the SAE2 E1-large subunit C-terminal tail [14]. Other members of the SUMOylation machinery are also localized to the nucleus, such as the SIZ1 E3 ligase that is present in the nucleoplasm and nuclear bodies [15]; the SUMO protease ESD4 that is enriched at the periphery of the nucleus [16]; and the SUMO proteases OST1 and OST2 that also localize to the nucleus, although OST1 is exclusively localized to the nucleoplasm while OST2 displays a nuclear punctuate pattern [17]. Other members of the SUMOylation machinery display a localization distributed among the nucleus and the cytosol such as SUMO1/2 [18], the E2 conjugating enzyme [18], and the E3 ligase MMS21 [19].

This protocol describes in detail a confocal image-based method to quantify and analyze the subcellular localization of some of the Arabidopsis SUMOylation machinery components. Specifically, we show that subcellular distribution of the SUMO E2-conjugating enzyme SCE1 is sensitive to its catalytic activity and to coexpression with SUMO1. We show that SCE1 was localized preferentially in the nucleus but could be also found in the cytosolic compartment. A point mutation in the SCE1 catalytic site, SCE1C94S, prevented efficient nuclear localization, suggesting a possible coupling of the catalytic activity to subcellular distribution. On the contrary, when SCE1 and SUMO1 were coexpressed, both proteins strongly colocalized in the nucleus and a significant signal reduction was observed in the cytosol. The quantitative analysis of the obtained confocal images allowed the statistical analysis of the observed subcellular protein dynamics. In this protocol, we describe the methods involving in vivo transient protein expression, image acquisition, quantification, and statistical analysis.

2 Materials

2.1 Vectors

All constructs were previously generated [18] and the map is shown in Fig. 1.

1. pWEN24 encoding ECFP.
2. pWEN25 encoding EYFP.
3. pWEN24 encoding the protein fusion ECFP:SUMO1 mature form (Met1-Gly93).
4. pWEN25 encoding the protein fusion EYFP:SCE1.
5. pWEN25 encoding the protein fusion EYFP:SCE1 catalytic inactive form C94S.

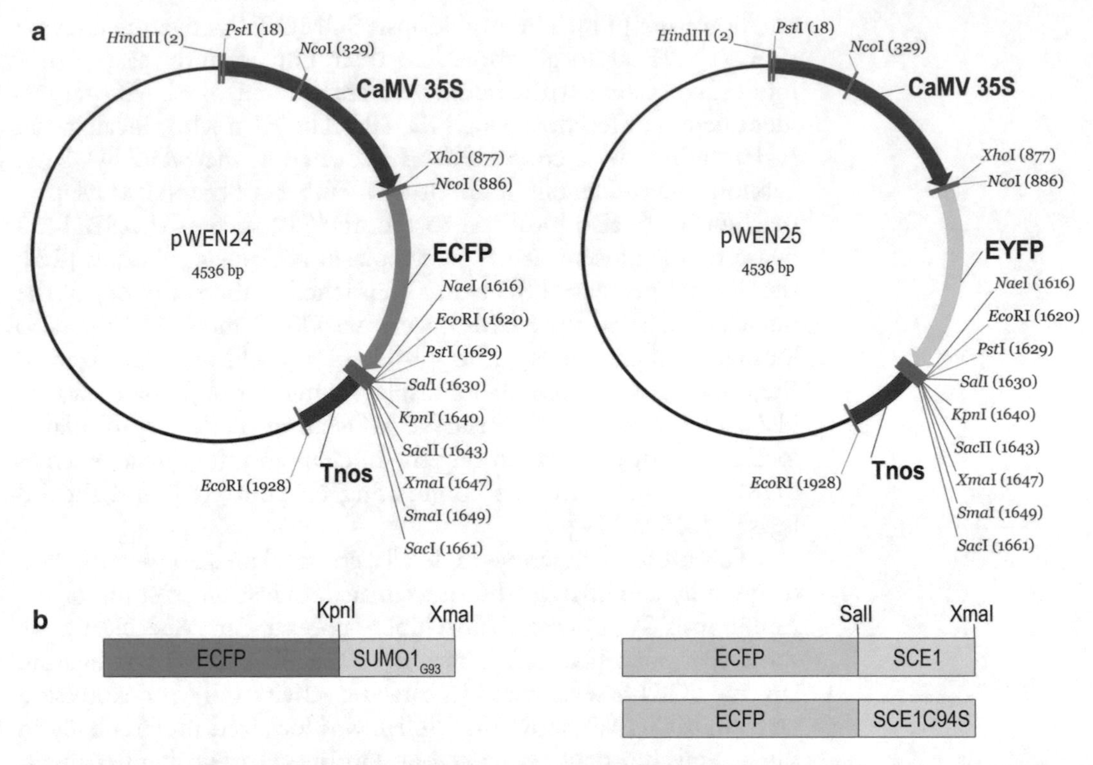

Fig. 1 Constructs used in this protocol for biolistic transient transformation. (**a**) pWEN24 (encoding the ECFP) and pWEN25 (encoding the EYFP) vectors were used as FP localization controls, and they were used for generating the ECFP::SUMO1, ECFP::SCE1, and ECFP::SCE1C94S protein fusion variants. The schematic representation of the protein fusions expressed in onion cells in this protocol are shown in panel (**b**)

2.2 Plant Tissue

Epidermis from inner onion leaves obtained at local stores (*see* **Note 1**).

2.3 Bombardment Equipment

1. PDS-1000/He System (BIO-RAD) Biolistic Particle Delivery System.
2. Macrocarriers Ref. 1652335, BIO-RAD.
3. Macrocarriers holders Ref. 1652322, BIO-RAD.
4. 1100 psi rupture disks Ref. 1652326, BIO-RAD.
5. Stopping screens Ref. 1652336, BIO-RAD.
6. Tungsten M17-Microcarriers Ref. 1652268, BIO-RAD.

2.4 Reagents

1. Pure Yield™ Plasmid Midiprep System (Promega) or similar.
2. Calcium Chloride Ref. C-4901 (Sigma Aldrich).
3. Spermidine Ref. S2626 (Sigma Aldrich).
4. Ethanol absolute, reagent grade ACS, ISO (Scharlau).

2.5 Microscopy Equipment

1. Surgical blades.
2. Microscope slides and cover slips.
3. Leica SPS confocal Laser Scanning Microscope.

2.6 Software

1. Leica SPS confocal software.
2. ImageJ freeware (http://rsbweb.nih.gov/ij/) and MS Excel.

3 Methods

3.1 Design and Generation of Fluorescence Chimeric Proteins

Choose a fluorescence protein for protein fusion chimera construction according to the available image acquisition equipment, biological sample restrictions, structural and functional organization of the protein of interest, and experimental design. The Green Fluorescence Protein (GFP) and its color genetic variants, as for example Yellow Fluoresce Protein (YFP), Cyan Fluorescence Protein (CFP), or Red Fluorescence Protein (RFP), are widely used in subcellular localization studies. Instruments with simple optical setup can easily distinguish between fluorescence proteins having none or minimal emission overlaps. We used as example (ECFP/EYFP) for the subcellular localization quantification of our proteins of interest as described in **item 2.1**, Subheading 2 (*see* **Note 2**).

Proteins of interest were fused at the C-terminus of fluorescent protein using standard molecular biology techniques. As for SUMO1, only N-terminal fusions (ECFP:SUMO) can be performed since the C-terminal fusion (SUMO:ECFP) would generate a nonconjugable SUMO form that could result in localization artifacts (*see* **Note 3**). Protein fusion expression was regulated by the strong and constitutive CaMV 35S promoter.

3.2 Biolistic Bombardment: Microcarrier Preparation

All steps were performed at room temperature and nonsterile conditions. Purity of used reagents meets the ACS reagent grade.

1. Weigh out 60 mg of microparticles into a 1.5 ml microfuge tube.
2. Add 1 ml of 70% ethanol (v/v).
3. Vortex vigorously for 3–5 min (a platform vortex is useful).
4. Allow the particles to soak in 70% ethanol for 15 min.
5. Pellet the microparticles by spinning for 5 s in a microfuge.
6. Remove and discard the supernatant.
7. Add 1 ml of autoclaved water in order to wash microparticles.
8. Vortex vigorously for 1 min.
9. Allow the particles to settle for 1 min.
10. Pellet the microparticles by briefly spinning in a microfuge.
11. Remove the liquid and discard.
12. Repeat **7–11** two additional times.
13. After the third wash, add 1 ml of sterile 50% glycerol to bring the microparticle concentration to 30 mg/ml (*see* **Note 4**).

3.3 Biolistic Bombardment: Coating Washed Microcarriers with DNA

1. Vortex prepared microcarriers for 5 min on a platform vortex to resuspend and disrupt agglomerated particles (*see* **Note 5**).
2. Transfer 12.5 μl of microcarriers to a 1.5 ml microcentrifuge tube.
3. Add 1–2 μg of DNA in a maximum volume of 2–3 μl (*see* **Note 6**).
4. Add the precipitation solution (12.5 μl 2.5 M $CaCl_2$ and 5 μl of 0.1 M spermidine) (*see* **Note 7**).
5. Vortex vigorously for 3 min.
6. Allow the particles to settle for 1 min.
7. Pellet the microcarriers by spinning 5 s in a microfuge.
8. Remove the supernatant and discard.
9. Add 200 μl of 70% ethanol.
10. Pellet the microcarriers by spinning 5 s in a microfuge.
11. Remove the supernatant and discard.
12. Add 200 μl of 100% ethanol.
13. Pellet the microcarriers by spinning 5 s in a microfuge.
14. Remove the supernatant and discard.
15. Add 20 μl of 100% ethanol.
16. Gently resuspend the pellet by tapping the side of the tube several times, followed by vortexing for 2–3 s (*see* **Note 8**).

3.4 Performing Bombardment

1. Prepare the onion samples by cutting the fresh inner leaves of the onion. Prepare three leaves for performing a triplicate transformation of each DNA sample.
2. Place the macrocarrier into the macrocarrier holder. Load 6 μl of microcarriers coated with DNA onto a macrocarrier. Prepare macrocarrier triplicates for each DNA sample (*see* **Note 9**).
3. Transfer selected macrocarriers to individual Petri dishes for easier handling.
4. Check helium supply, 200 psi in excess of desired rupture pressure (*see* **Note 10**).
5. Turn on the vacuum source and power ON the PDS-1000/He unit (*see* **Note 11**).
6. Load the rupture disk into retaining cap and tighten with torque wrench.
7. Load macrocarrier and stopping screen into the microcarrier launch assembly.
8. Place microcarrier launch assembly and target tissue in chamber and close door (*see* **Note 12**).
9. Generate vacuum in the chamber until a 27-mmHg (0.063 atm) pressure is reached and hold it (*see* **Note 13**).

10. Fire button continuously depressed until rupture disk burst and release Fire button (*see* **Note 14**).
11. Release vacuum from chamber.
12. Remove target tissue from chamber and unload macrocarrier, stopping screen from microcarrier launch assembly and broken rupture disk.
13. Repeat **steps 6–12** for each replicate.
14. Remove helium pressure from the system (after all experiments are completed).
15. Place the onion leaves over filter paper soaked in water and wrap in aluminum foil. Leave it at room temperature in the dark.

3.5 Fluorescence Protein Detection and Imaging

1. Screen plant samples under a fluorescence stereomicroscope to check if transient expression was successful 16 h after bombarding.
2. Cut with surgical blades an appropriate onion leaf piece containing cells exhibiting strong fluorescence, as a result of having a good transformation rate and expression level.
3. Remove the epidermal cell layer of the selected onion leaf area and place it on a microscopic slide containing a drop of water. Cover with a cover slip.
4. Set up all the hardware parameters and imaging settings of your confocal laser scanning microscope, and activate the sequential mode imaging in order to collect the fluorescence of coexpressed fluorophores independently (*see* **Note 15**).
5. Place the microscopic slide under a 20× objective in the confocal microscope in order to observe complete single onion cells.
6. Take a z-stack of a cell fixing the upper and lower limits of the z-series with a step size of 1 μm to reach the maximum cell depth (Fig. 2a) The maximum cell depth of the z-series is defined as the depth necessary for covering the maximum cell area and the whole nuclear volume (*see* **Note 16**).
7. Monitor image saturation degree under the imaging settings selected for EYFP imaging. Select HiLo Lut mode and scan the defined maximum cell depth for detecting saturated pixels, which appear highlighted on the screen.
8. Adjust gain parameter for generating an image displaying the minimum saturated pixels that ensure the full range quantification from 0 to 65553 in a 16 bit color depth. The presence of a portion of saturated pixels is necessary when comparing cell compartments displaying large differences in fluorescence intensities, such as nucleus versus cytosol, in order to measure significant fluorescence signal from the compartment exhibiting less intensity (the cytosol in this case).

9. Repeat **steps 7–8** for ECFP imaging.
10. Collect the z-stack series for EYFP, ECFP, and transmission light.

3.6 Average Fluorescence Intensity Measurements

1. Open collected z-stack image series with ImageJ software using split channel mode (http://rsbweb.nih.gov/ij/).
2. In the main menu, select the *image/stacks/z-projection/max intensity* option for generating the maximum projection of the z-series corresponding to the cell being analyzed (Fig. 2b) (*see* **Note 17**).
3. Check the nucleus saturation pixel degree: draw a circular region of interest (ROI) comprising the whole nucleus by using the freehand selection tool, open the *analyze/histogram* window (Ctrl + H) and determine the portion of saturated pixels contained in the ROI (Fig. 2c) We used as a criterion not analyzing images displaying fluorescence saturation for more than 20% of captured pixels (we estimate that pixels contained in the final upper bin of the histogram display fluorescence saturation or are very close to it) (Fig. 2d) (*see* **step 8** in Subheading 3.3).
4. With the *drawing/freehand selection tool*, draw in the maximum intensity projection image a first region of interest (ROI1) following the perimeter of the nucleus and click on the *Add button* on the *ROI manager* window. Then make a ROI2 enclosing cytosol but excluding the nucleus and click on the *Add button* on the *ROI manager* window. Finally, make a third ROI outside of the cell as a control of the background, for which we use the same area as the cytosol (Fig. 3a), click on the *Add button* on the *ROI manager* window. To analyze ROIs of the same size, for example cytosol and background, the selected ROI can be dragged with the cursor to other region of interest.
5. From the main menu open *Analyze/Set measurements* window and select *Area* and *Mean Gray Value* in the check box list (Fig. 3b). Next, open the ROI manager window (main menu\analyze\tools) and select both check boxes (show all and labels) (Fig. 3c) (*see* **Note 18**).
6. On the *ROI manager* window, select all generated ROIs and click on the *Measure button*. The Results window containing the information regarding the Areas and Average intensities for the selected ROIs will open (Fig. 3d).

3.7 Statistical Analysis of Average Fluorescence Intensity

The average fluorescence intensity in specific cellular compartments such as nucleus and cytosol must be quantified for each cell as follows:

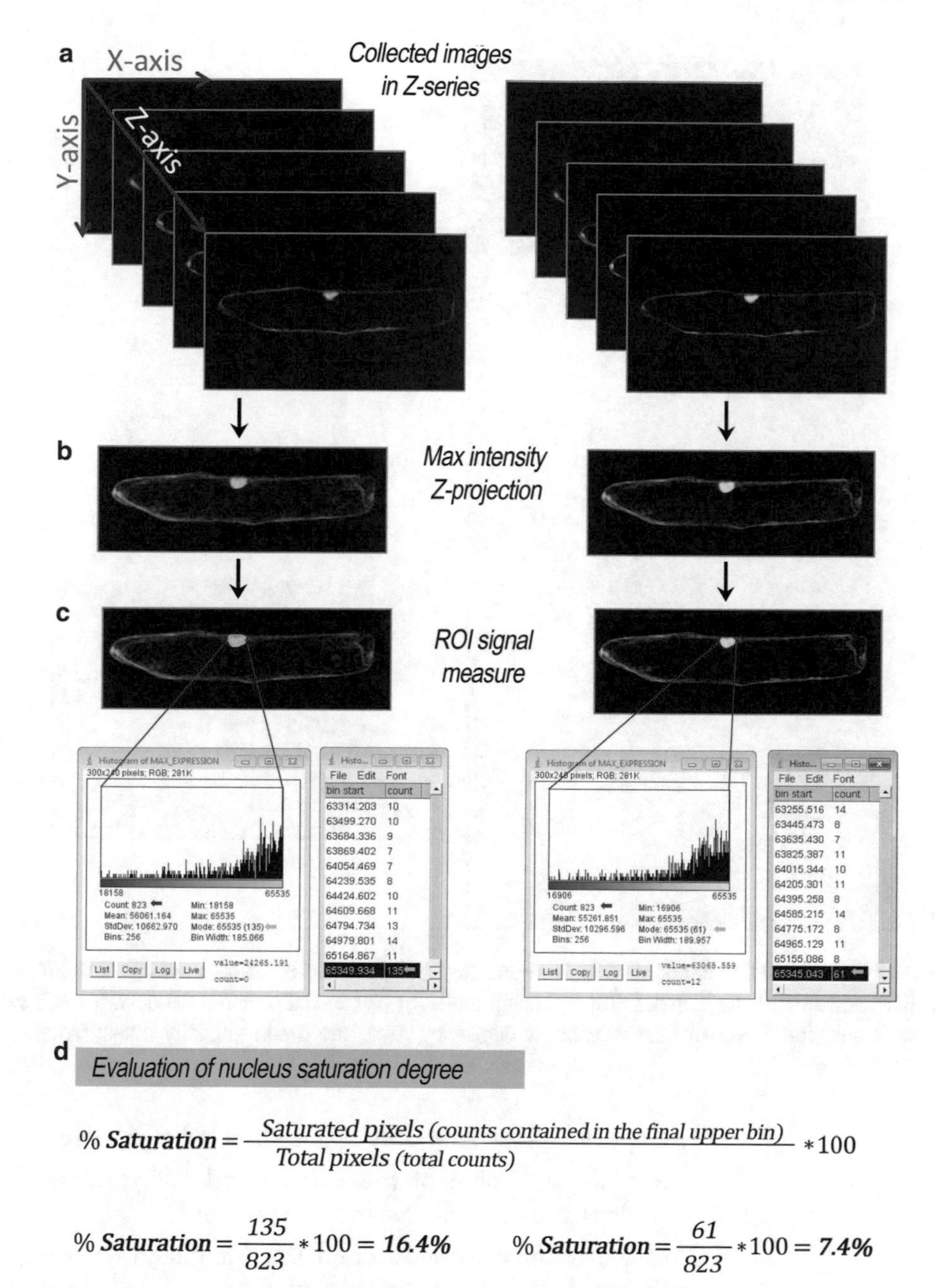

Fig. 2 Imaging of fluorescence protein detection. (**a**) Collected images in Z-series. (**b**) Maximum intensity Z-projection of the image stack. (**c**) The evaluation of nucleous saturation degree is estimated using the histogram tool. The histogram is built counting 823 pixels (Count) distributed among 256 bins. (**d**) The saturation degree is calculated as the relation between the number of saturated pixels (pixels contained in the final upper bin; *green arrow*) and the total pixels (*red arrow*). For instance, in the case of the ECFP, 135 pixels displays intensities between 65349 and the upper limit 65535, comprising the 17.2 % of the total pixels

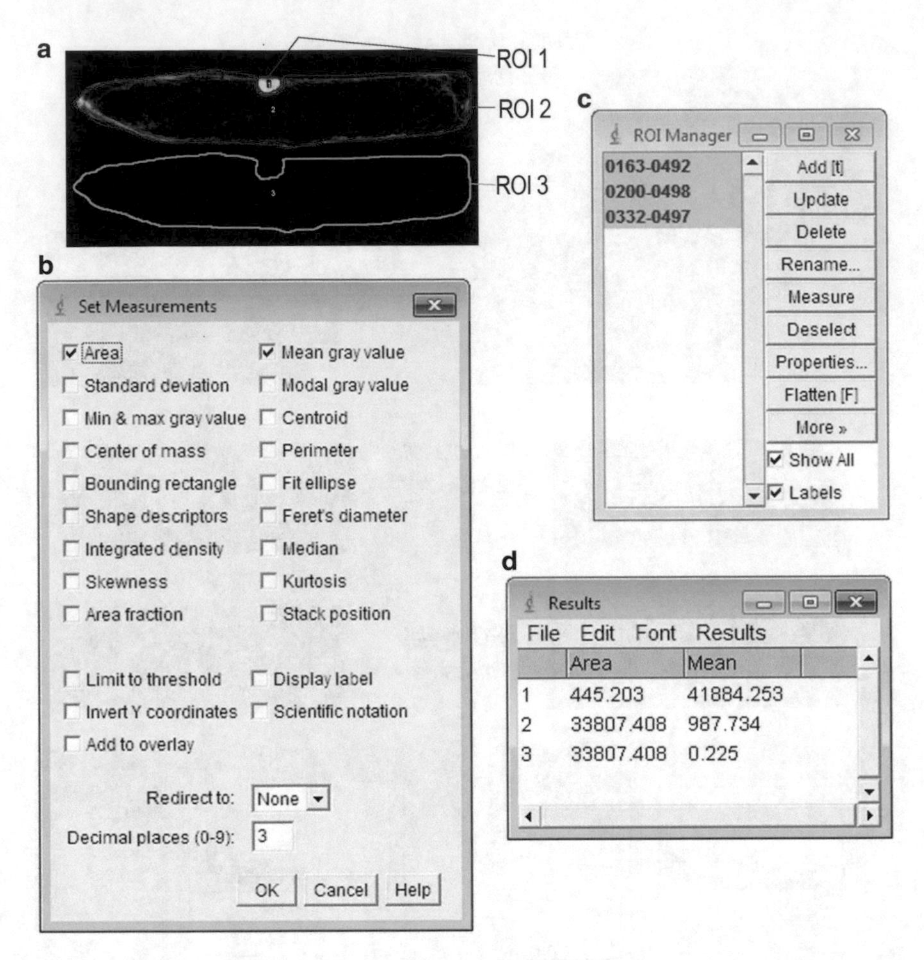

Fig. 3 Average fluorescence intensity measurement. (**a**) Multi ROI fluorescence intensity measurements by ImageJ. ROI1, nucleus. ROI2, cytosol. ROI 3, background. (**b**) *Set measurements* window. (**c**) *ROI manager* window with selected ROIs. (**d**) *Results* window displaying Area and mean intensity measurements of the selected ROIs

1. Copy data from **step 6** in Subheading 3.4 to an Excel file.
2. Remove the value of the background (BG) average to the nucleus and cytosolic mean intensity value.
3. Calculate the cytosolic and nuclear Integrated Density (ID) as the product of the cytosol or nuclear Area and the corrected Mean intensity without the background (*see* **Note 19**).

$$ID_{nucleus} = \text{ROI1 area} \times (\text{ROI1 mean intensity} - \text{ROI3 background intensity})$$

$$ID_{cytosol} = \text{ROI2 area} \times (\text{ROI2 mean intensity} - \text{ROI3 background intensity})$$

4. Calculate the Cytosol Fluorescence Ratio. In order to compare between different transformed cells, the Cytosol Fluorescence Ratio is calculated as a measure of the cytosolic signal enrichment.

$$\text{Cytosol Fluorescence Ratio} = \frac{\text{ID}_{\text{cytosol}}}{\text{ID}_{\text{nucleus}} + \text{ID}_{\text{cytosol}}}$$

At least seven cells must be analyzed in each transformation experiment.

5. Repeat **steps 1–3** for each fluorescence channel analyzed.
6. Calculate and plot the average of all obtained ratios and the corresponding standard errors (*see* **Note 20**).

As a practical example for the present protocol, we have analyzed the quantitative subcellular distribution of the SUMO E2 conjugating enzyme, SCE1, and evaluated the effect of SUMO coexpression and/or its catalytic activity on its localization. In Fig. 4, we show that SCE1 was localized preferentially in the nucleus but could also be found in the cytosolic compartment, consistent with a potential role for SCE1 in SUMOylating cytoplasmic proteins. A point mutation in the SCE1 catalytic site, SCE1C94S, prevented efficient nuclear localization, suggesting a possible coupling of catalytic activity to cellular localization. On the contrary, when SCE1a was coexpressed with SUMO1, both proteins colocalized strongly in the nucleus, with little signal detected in the cytoplasm. The effect of SUMO1 on SCE1 nuclear enrichment is not observed when the activity mutant SCE1C94S was coexpressed with SUMO, supporting a potential coupling of the catalytic activity to cellular localization (Fig. 4a). These observations were supported by quantitative data obtained applying the present protocol (Fig. 4b).

4 Notes

1. For biolistic bombardment assay we recommend to used fresh inner onion leaves as plant tissue because it is easy to obtain and, after peeling, it provides living cells in a monolayer, which facilitates confocal microscopy imaging. The cells of this tissue can be efficiently transformed since the microcarriers bombardment can be spread over a large homogenous area, without nonoverlying cell layers intercepting some of the particles delivered. Moreover, this tissue consists of large cells containing big nucleus and cytosol and, more interestingly, no chlorophyll interference, which make them easy to analyze. Alternatively, *Arabidopsis* roots are also suitable for this technique.
2. Optimal transformation and expression efficiency is obtained by using small plasmids such as the ones proposed in this protocol.
3. In absence of information about structural and functional protein organization, protein fusions should be performed at FP

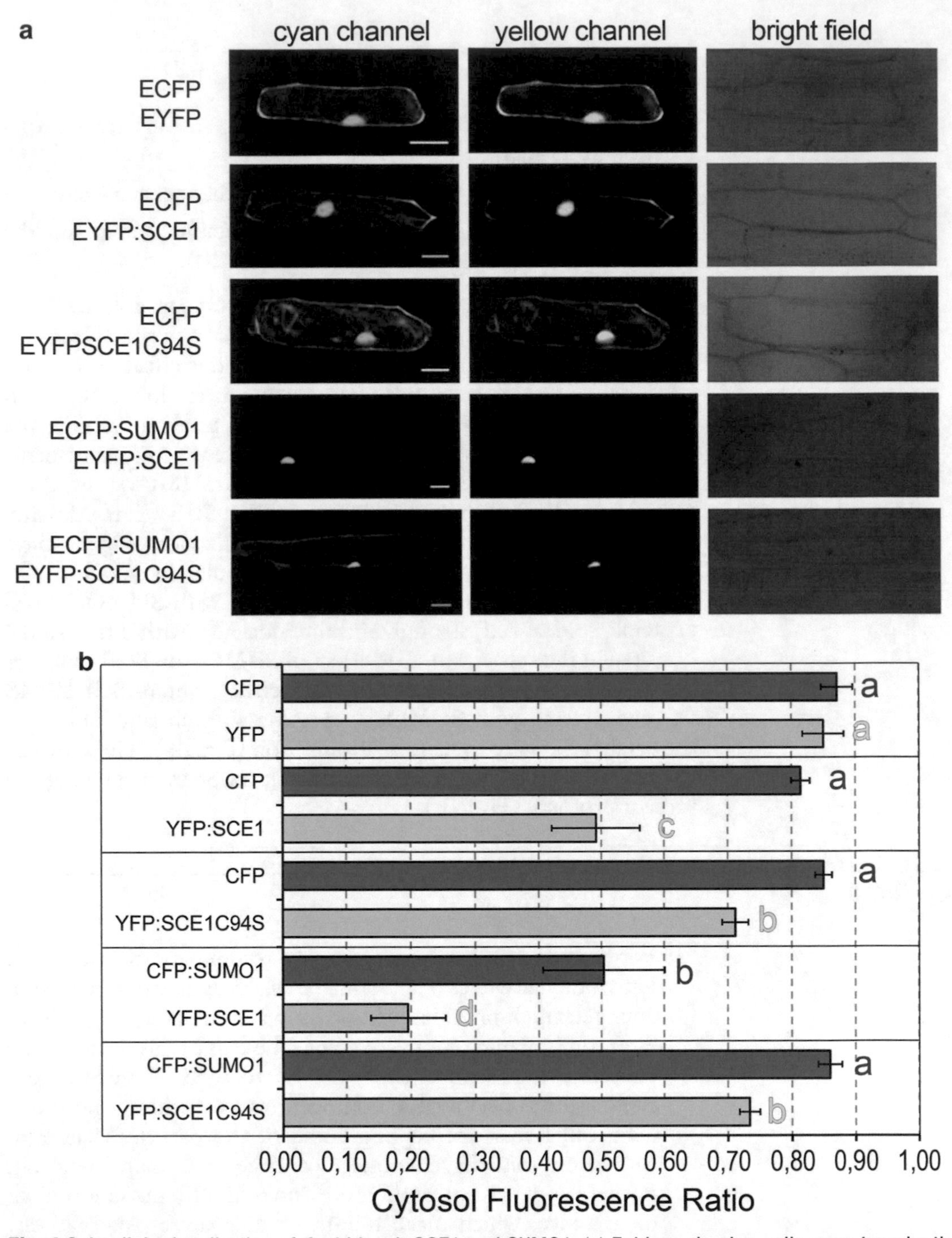

Fig. 4 Subcellular localization of *Arabidopsis* SCE1 and SUMO1. (**a**) Epidermal onion cells were transiently transformed with vectors expressing the following fluorescence proteins as indicated on the left: EYFP + ECFP, ECFP + EYFP:SCE1, ECFP + EYFP:SCEC94S catalytic mutant, ECFP:SUMO1 + EYFP:SCE1, and ECFP:SUMO1 + EYFP:SCE1C94S catalytic mutant. Bars = 50 μm. (**b**) Cytosol Fluorescence Ratio was measure for at least seven cells in each experiment as indicated in the present protocol. Average values and standard error are shown in the plot. *T*-test was performed for each fluorophore and *letters* next to the bars indicate those proteins displaying a significant distinct subcellular localization ($p \leq 0.02$)

C-terminus and N-terminus to corroborate that localization is not affected by the position of the fluorescent protein.

4. Tungsten aliquots should be stored at −20 °C to prevent oxidation. Avoid using old aliquots which will reduce the transformation efficiency of the assay.
5. When removing aliquots of microcarriers, it is important to continuously vortex the tube containing the microcarriers to maximize uniform sampling. When pipetting aliquots, hold the microcentrifuge tube firmly at the top while continually vortexing the base of the tube.
6. In the case of a cotransformation, for allowing equal transformation efficiency, both plasmids must be mixed before adding the microcarriers to the DNA sample. It is also desirable to use plasmids of similar size.
7. It is highly recommended to handle spermidine in one-use aliquots since freezing can affect its stability as well as transformation efficiency of the assay.
8. The DNA-coated microcarriers can be stored at −20 °C for few days, although is better to use it immediately.
9. The edge of the macrocarrier should be securely inserted under the lip of the macrocarrier holder. In case that there are not enough macrocarrier holders for all the samples, DNA-coated microcarriers can be loaded on the macrocarrier and transfer to the holder before performing the bombardment (in this case, we keep the prepared macrocarriers in Petri dishes labeled according to the DNA construct used).
10. 1100 psi rupture disks are recommended for plant tissues so the helium supply should have a pressure of 1300 psi.
11. It is recommended that vacuum generation and release are performed at the highest speed.
12. One of the most important parameters to optimize is target shelf placement within the bombardment chamber. This placement directly affects the distance that the microcarriers travel to the target cells for microcarrier penetration and transformation. We recommend starting with the closest second position to the stopping screen.
13. Set the vacuum switch on the PDS-1000/He (middle red control switch) to VAC position. When the desired vacuum level is reached, hold the camber vacuum at that level by quickly pressing the vacuum control switch through the middle VENT position to the bottom HOLD position.
14. With the vacuum level in the bombardment chamber stabilized, press and hold the FIRE switch to allow helium pressure to build inside the gas acceleration tube that is sealed by a selected rupture disk. A small pop will be heard when the rupture

disk burst. The rupture disk should burst within 10% of the indicated rupture pressure and within 11–13 s. Release the FIRE switch immediately after the disk ruptures to avoid wasting helium.

15. For accurate quantitative evaluation and comparison of average fluorescence intensities, we recommend using a confocal microscopy with high chromatic resolution with 16 Bits and 65535 grey levels. It is also highly recommended to take images with the same hardware parameters such as objective, laser power, pinhole opening, gain, offset values, and zoom factor as well as to prepare and analyze all experimental variants at the same time under the same conditions. All of this will allow reducing the influence of experimental conditions on the fluorescence intensity measurement and quantification. Regarding to laser power, adjust the intensity in order to avoid photodamage and photobleaching of the fluorescence. Try to use the minimum amount of laser power to get sufficient signal at gain levels that not result in too much background (700–800av). To enhance the quality of your image acquisition, a double scan or an average line of two from the image acquisition set up is recommended since it will diminish the background. The pinhole aperture can be increased if photodamage is observed due to laser illumination or if electronic noise occurs when the photomultiplied gain is increased. Take into account that the more you open the pinhole the more noise fluorescence you have, loosing confocality.

16. To eliminate the influence of the imaging depth on the fluorescence intensity, avoid plant cells with the nucleus located deeper, and start the z-series from the surface of the cell, otherwise the quality of images collected from deeper layers is worse due to the dispersion of laser light and the quantification and comparison of fluorescence intensity will be not appropriate. We have consider the maximum cell depth of the z-series as the depth necessary for covering the maximum cell area and the whole nuclear volume since we assume that half of the cell is more or less symmetrical to the other half. The main advantage of this maximum cell depth set up consists in a reduction of layer number in the z-series, which translates into shorter acquisition time and fluorescence photobleaching decrease.

17. In this method we perform the fluorescence intensity quantification in a maximum intensity projection, which is defined as an output image each of whose pixels contains the maximum value over all images in the stack at the particular pixel location.

18. The area is defined as the area of selection in square pixels. The Mean Grey Value, or average intensity, is the sum of the gray values of all the pixels in the selection divided by the number of pixels.

19. Integrated Density (ID) is an appropriate descriptor that allows comparing the cytosolic and nucleus intensity within cells with different size.
20. Perform statistical analysis applying the T-test (significant differences are considered when $p \leq 0.02$).

Acknowledgments

This work was supported by the European Research Council (grant ERC-2007-StG-205927) and Departament d'Innovació, Universitats i Empresa from the Generalitat de Catalunya (Xarxa de Referència en Biotecnologia and 2014SGR447). A.M. was supported by predoctoral fellowships FPU12/05292. This article is based upon work from COST Action (PROTEOSTASIS BM1307), supported by COST (European Cooperation in Science and Technology). We thank Reyes Benlloch for critical reading.

References

1. Hung MC, Link W (2011) Protein localization in disease and therapy. J Cell Sci 124(Pt 20):3381–3392. doi:10.1242/jcs.089110
2. Waters JC (2009) Accuracy and precision in quantitative fluorescence microscopy. J Cell Biol 185(7):1135–1148. doi:10.1083/jcb.200903097
3. Hanson MR, Kohler RH (2001) GFP imaging: methodology and application to investigate cellular compartmentation in plants. J Exp Bot 52(356):529–539
4. Lichocka M, Schmelzer E (2014) Subcellular localization experiments and FRET-FLIM measurements in plants. Bio Protoc 4(1):e1018, http://www.bio-protocol.org/e1018
5. Ulrich H (2009) The SUMO system: an overview. In: Ulrich H (ed) SUMO protocols, vol 497, Methods in molecular biology. Humana Press, Totowa, NJ, pp 3–16. doi:10.1007/978-1-59745-566-4_1
6. Wilkinson KA, Henley JM (2010) Mechanisms, regulation and consequences of protein SUMOylation. Biochem J 428(2):133–145
7. Wasik U, Filipek A (2014) Non-nuclear function of sumoylated proteins. Biochim Biophys Acta 1843(12):2878–2885. doi:10.1016/j.bbamcr.2014.07.018
8. Moutty MC, Sakin V, Melchior F (2011) Importin alpha/beta mediates nuclear import of individual SUMO E1 subunits and of the holo-enzyme. Mol Biol Cell 22(5):652–660
9. Truong K, Lee TD, Li BZ, Chen Y (2012) Sumoylation of SAE2 C terminus regulates SAE nuclear localization. J Biol Chem 287(51):42611–42619. doi:10.1074/jbc.M112.420877
10. Shen TH, Lin HK, Scaglioni PP, Yung TM, Pandolfi PP (2006) The mechanisms of PML-nuclear body formation. Mol Cell 24(3):331–339. doi:10.1016/j.molcel.2006.09.013
11. Lois LM (2010) Diversity of the SUMOylation machinery in plants. Biochem Soc Trans 38(Pt 1):60–64. doi:10.1042/BST0380060
12. Miller MJ, Scalf M, Rytz TC, Hubler SL, Smith LM, Vierstra RD (2013) Quantitative proteomics reveals factors regulating RNA biology as dynamic targets of stress-induced SUMOylation in Arabidopsis. Mol Cell Proteomics 12(2):449–463
13. Miller MJ, Barrett-Wilt GA, Hua Z, Vierstra RD (2010) Proteomic analyses identify a diverse array of nuclear processes affected by small ubiquitin-like modifier conjugation in Arabidopsis. Proc Natl Acad Sci U S A 107:16512
14. Castaño-Miquel L, Seguí J, Manrique S, Teixeira I, Carretero-Paulet L, Atencio F, Lois LM (2013) Diversification of SUMO-activating enzyme in arabidopsis: implications in SUMO conjugation. Mol Plant 6(5):1646–1660. doi:10.1093/mp/sst049
15. Miura K, Rus A, Sharkhuu A, Yokoi S, Karthikeyan AS, Raghothama KG, Baek D, Koo YD, Jin JB, Bressan RA, Yun DJ, Hasegawa PM (2005) The Arabidopsis SUMO E3 ligase SIZ1 controls phosphate deficiency responses.

Proc Natl Acad Sci U S A 102(21): 7760–7765

16. Murtas G, Reeves PH, Fu YF, Bancroft I, Dean C, Coupland G (2003) A nuclear protease required for flowering-time regulation in Arabidopsis reduces the abundance of SMALL UBIQUITIN-RELATED MODIFIER conjugates. Plant Cell 15(10):2308–2319
17. Conti L, Price G, O'Donnell E, Schwessinger B, Dominy P, Sadanandom A (2008) Small ubiquitin-like modifier proteases OVERLY TOLERANT TO SALT1 and -2 regulate salt stress responses in Arabidopsis. Plant Cell 20(10):2894–2908
18. Lois LM, Lima CD, Chua NH (2003) Small ubiquitin-like modifier modulates abscisic acid signaling in Arabidopsis. Plant Cell 15(6): 1347–1359
19. Huang L, Yang S, Zhang S, Liu M, Lai J, Qi Y, Shi S, Wang J, Wang Y, Xie Q, Yang C (2009) The Arabidopsis SUMO E3 ligase AtMMS21, a homologue of NSE2/MMS21, regulates cell proliferation in the root. Plant J 60(999A): 666–678

Chapter 12

Biochemical Analysis of Autophagy in Algae and Plants by Monitoring the Electrophoretic Mobility of ATG8

María Esther Pérez-Pérez, Ascensión Andrés-Garrido, and José L. Crespo

Abstract

Identification of specific autophagy markers has been fundamental to investigate autophagy as catabolic process. Among them, the ATG8 protein turned out to be one of the most widely used and specific molecular markers of autophagy both in higher and lower eukaryotes. Here, we describe how ATG8 can be used to monitor autophagy in Chlamydomonas and Arabidopsis by western blot analysis.

Key words Autophagy, ATG8, SDS-PAGE, Western blot, Chlamydomonas, Alga

1 Introduction

Autophagy is a widely conserved catabolic process by which eukaryotic cells degrade and recycle intracellular material or clear damaged organelles. Autophagy is characterized by the formation of double-membrane vesicles known as autophagosomes that engulf, in bulk or selectively, cellular components for degradation via fusion with the vacuole or lysosome. The autophagy machinery is composed of conserved autophagy-related (ATG) proteins that mediate the formation of the autophagosome [1, 2]. Among these core ATG proteins, ATG8 plays an essential role in autophagy and has been widely used to monitor this degradative process in multiple systems including plants and algae [3, 4]. Unlike other ATG proteins, ATG8 associates with both inner and outer membranes of the autophagosome by covalent binding to the lipid molecule phosphatidylethanolamine (PE) at a highly conserved glycine residue exposed at the C terminus of the protein [1]. ATG8 lipidation occurs through the sequential action of other highly conserved ATG proteins that constitute the ATG8 conjugation system. This system is composed by the ATG4 cysteine protease that cleaves ATG8 at the C terminus to expose the reactive glycine

L. Maria Lois and Rune Matthiesen (eds.), *Plant Proteostasis: Methods and Protocols*, Methods in Molecular Biology, vol. 1450,
DOI 10.1007/978-1-4939-3759-2_12,

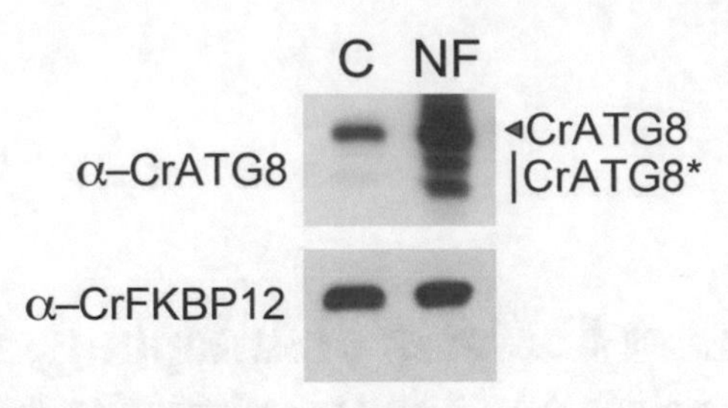

Fig. 1 Western blot analysis of CrATG8 in Chlamydomonas cells treated with 20 μM norflurazon (NF) for 24 h. Untreated cells were used as a negative control (C) of autophagy activation for this experiment. Growth conditions and NF treatment were as described in Pérez-Pérez et al. [12]. Modified CrATG8 forms migrate faster than unmodified CrATG8 and are indicated with an *asterisk*. Optimal resolution of total extract proteins by 15 % SDS-PAGE allows detection of several bands corresponding to modified CrATG8. Western blot analysis with anti-CrFKBP12 was used as loading control

residue, the activating E1-like enzyme ATG7, and the conjugating E2-like enzyme ATG3 that binds PE to ATG8. Efficient lipidation of ATG8 also requires the participation of the E3-like system composed by ATG5, ATG10, ATG12, and ATG16.

Initial biochemical studies on yeasts ATG8 revealed that the modified form of this protein, which is conjugated with PE, migrates faster compared to unmodified ATG8 on SDS gels [5]. This biochemical feature of lipidated ATG8 has been used to investigate autophagy since modified ATG8 accumulates under conditions that activate this process. Detection of lipidated ATG8 by western blot techniques has been proved to be an effective method to monitor autophagy. In yeasts, algae, and plants accumulation of modified ATG8 forms have been reported upon autophagy activation [6]. For instance, in exponentially growing Chlamydomonas cells, a single band corresponding to unmodified CrATG8 can be detected by western blot analysis. By contrast, when cells are subjected to autophagy-activating conditions, lower apparent molecular mass bands corresponding to modified CrATG8 can be also detected (Fig. 1) [4]. Moreover, it was shown in Chlamydomonas that the overall protein abundance of CrATG8 also increases in response to autophagy activation [4]. The localization of ATG8 in the cell either by immunodetection of the endogenous protein or using GFP-ATG8 fusion proteins has also been used as a specific autophagy marker in different organisms [6]. In yeast cells with active autophagy, GFP-ATG8 is recruited together with other ATG proteins to the site of autophagosome formation, resulting in the detection of the fusion protein in spots that can be easily observed by fluorescence microscopy. GFP-ATG8 has been used in plants to monitor autophagy by labeling the accumulation of autophagic bodies inside the vacuole. In Chlamydomonas, the subcellular localization of endogenous CrATG8 has been analyzed by indirect immunofluorescence microscopy, revealing that this protein localizes to punctate structures upon autophagy activation. Here, we describe how ATG8 can be used to monitor autophagy in Chlamydomonas and Arabidopsis by western blot analysis.

2 Materials

Prepare all solutions using ultrapure water quality (obtained by purifying deionized water to reach a sensitivity of 18.2 MΩ cm at 25 °C). Prepare and store all reagents at the indicated temperature (unless indicated otherwise).

2.1 Growth Media and Components

Prepare, sterilize, and store stock solutions:

1. Tris-acetate phosphate (TAP) medium: add about 900 ml water to a 1 l graduated cylinder. Add 10 ml Tris-Acetate (100×), 25 ml Beij solution (40×), 1 ml 1 M phosphate potassium buffer (pH 7.0), and 1 ml Mineral Traces. Mix and check pH is 7.0. Make up to 1 l with water. Sterilize and store at room temperature (*see* **Note 1**).
2. Tris-Acetate (100×): dissolve 242 g Tris in 900 ml water, then add 100 ml glacial acetic acid.
3. 1 M Phosphate potassium buffer (pH 7.0): mix 250 ml 1 M K_2HPO_4 with 170 ml 1 M KH_2PO_4.
4. Beij solution (40×): add about 400 ml water to a 500 ml graduated glass beaker. Weigh and mix 2 g $CaCl_2 \cdot 2H_2O$. Add about 400 ml water to another 500 ml graduated glass beaker. Weigh and mix 16 g NH_4Cl and 4 g $MgSO_4 \cdot 7H_2O$. Transfer everything to a 1-l graduated cylinder, mix and make up to 1 l with water. Sterilize and store at room temperature (*see* **Note 2**).
5. Mineral traces:

   Prepare solution 1: dissolve in 550 ml water in the order indicated below, then heat at 100 °C:

   (a) 11.4 g H_3BO_3.

   (b) 22 g $ZnSO_4 \cdot 7H_2O$.

   (c) 5.06 g $MnCl_2 \cdot 4H_2O$.

   (d) 4.99 g $FeSO_4 \cdot 7H_2O$.

   (e) 1.61 g $CoCl_2 \cdot 6H_2O$.

   (f) 1.57 g $CuSO_4 \cdot 5H_2O$.

   (g) 1.1 g $(NH_4)_6Mo_7O_{24} \cdot 4H_2O$.

   Prepare solution 2: dissolve 50 g Na_2EDTA in 250 ml water by heating and add to the solution 1 at 100 °C.

   Heat the combined solutions to 100 °C, cool to 80–90 °C, and adjust to pH 6.5–6.8 with 20% KOH. The pH meter should first be calibrated at 75 °C; the temperature should remain above 70 °C. Adjust to 1 l, and allow a rust-colored precipitate to form, during 2 weeks at room temperature, in a 2 l Erlenmeyer flask loosely stoppered with cotton. The solution will change from green to purple. Then, the solution is filtered several times through three layers of Whatman paper, and a clear purple solution is obtained. Finally, the mineral traces are aliquoted in 50 ml tubes and stored at –80 °C. This protocol is based on Hutner et al. [7].

2.2 SDS-PAGE Components

1. Resolving gel buffer: 1.5 M Tris–HCl, pH 8.8.
   Add about 400 ml water to a 500 ml glass beaker. Weigh 90.9 g Tris and dissolve it in the water. Mix and adjust pH with HCl 37% (v/v). Transfer to a 500 ml graduated cylinder and make up to 500 ml with water. Sterilize and store at 4 °C.
2. Stacking gel buffer: 0.5 M Tris–HCl, pH 6.8.
   Weigh 30.3 g Tris and prepare a 500 ml solution as described above for the resolving buffer. Sterilize and store at 4 °C.
3. Acrylamide 40% (w/v) solution (Acrylamide:Bis-acrylamide, 29:1), electrophoresis grade (Fisher Scientific). Store at 4 °C.
4. *N*,*N*,*N*,*N*′-tetramethyl-ethylenediamine (TEMED) (AppliChem). Store at 4 °C.
5. 10% ammonium persulfate (APS) (w/v) solution in water. Prepare 10 ml, aliquot in 0.5 ml tubes and store at –20 °C.
6. SDS-PAGE running buffer: 0.025 M Tris, 0.192 M glycine and 0.1% (w/v) SDS (*see* **Note 3**).
7. Loading buffer: 0.125 M Tris–HCl (pH 6.8), 4% (w/v) SDS, 10% (v/v) β-mercaptoethanol, 0.025% (w/v) bromophenol blue, 20% (v/v) glycerol. This is a 4× stock solution; prepare 5 ml, aliquot in 1 ml tubes, and store at –20 °C. Use it as 1× solution with the protein sample.
8. Bromophenol blue solution: prepare a 0.1% (w/v) solution in water.
9. Electrophoresis unit: SE260 Mighty Small II (GE Healthcare) (*see* **Note 4**).
10. Protein assay dye reagent (Bio-Rad) was used to determine protein concentration of samples as indicated by the manufacturer.

2.3 Immunoblot Components

1. Blotter: TE 77 PWR Semi-Dry transfer unit (GE Healthcare). Follow the manufacturer's instructions.
2. Nitrocellulose membranes: Hybridization nitrocellulose filter (0.45 μm HATF) provided by Millipore.
3. Blotting paper: Grade 3MM Chr cellulose chromatography papers (GE Healthcare).

4. Western blot transfer buffer: 0.025 M Tris, 0.192 M glycine, 20% (v/v) methanol, and 3.75% (w/v) SDS. Prepare and store at 4 °C.
5. Phosphate buffered saline (PBS): 0.136 M NaCl, 0.027 M KCl, 0.010 M Na_2HPO_4, 0.009 M KH_2PO_4 at pH 7.4 (*see* **Note 5**).
6. PBS (1×) containing 0.1% (v/v) Tween-20 (PBS-T): Prepare PBS (1×) from PBS (10×) and add 0.1% (v/v) Tween-20 (Sigma) (*see* **Note 6**).
7. Blocking solution: PBS-T with 5% (w/v) milk powder (*see* **Note 7**).
8. Container: we use square, plastic Petri dishes (120 mm × 120 mm) to incubate and wash nitrocellulose membranes.
9. Anti-CrAtg8 polyclonal antibody: the antibody was produced as described in Pérez-Pérez et al. [4]. Dilute to a final concentration of 1:2500 in blocking solution (*see* **Note 8**).
10. Secondary antibody: horseradish peroxidase (HRP)-conjugated anti-rabbit antibody (Sigma) was used to a final dilution of 1:10,000 in blocking solution.

3 Methods

3.1 Preparation of Proteins from Chlamydomonas

1. Chlamydomonas cells were grown under continuous illumination (20–30 μE/m²s) at 100 rpm at 25 °C in TAP liquid medium. Typically, a 250-ml flask containing 50 ml of TAP medium was inoculated with Chlamydomonas cells to a final density of 10^5 cells/ml. Allow cells to grow to a cell density of 1–2 × 10^6 cells/ml, usually this takes about 24 h (*see* **Note 9**). Cell density was measured by using a Scepter cell counter (Millipore) with a 40 μm sensor.
2. Log phase cells (1–2 × 10^6 cells/ml) were then subjected to autophagy activation (*see* **Note 10**).
3. Cells were collected by centrifugation (4000 × *g*, 5 min), washed once in 50 mM Tris–HCl (pH 7.5), and resuspended in a minimal volume of the same solution (lysis buffer) (*see* **Note 11**).
4. Cells were lysed by two cycles of slow freezing (samples were introduced in a −80 °C freezer for at least 1 h) followed by thawing at room temperature. The soluble cell extract was separated from the insoluble fraction by centrifugation (15,000 × *g*, 20 min) in a microcentrifuge at 4 °C (*see* **Note 12**).

3.2 Preparation of Proteins from Arabidopsis

1. Plants were first subjected to autophagy-activating conditions such as nutrient (nitrogen or carbon) limitation or oxidative stress (hydrogen peroxide or methyl viologen treatment) as

previously described [8–10] in order to activate this degradative process.

2. Total protein extracts were obtained as described by Yoshimoto et al. [10] and modified by Alvarez et al. [11]. Essentially, about 200 mg leaves were pestled in liquid nitrogen with a minimal volume of extraction buffer (100 mM Tris–HCl pH 7.5, 400 mM sucrose, 1 mM EDTA, 0.1 mM phenylmethylsulfonyl fluoride (PMSF), 10 mg/ml sodium deoxycholate, 10 μg/ml of leupeptin, 10 μg/ml of pepstatin A, 4% (v/v) protease inhibitor cocktail from Roche) using a mortar.
3. Soluble material was obtained by centrifuging at 500×*g* for 10 min at 4 °C to remove cell debris.
4. The supernatant fraction containing soluble and membrane-bound proteins was analyzed by SDS-PAGE and immunoblot using anti-CrATG8 antibody.

3.3 Separation and Analysis of Proteins by Electrophoresis

1. Proteins from total extracts were quantified with the Bio-Rad protein assay dye reagent as described by the manufacturer.
2. Assemble the glass plates of the SE260 Mighty Small system following the manufacturer's instructions (GE Healthcare).
3. Prepare 15% acrylamide resolving gel (*see* **Note 13**). Leave sufficient space for the stacking gel and carefully overlay the acrylamide solution with water or isopropanol.
4. After polymerization is complete, pour the 5% acrylamide stacking gel onto the surface of the polymerized resolving gel. Immediately insert a Teflon comb, being careful to avoid trapping air bubbles.
5. Prepare samples for SDS-PAGE analysis. Typically, 15 μg of protein in a total volume of 20 μl were mixed with 5 μl of sample loading buffer. Samples were heated at 65 °C for 5 min before loading. Load prestained protein standards (*see* **Note 14**).
6. Apply a voltage of 150 V to the gel and let the dye front run out of the gel. Stop electrophoresis when the 6 kDa marker reaches the bottom of the gel (*see* **Note 15**).

3.4 Western Blotting and ATG8 Protein Detection

1. After SDS-PAGE, prepare the transfer unit with three pieces of 10×10 cm blotting paper and a nitrocellulose membrane of the same size, previously humidified with transfer buffer.
2. Gently lay the gel on the nitrocellulose membrane and place three pieces of humidified blotting paper on the gel.
3. Electrotransfer proteins from the gel to the membrane by applying a maximum current of 1 mA/cm^2 of the gel surface during 75 min.
4. After transfer, submerge the membrane in blocking solution for 60 min.

5. Add anti-CrATG8 antibody to the blocking solution to a final dilution of 1:2500 and incubate 3 h at room temperature or over night at 4 °C.
6. Wash the membrane four times with PBS-T, 5 min each.
7. Incubate the membrane with the secondary antibody (1:10,000 final dilution) in blocking solution at room temperature for 1 h.
8. Wash the membrane four times with PBS-T, 5 min each.
9. Signal was developed using the Luminata Crescendo Western HRP Substrate (Millipore) according to manufacturer's instructions.
10. Remove excess reagent and cover the membrane in transparent plastic wrap.
11. Acquire image using darkroom development techniques or a scanner for chemiluminiscence signal (*see* **Note 16**).

4 Notes

1. Add 1.2 % of agar to prepare TAP plates. When required, antibiotics, vitamins, amino acids, or drugs can be added before plating.
2. To prepare Beij solution, dilute $CaCl_2$ and $MgSO_4$–NH_4Cl separately.
3. A simple and useful method of preparing running buffer is to prepare a 10× stock solution buffer (0.25 M Tris, 1.92 M glycine, 1 % SDS). Add about 900 ml water to a 1 l graduated cylinder. Weigh and add 30.3 g Tris and 144 g glycine, mix and make up to 950 ml with water. Add 50 ml of SDS 20 % (Sigma). SDS should be added last since it is a detergent and makes bubbles. Store at room temperature. Use this stock solution to prepare a 1× running buffer solution in water when required.
4. The SE260 system from GE Healthcare allows running of 10 × 10.5 cm gels, which compared to standard 10 × 8 cm minigels gives more than 25 % higher resolution, specially for low molecular weight proteins.
5. A simple and useful method of preparing PBS is to prepare a 10× stock solution buffer (1.36 M NaCl, 0.27 M KCl, 0.10 M Na_2HPO_4, 0.09 M KH_2PO_4). Add about 900 ml water to a 1 l graduated cylinder. Weigh and add 80 g NaCl, 2 g KCl, 36.3 g Na_2HPO_4, 2.4 g KH_2PO_4. Mix and adjust pH with KOH 10 M. Sterilize and store at room temperature. Use this stock solution to prepare a 1× PBS solution in water when required.
6. Due to the high density of Tween 20 it is better to prepare a 50 % (v/v) stock solution in water and use it to prepare PBS-T.

7. Mix properly before using. Store at 4 °C, no longer than 1 day.
8. After using it, the blocking solution containing the CrATG8 antibody can be stored at −20 °C and reused several times.
9. It is very important cells do not enter stationary growth phase ($4–6 \times 10^6$ cells/ml) before applying any treatment or stress condition because CrATG8 protein abundance is upregulated when cells reach stationary growth [4].
10. Autophagy can be triggered in Chlamydomonas cell cultures by adding different drugs and compounds or by shifting cells to physiological stress conditions. Chemical induction of autophagy can be achieved by treating cells with any of the following compounds:
    - Rapamycin (Rap): 500 nM rapamycin (LC Laboratories; prepare a stock solution of 4 mM rapamycin in 90% ethanol and 10% Tween-20).
    - Hydrogen peroxide (H_2O_2): 1 mM H_2O_2 (prepare a 1 M stock solution by mixing 105 μl 9.6 M H_2O_2 (30% w/v, Fisher Chemical) and 895 μl ultrapure water).
    - Tunicamycin (Tun): 5 μg/ml tunicamycin (Calbiochem; prepare a 5 mg/ml stock solution in dimethyl formamide).
    - Norflurazon (NF): 20 μM norflurazon (Supelco Analytical; prepare a 10 mM stock solution in methanol).
    - Methyl viologen (MV): 1 μM methyl viologen (Sigma; prepare a 10 mM stock solution in water).
    - Dithiothreitol (DTT): 2.5 mM dithiothreitol (Sigma; prepare a 1 M stock solution in water).

    Autophagy can also be induced by shifting Chlamydomonas cells to nitrogen-free medium or to darkness in acetate-free medium or by exposing cells to high light stress. Detailed information about all these treatments and stress conditions can be found in [4, 12].
11. Typically, 50 ml cells to a density of $2–4 \times 10^6$ cells/ml are resuspended in 400–500 μl of lysis buffer. Optionally, a cocktail of protease inhibitors can be added to the lysis buffer. We observed that the presence of protease inhibitors has no effect on the detection of CrATG8 by western blot.
12. Total soluble extracts obtained by this method usually are colorless or display a pale yellow color.
13. We have experimentally determined that 15% acrylamide gels provide optimal resolution and separation of modified and unmodified forms of CrATG8 protein.
14. We use SeeBlue Pre-Stained standard (Novex) that contains low molecular weight markers of 14 kDa, 6 kDa and 3 kDa.

15. To get a better resolution of modified and unmodified CrATG8 forms, we let the prestained 6 and 3 kDa markers to run out of the gel.
16. We use Hyperfilm ECL (GE Healthcare) to get a high sensitivity.

Acknowledgments

MEPP, AAG, and JLC are supported by the Spanish Ministry of Economy and Competitiveness (grant BFU2012-35913 to JLC), and by Junta de Andalucía (grant CVI-7336 to JLC).

References

1. Nakatogawa H, Suzuki K, Kamada Y, Ohsumi Y (2009) Dynamics and diversity in autophagy mechanisms: lessons from yeast. Nat Rev Mol Cell Biol 10:458–467
2. Xie Z, Klionsky DJ (2007) Autophagosome formation: core machinery and adaptations. Nat Cell Biol 9:1102–1109
3. Liu Y, Bassham DC (2012) Autophagy: pathways for self-eating in plant cells. Annu Rev Plant Biol 63:215–237
4. Perez-Perez ME, Florencio FJ, Crespo JL (2010) Inhibition of target of rapamycin signaling and stress activate autophagy in *Chlamydomonas reinhardtii*. Plant Physiol 152:1874–1888
5. Kirisako T, Ichimura Y, Okada H, Kabeya Y, Mizushima N, Yoshimori T, Ohsumi M, Takao T, Noda T, Ohsumi Y (2000) The reversible modification regulates the membrane-binding state of Apg8/Aut7 essential for autophagy and the cytoplasm to vacuole targeting pathway. J Cell Biol 151:263–276
6. Klionsky DJ, Abdalla FC, Abeliovich H, Abraham RT, Acevedo-Arozena A, Adeli K, Agholme L, Agnello M, Agostinis P, Aguirre-Ghiso JA, Ahn HJ, Ait-Mohamed O, Ait-Si-Ali S, Akematsu T, Akira S, Al-Younes HM, Al-Zeer MA, Albert ML, Albin RL, Alegre-Abarrategui J, Aleo MF, Alirezaei M, Almasan A, Almonte-Becerril M, Amano A, Amaravadi R, Amarnath S, Amer AO, Andrieu-Abadie N, Anantharam V, Ann DK, Anoopkumar-Dukie S, Aoki H, Apostolova N, Arancia G, Aris JP, Asanuma K, Asare NY, Ashida H, Askanas V, Askew DS, Auberger P, Baba M, Backues SK, Baehrecke EH, Bahr BA, Bai XY, Bailly Y, Baiocchi R, Baldini G et al (2012) Guidelines for the use and interpretation of assays for monitoring autophagy. Autophagy 8:445–544
7. Hutner SH, Provasoli L, Schatz A, Haskins CP (1950) Some approaches to the study of the role of metals in the metabolism of microorganisms. Proc Am Philos Soc 94:152–170
8. Thompson AR, Doelling JH, Suttangkakul A, Vierstra RD (2005) Autophagic nutrient recycling in Arabidopsis directed by the ATG8 and ATG12 conjugation pathways. Plant Physiol 138:2097–2110
9. Xiong Y, Contento AL, Nguyen PQ, Bassham DC (2007) Degradation of oxidized proteins by autophagy during oxidative stress in Arabidopsis. Plant Physiol 143:291–299
10. Yoshimoto K, Hanaoka H, Sato S, Kato T, Tabata S, Noda T, Ohsumi Y (2004) Processing of ATG8s, ubiquitin-like proteins, and their deconjugation by ATG4s are essential for plant autophagy. Plant Cell 16:2967–2983
11. Alvarez C, Garcia I, Moreno I, Perez-Perez ME, Crespo JL, Romero LC, Gotor C (2012) Cysteine-generated sulfide in the cytosol negatively regulates autophagy and modulates the transcriptional profile in Arabidopsis. Plant Cell 24:4621–4634
12. Perez-Perez ME, Couso I, Crespo JL (2012) Carotenoid deficiency triggers autophagy in the model green alga *Chlamydomonas reinhardtii*. Autophagy 8:376–388

Chapter 13

Detection of Autophagy in Plants by Fluorescence Microscopy

Yunting Pu and Diane C. Bassham

Abstract

Autophagy is a key process for degradation and recycling of proteins or organelles in eukaryotes. Autophagy in plants has been shown to function in stress responses, pathogen immunity, and senescence, while a basal level of autophagy plays a housekeeping role in cells. Upon activation of autophagy, vesicles termed autophagosomes are formed to deliver proteins or organelles to the vacuole for degradation. The number of autophagosomes can thus be used to indicate the level of autophagy. Here, we describe two common methods used for detection of autophagosomes, staining of autophagosomes with the fluorescent dye monodansylcadaverine, and expression of a fusion between GFP and the autophagosomal membrane protein ATG8.

Key words Autophagy, *Arabidopsis thaliana*, Autophagosome, Monodansylcadaverine, GFP-ATG8, Vacuole

1 Introduction

Autophagy is an important process for delivering macromolecules or organelles to be degraded and recycled in animal and plant cells. Three types of autophagic pathway have been identified with distinct mechanisms: macroautophagy, microautophagy, and chaperone-mediated autophagy [1–3]. In plants, although both macroautophagy and microautophagy have been identified, the function and mechanism of macroautophagy is better studied. In this protocol, autophagy hereafter refers to macroautophagy. Upon activation of autophagy, a cup-shaped double-membrane vesicle called a phagophore forms to engulf cargo that will be degraded. Expansion and closure of the phagophore leads to the formation of an intact vesicle called an autophagosome. Autophagosomes then deliver the cargo into the lysosome in animal cells or the vacuole in plant cells for degradation. In plant cells, the outer membrane of the autophagosome fuses with the tonoplast, or vacuole membrane, while the inner membrane along with the cargo is delivered into the vacuole as an

L. Maria Lois and Rune Matthiesen (eds.), *Plant Proteostasis: Methods and Protocols*, Methods in Molecular Biology, vol. 1450, DOI 10.1007/978-1-4939-3759-2_13,

autophagic body and degraded by lytic enzymes. The products of degradation are released into the cytoplasm for reuse [1].

Initiation and formation of autophagosomes involves a series of autophagy-related (*ATG*) genes. The ATG8-PE conjugation system plays a key role in autophagosome formation [4]. In *Arabidopsis thaliana*, nine isoforms within an *ATG8* gene family have been identified, *AtATG8a–AtATG8i* [5]. Upon induction of autophagy, the C-terminus of ATG8 is cleaved by the ATG4 protease and eventually conjugated to the membrane lipid phosphatidylethanolamine (PE). ATG8 is therefore attached to the autophagosome membrane, enabling it to participate in the formation of autophagosomes [4, 6]. The conjugation is reversible via cleavage by ATG4 for ATG8 recycling [7]. ATG8 has also been characterized in other photosynthetic organisms, including Chlamydomonas, rice, and maize, where it acts via a similar mechanism [8–10].

Autophagy is maintained at a basal level in cells as a housekeeping process. It is induced in both biotic and abiotic stress conditions, including nutrient starvation, salt and drought stress, oxidative stress, endoplasmic reticulum (ER) stress, pathogen infection, and senescence [1, 11]. Several assays have been established to monitor autophagy in plant cells, such as detection of ATG8 lipidation by immunoblot with ATG8 antibodies, visualization of autophagosomes using a GFP-ATG8 fusion protein, and staining of autophagosomes with acidotropic dyes such as LysoTracker Red and monodansylcadaverine (MDC), followed by fluorescence microscopy [12]. In this protocol, we describe two methods to monitor autophagy in vivo by fluorescence microscopy. ATG8 decorates both the outer and inner membranes of autophagosomes through ATG8-PE adduct formation via lipidation, and remains on the inner membrane upon its delivery into the vacuole as an autophagic body. Therefore, fluorescent protein-fused ATG8 can be used as a marker both of autophagosomes and of autophagic bodies inside the vacuole [4, 13]. Due to the rapid degradation of autophagic bodies, their visualization is sometimes facilitated by incubation with degradation inhibitors such as Concanamycin A, a V-ATPase inhibitor that blocks hydrolase activity by elevating vacuolar pH [4, 14]. MDC is an acidotropic dye that stains acidic cellular compartments, including autophagosomes [13]. Although other acidic vesicles might also be stained by MDC and thus be confused with autophagosomes, the simplicity and time-saving advantages make it a good method for rapid autophagy detection when combined with other approaches [15].

2 Materials

2.1 Plant Growth Materials

1. Seed sterilizing solution: 33% (v/v) Bleach and 0.1% (v/v) Triton X-100 (Sigma, St. Louis, MO, USA). Mix 1 mL Triton X-100 with 9 mL water to prepare a 10% stock solution.

Add 30 μL 10% Triton X-100 solution and 10 mL household bleach to a tube containing 20 mL water. Store at room temperature.

2. Solid half-strength MS medium with sucrose: 0.22% (w/v) Murashige–Skoog vitamin mixture (Caisson Laboratories, North Logan, UT, USA), 2.4 mM 2-morphinolino-ethanesulfonic acid (MES) (Sigma), 0.6% (w/v) Phytoblend agar (Caisson Laboratories), 1% sucrose (Sigma). Adjust pH to 5.7 with KOH. Autoclave the medium at 121 °C for 20 min. Allow the medium to cool to 45–50 °C. Pour the medium into petri dishes in a laminar flow hood to approximately half the depth of the plate. Allow the medium to cool to room temperature for about an hour to solidify. Store plates at 4 °C (*see* **Note 1**).
3. Solid half-strength MS medium without nitrogen: 5% (v/v) Murashige–Skoog basal salt micronutrient solution (10×) (Sigma), 1.5 mM $CaCl_2$, 0.75 mM $MgSO_4$, 0.625 mM KH_2PO_4, 2.5 mM KCl, 2 mM MES, 1% (w/v) sucrose. Adjust pH to 5.7 with KOH. Autoclave and pour the medium as for solid half-strength MS medium with sucrose.
4. 0.1% Agarose: 0.1% (w/v) Agarose (Fisher Scientific, Dallas, TX, USA) in water. Autoclave at 121 °C for 20 min. Store at room temperature.
5. Petri dishes (100 mm × 20 mm) (Fisher Scientific).

2.2 Seedling Treatment and Staining

1. Phosphate-buffered saline (PBS, 10×): 8% (w/v) NaCl, 0.2% (w/v) KCl, 1.4% (w/v) Na_2HPO_4, 0.24% (w/v) KH_2PO_4, pH 7.4. Store at room temperature.
2. 20× MDC stock solution: 1 mM dansylcadaverine (Sigma). Aliquot 500 μL or 1 mL into microcentrifuge tubes. Store at −20 °C in the dark (*see* **Note 2**).
3. 6-well plates (BioExpress, Kaysville, UT, USA).
4. Mannitol (Sigma).
5. Dithiothreitol (DTT, 100×): 0.2 M DTT (Fisher Scientific) in water. Store at −20 °C.
6. Tunicamycin (200×): 1 g/mL tunicamycin (Sigma) in dimethyl sulfoxide (DMSO). Store at 4 °C.
7. Hydrogen peroxide (H_2O_2) 30% (w/w) (Sigma).
8. Methyl viologen (1000×): 10 mM methyl viologen dichloride hydrate (Sigma) in water. Filter the solution through a 0.22 μm sterile syringe filter (VWR, Radnor, PA, USA). Store at 4 °C.
9. Concanamycin A (1000×): 1 mM Concanamycin A (Sigma) in DMSO. Store at −20 °C.
10. Aluminum foil.

2.3 Fluorescence Microscopy

1. Glass slides (Fisher Scientific).
2. Glass cover slips (22 × 22 mm) (Fisher Scientific).
3. Light microscope: Zeiss Axio Imager.A2 upright microscope (Zeiss, Jena, Germany) with Zeiss Axiocam BW/color digital cameras (*see* **Note 3**).
4. Confocal microscope: Leica SP5 × MP confocal/multiphoton microscope system (Leica, Wetzlar, Germany) (*see* **Note 4**).
5. ZEN 2012 (blue edition) (Zeiss) (*see* **Note 5**).
6. Leica Application Suite (Leica) (*see* **Note 6**).

3 Methods

3.1 Plant Materials and Growth Conditions

1. Sterilize Arabidopsis seeds with seed sterilizing solution for 20 min with agitation or rocking, followed by five washes with sterile water for 5 min each. Sterilized seeds are stored at 4 °C in the dark for at least 2 days before plating to allow stratification.
2. Plate seeds in lines on solid half-strength MS medium with sucrose. Suspend seeds in a tube containing 0.1% agarose. Use a pipette to spot 10–13 seeds per line and at most 8 lines per plate to allow sufficient growth space. Keep the plates vertically under long day conditions (16 h light) at 22 °C for 7 days.

3.2 Autophagy Activation in Seedlings by Abiotic Stresses (*See* Note 7)

1. To induce autophagy by starvation, transfer 7-day-old seedlings onto solid half-strength MS medium without sucrose for carbon starvation, or solid half-strength MS medium without nitrogen for nitrogen starvation. Meanwhile, transfer seedlings onto half-strength MS medium with sucrose as a control. Wrap the plates for sucrose starvation with aluminum foil to maintain darkness. Grow the transferred seedlings on sucrose starvation plates in the dark, and seedlings on control or nitrogen starvation plates in the light for an additional 2–4 days.
2. To induce autophagy by salt or osmotic stress, immerse 5–10 7-day-old seedlings in 2 mL liquid half-strength MS medium with sucrose plus 0.16 M NaCl or 0.35 M mannitol in a 6-well plate. To induce autophagy by ER stress, immerse 7-day-old seedlings in 2 mL liquid half-strength MS medium with sucrose plus 2 mM DTT or 5 μg/mL tunicamycin in a 6-well plate. When inducing ER stress with tunicamycin, add an equal volume of DMSO into liquid medium for a control treatment. Wrap the 6-well plate with aluminum foil and gently shake for 6–8 h (*see* **Note 8**).
3. To induce autophagy with oxidative stress, immerse 5–10 7-day-old seedlings in 2 mL liquid half-strength MS medium with sucrose plus 5 mM H_2O_2 or 10 μM methyl viologen in a 6-well plate. Wrap the 6-well plate with aluminum foil and gently shake for 1–2 h.

3.3 Labeling of Autophagosomes in Seedlings by MDC Staining

1. Dilute the 20× MDC stock solution to 1× MDC solution with PBS buffer.
2. Carefully transfer 5–10 seedlings from the solid medium to 6-well plates with 2 mL MDC solution in each well. If seedlings are in 6-well plates with liquid medium, carefully remove liquid medium from seedlings by pipetting. Gently dispense 2 mL MDC solution into each well. Immerse seedlings in the MDC solution and shake gently for 10 min in the dark.
3. Wash the seedlings twice with PBS buffer for 5 min each. Be sure to remove any visible remains of MDC solution. Leave the seedlings in PBS buffer and wrap the plate with aluminum foil until observation by microscopy (*see* **Note 9**).

3.4 Labeling of Autophagosomes in Seedlings with GFP-ATG8 Fusion Protein (*See* Note 10)

1. To detect autophagy using GFP-ATG8 fusion proteins, grow GFP-ATG8e transgenic seedlings under the same conditions as described in Subheading 3.1, and induce autophagy using stress conditions as in Subheading 3.2 (*see* **Note 11**).
2. To detect autophagy using a GFP-ATG8 fusion protein in mutant lines or other genetic backgrounds, cross the desired lines with GFP-ATG8e transgenic plants or use Agrobacterium-mediated transformation to generate transgenic plants with both the desired genotype and GFP-ATG8e.
3. To facilitate visualization of autophagosomes by inhibiting autophagic body degradation with Concanamycin A (optional), transfer 5–10 seedlings to a 6-well plate with 2 mL liquid half-strength MS medium with sucrose plus 1 μM Concanamycin A. Incubate the plate with shaking for 6–8 h (*see* **Note 12**).

3.5 Visualization of MDC-Stained or GFP-Labeled Autophagosomes by Fluorescence Microscopy

Procedure described is for both epifluorescence microscopy and confocal microscopy unless otherwise specified.

1. Add a drop of PBS buffer to a slide and gently lay out 5–8 seedlings onto the slide with roots submerged in buffer. Cover the roots with a cover slip. Carefully place the slide onto the stage of the microscope.
2. Adjust the focus of the eyepiece (10×). Set the objective lens to 10× or 20× to find the root tips under bright field illumination. From the root tips, move up along the root to the elongation zone, where autophagy can be observed most easily.
3. Switch the objective lens to 40×/0.75 for epifluorescence microscopy or 63×/1.4 oil for confocal microscopy and adjust focus.
4. For MDC detection, select filter sets specific for imaging DAPI, UV, or with an excitation wavelength of 335 nm and emission wavelength of 508 nm. For GFP fluorescence detection,

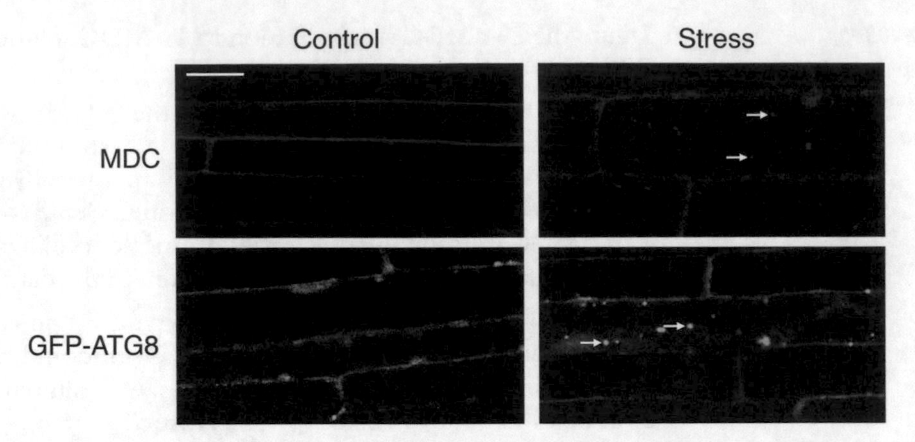

Fig. 1 Imaging of autophagosomes labeled with MDC or GFP-ATG8 in Arabidopsis root cells. Arabidopsis seedlings stained with MDC (*upper panels*) or expressing GFP-ATG8e (*lower panels*) were incubated in stress or control conditions as described in this protocol. Root cells in the elongation zone were observed using confocal microscopy. MDC-stained or GFP-ATG8e labeled autophagosomes are indicated by *white arrows*, showing autophagy induction in Arabidopsis root cells upon stress treatment. Scale bar = 20 μm

select filter sets specific for imaging Fluorescein Isothiocyanate (FITC), GFP, or with excitation wavelength of 488 nm and emission wavelength of 525 nm.

5. Observe the elongation zone, focusing on different layers of the root to get an initial overview of the level of autophagy in the seedling. In seedlings without stress treatment, GFP fluorescence signal should be diffuse in the cytoplasm, and MDC weakly stains cell walls. Upon stress treatment, small spherical puncta form and move around rapidly in the cytoplasm, indicating autophagosome accumulation due to autophagy activation (Fig. 1). If Concanamycin A is used, the majority of GFP fluorescence will be associated with autophagic bodies inside the vacuole (*see* **Note 13**).
6. For quantification, epifluorescence microscopy is most convenient for imaging large numbers of autophagosomes. Take 2–3 representative images for each seedling at different places of the elongation zone of the root, with at least ten images for all seedlings of a certain genotype or treatment. The corresponding bright field images are used as a reference. A differential interference contrast (DIC) filter may also be used for a bright field reference as it provides better contrast. Save and export the images including the scale bar using ZEN or other appropriate software for future quantification and statistical analysis.

   For higher quality images, take representative images of autophagosomes in each seedling using confocal microscopy, again with bright field images as reference. Save and export the fluorescence images, bright field images, and merged images for qualitative presentation of autophagy in a certain genotype or treatment (*see* **Note 14**).

7. For quantitative analysis of autophagy, count the number of autophagosomes in each frame and calculate the average number of autophagosomes per frame for all images for each genotype or treatment. The average number of autophagosomes in each image indicates the level of autophagy. At a minimum, calculate the standard deviation to indicate the variation for each data set, and determine the statistical significance of any differences seen using a Student's *t*-test or other appropriate analysis. Statistical analysis can be performed using EXCEL (Microsoft Corporation, Redmond, WA, USA), SAS (SAS Institute Inc. Cary, NC, USA), JMP (JMP Group Inc. San Francisco, CA, USA), or other similar software.

4 Notes

1. For sucrose starvation induction of autophagy, prepare standard half-strength MS medium, but do not add sucrose. For liquid medium, prepare as for solid medium but without addition of Phytoblend agar. Store liquid medium at room temperature after autoclaving. Addition of other chemicals to the medium should be performed when the medium has cooled to 45–50 °C (when the bottle can be held with hands). Chemicals should be dissolved into solution and sterilized by autoclaving or using a 0.22 μm syringe filter before addition to the medium.
2. MDC is difficult to dissolve into a 20× stock solution and precipitation may occur in the stock solution. Use pipet tips to grind the MDC powder in PBS buffer, and aliquot to 1 mL or 500 μL with regular vortexing to assure equal distribution. Since MDC is light-sensitive, fast preparation and dark storage is necessary to maintain its activity.
3. Any microscope with fluorescence capability and attached camera can be used for basic detection and imaging of autophagosomes.
4. Any confocal microscopy system with fluorescence capability and attached camera can be used for detection and imaging of autophagosomes.
5. The ZEN 2012 software is used for image analysis and export of the original images taken from the Zeiss Axio Imager.A2 upright microscope. The software package used will depend on the microscope being used.
6. The Leica Application Suite software is used for image analysis and export from the Leica Sp5 × MP confocal/multiphoton

microscope. The software package used will depend on the microscope being used.

7. Besides abiotic stresses as discussed here, autophagy can also be induced by biotic stresses including pathogen infection, and is activated during leaf senescence.
8. When immersing seedlings in liquid medium, 2 mL medium is typically sufficient for 5–10 7-day-old seedlings. For treatment of more seedlings, increase the volume accordingly to allow all seedlings to be fully immersed in liquid. The speed of shaking should be around 50–100 rpm to avoid root damage caused by excessive agitation.
9. Detection of autophagy by microscopy should begin immediately after sample preparation and staining. In the interval between staining and microscopy, avoid exposure of the seedlings to high temperatures, as heat stress may induce autophagy.
10. Fluorescent proteins other than GFP can also be used for detection of autophagosomes by fusion with ATG8. Fluorescent protein fusions should be designed with the fluorescent protein at the N-terminus of ATG8, as ATG8 lipidation occurs at the C-terminus during autophagosome formation. The fusion protein can be expressed either transiently or in transgenic plants. In this protocol, only detection of autophagosomes in transgenic plants is discussed.
11. The GFP-ATG8e transgenic plants and constructs are as described in Contento et al. [13] and can be obtained from the Arabidopsis Biological Resource Center (stock # CS66943). GFP-ATG8a transgenic Arabidopsis lines have also been commonly used to detect autophagosomes (ABRC stock # CS39996) [16]. There are nine isoforms of ATG8 in *Arabidopsis thaliana*, all of which are thought to associate with the autophagosome membrane. The detection of autophagosomes can therefore be performed by fusion of GFP with any isoform of ATG8.
12. The addition of Concanamycin A is optional, for the purpose of increasing the number of autophagosomes to be visualized, as it inhibits the degradation of autophagic bodies in the vacuole by raising the vacuolar pH. However, this precludes its use with acidotropic dye staining methods such as MDC staining. Be careful and use personal protective equipment when dealing with Concanamycin A or treated samples since Concanamycin A is a carcinogen.
13. Occasionally, small puncta are also visible in seedlings without stress treatment, suggesting the presence of a basal level of autophagy functioning as a housekeeping mechanism.

14. Representative images should be taken randomly in different regions of the elongation zone with similar exposure times. The exposure time, cell layer, and region of the root should be kept constant between samples to ensure validity of the results.

Acknowledgments

This work was supported by grant no. IOS-1353867 from the National Science Foundation to D.C.B. We thank Xiaochen Yang for providing GFP images for Fig. 1.

References

1. Liu Y, Bassham DC (2012) Autophagy: pathways for self-eating in plant cells. Annu Rev Plant Biol 63:215–237. doi:10.1146/annurev-arplant-042811-105441
2. Mijaljica D, Prescott M, Devenish RJ (2011) Microautophagy in mammalian cells: revisiting a 40-year-old conundrum. Autophagy 7(7):673–682. doi: 10.4161/auto.7.7.14733
3. Orenstein SJ, Cuervo AM (2010) Chaperone-mediated autophagy: molecular mechanisms and physiological relevance. Semin Cell Dev Biol 21(7):719–726. doi:10.1016/j.semcdb.2010.02.005
4. Yoshimoto K, Hanaoka H, Sato S, Kato T, Tabata S, Noda T, Ohsumi Y (2004) Processing of ATG8s, ubiquitin-like proteins, and their deconjugation by ATG4s are essential for plant autophagy. Plant Cell 16(11):2967–2983. doi:10.1105/tpc.104.025395
5. Doelling JH, Walker JM, Friedman EM, Thompson AR, Vierstra RD (2002) The APG8/12-activating enzyme APG7 is required for proper nutrient recycling and senescence in Arabidopsis thaliana. J Biol Chem 277(36):33105–33114. doi:10.1074/jbc.M204630200
6. Woo J, Park E, Dinesh-Kumar SP (2014) Differential processing of Arabidopsis ubiquitin-like Atg8 autophagy proteins by Atg4 cysteine proteases. Proc Natl Acad Sci U S A 111(2):863–868. doi:10.1073/pnas.1318207111
7. Kirisako T, Ichimura Y, Okada H, Kabeya Y, Mizushima N, Yoshimori T, Ohsumi M, Takao T, Noda T, Ohsumi Y (2000) The reversible modification regulates the membrane-binding state of Apg8/Aut7 essential for autophagy and the cytoplasm to vacuole targeting pathway. J Cell Biol 151(2):263–276. doi:10.1083/jcb.151.2.263
8. Pérez-Pérez ME, Florencio FJ, Crespo JL (2010) Inhibition of target of rapamycin signaling and stress activate autophagy in Chlamydomonas reinhardtii. Plant Physiol 152(4):1874–1888. doi:10.1104/pp.109.152520
9. Su W, Ma H, Liu C, Wu J, Yang J (2006) Identification and characterization of two rice autophagy associated genes, OsAtg8 and OsAtg4. Mol Biol Rep 33(4):273–278. doi:10.1007/s11033-006-9011-0
10. Chung T, Suttangkakul A, Vierstra RD (2009) The ATG autophagic conjugation system in maize: ATG transcripts and abundance of the ATG8-lipid adduct are regulated by development and nutrient availability. Plant Physiol 149(1):220–234. doi:10.1104/pp.108.126714
11. Li F, Vierstra RD (2012) Autophagy: a multifaceted intracellular system for bulk and selective recycling. Trends Plant Sci 17(9):526–537. doi:10.1016/j.tplants.2012.05.006
12. Bassham DC (2015) Methods for analysis of autophagy in plants. Methods 75:181–188. doi:10.1016/j.ymeth.2014.09.003
13. Contento AL, Xiong Y, Bassham DC (2005) Visualization of autophagy in Arabidopsis using the fluorescent dye monodansylcadaverine and a GFP-AtATG8e fusion protein. Plant J 42(4):598–608. doi:10.1111/j.1365-313X.2005.02396.x
14. Matsuoka K, Higuchi T, Maeshima M, Nakamura K (1997) A vacuolar-type H+-ATPase in a nonvacuolar organelle is required for the sorting of soluble vacuolar protein precursors in tobacco cells. Plant Cell 9(4):533–546. doi:10.1105/tpc.9.4.533
15. Klionsky DJ, Abdalla FC, Abeliovich H, Abraham RT, Acevedo-Arozena A, Adeli K, Agholme L, Agnello M, Agostinis P, Aguirre-Ghiso JA, Ahn HJ, Ait-Mohamed O, Ait-Si-Ali S, Akematsu T, Akira S, Al-Younes HM, Al-Zeer MA, Albert ML, Albin RL, Alegre-Abarrategui J, Aleo MF, Alirezaei M, Almasan A, Almonte-Becerril M, Amano A, Amaravadi R, Amarnath S, Amer AO, Andrieu-Abadie N, Anantharam V, Ann DK, Anoopkumar-Dukie

S, Aoki H, Apostolova N, Arancia G, Aris JP, Asanuma K, Asare NY, Ashida H, Askanas V, Askew DS, Auberger P, Baba M, Backues SK, Baehrecke EH, Bahr BA, Bai XY, Bailly Y, Baiocchi R, Baldini G, Balduini W, Ballabio A, Bamber BA, Bampton ET, Banhegyi G, Bartholomew CR, Bassham DC, Bast RC, Jr., Batoko H, Bay BH, Beau I, Bechet DM, Begley TJ, Behl C, Behrends C, Bekri S, Bellaire B, Bendall LJ, Benetti L, Berliocchi L, Bernardi H, Bernassola F, Besteiro S, Bhatia-Kissova I, Bi X, Biard-Piechaczyk M, Blum JS, Boise LH, Bonaldo P, Boone DL, Bornhauser BC, Bortoluci KR, Bossis I, Bost F, Bourquin JP, Boya P, Boyer-Guittaut M, Bozhkov PV, Brady NR, Brancolini C, Brech A, Brenman JE, Brennand A, Bresnick EH, Brest P, Bridges D, Bristol ML, Brookes PS, Brown EJ, Brumell JH, Brunetti-Pierri N, Brunk UT, Bulman DE, Bultman SJ, Bultynck G, Burbulla LF, Bursch W, Butchar JP, Buzgariu W, Bydlowski SP, Cadwell K, Cahova M, Cai D, Cai J, Cai Q, Calabretta B, Calvo-Garrido J, Camougrand N, Campanella M, Campos-Salinas J, Candi E, Cao L, Caplan AB, Carding SR, Cardoso SM, Carew JS, Carlin CR, Carmignac V, Carneiro LA, Carra S, Caruso RA, Casari G, Casas C, Castino R, Cebollero E, Cecconi F, Celli J, Chaachouay H, Chae HJ, Chai CY, Chan DC, Chan EY, Chang RC, Che CM, Chen CC, Chen GC, Chen GQ, Chen M, Chen Q, Chen SS, Chen W, Chen X, Chen X, Chen X, Chen YG, Chen Y, Chen Y, Chen YJ, Chen Z, Cheng A, Cheng CH, Cheng Y, Cheong H, Cheong JH, Cherry S, Chess-Williams R, Cheung ZH, Chevet E, Chiang HL, Chiarelli R, Chiba T, Chin LS, Chiou SH, Chisari FV, Cho CH, Cho DH, Choi AM, Choi D, Choi KS, Choi ME, Chouaib S, Choubey D, Choubey V, Chu CT, Chuang TH, Chueh SH, Chun T, Chwae YJ, Chye ML, Ciarcia R, Ciriolo MR, Clague MJ, Clark RS, Clarke PG, Clarke R, Codogno P, Coller HA, Colombo MI, Comincini S, Condello M, Condorelli F, Cookson MR, Coombs GH, Coppens I, Corbalan R, Cossart P, Costelli P, Costes S, Coto-Montes A, Couve E, Coxon FP, Cregg JM, Crespo JL, Cronje MJ, Cuervo AM, Cullen JJ, Czaja MJ, D'Amelio M, Darfeuille-Michaud A, Davids LM, Davies FE, De Felici M, de Groot JF, de Haan CA, De Martino L, De Milito A, De Tata V, Debnath J, Degterev A, Dehay B, Delbridge LM, Demarchi F, Deng YZ, Dengjel J, Dent P, Denton D, Deretic V, Desai SD, Devenish RJ, Di Gioacchino M, Di Paolo G, Di Pietro C, Diaz-Araya G, Diaz-Laviada I, Diaz-Meco MT, Diaz-Nido J, Dikic I, Dinesh-Kumar SP, Ding WX, Distelhorst CW, Diwan A, Djavaheri-Mergny M, Dokudovskaya S, Dong Z, Dorsey FC, Dosenko V, Dowling JJ, Doxsey S, Dreux M, Drew ME, Duan Q, Duchosal MA, Duff K, Dugail I, Durbeej M, Duszenko M, Edelstein CL, Edinger AL, Egea G, Eichinger L, Eissa NT, Ekmekcioglu S, El-Deiry WS, Elazar Z, Elgendy M, Ellerby LM, Eng KE, Engelbrecht AM, Engelender S, Erenpreisa J, Escalante R, Esclatine A, Eskelinen EL, Espert L, Espina V, Fan H, Fan J, Fan QW, Fan Z, Fang S, Fang Y, Fanto M, Fanzani A, Farkas T, Farre JC, Faure M, Fechheimer M, Feng CG, Feng J, Feng Q, Feng Y, Fesus L, Feuer R, Figueiredo-Pereira ME, Fimia GM, Fingar DC, Finkbeiner S, Finkel T, Finley KD, Fiorito F, Fisher EA, Fisher PB, Flajolet M, Florez-McClure ML, Florio S, Fon EA, Fornai F, Fortunato F, Fotedar R, Fowler DH, Fox HS, Franco R, Frankel LB, Fransen M, Fuentes JM, Fueyo J, Fujii J, Fujisaki K, Fujita E, Fukuda M, Furukawa RH, Gaestel M, Gailly P, Gajewska M, Galliot B, Galy V, Ganesh S, Ganetzky B, Ganley IG, Gao FB, Gao GF, Gao J, Garcia L, Garcia-Manero G, Garcia-Marcos M, Garmyn M, Gartel AL, Gatti E, Gautel M, Gawriluk TR, Gegg ME, Geng J, Germain M, Gestwicki JE, Gewirtz DA, Ghavami S, Ghosh P, Giammarioli AM, Giatromanolaki AN, Gibson SB, Gilkerson RW, Ginger ML, Ginsberg HN, Golab J, Goligorsky MS, Golstein P, Gomez-Manzano C, Goncu E, Gongora C, Gonzalez CD, Gonzalez R, Gonzalez-Estevez C, Gonzalez-Polo RA, Gonzalez-Rey E, Gorbunov NV, Gorski S, Goruppi S, Gottlieb RA, Gozuacik D, Granato GE, Grant GD, Green KN, Gregorc A, Gros F, Grose C, Grunt TW, Gual P, Guan JL, Guan KL, Guichard SM, Gukovskaya AS, Gukovsky I, Gunst J, Gustafsson AB, Halayko AJ, Hale AN, Halonen SK, Hamasaki M, Han F, Han T, Hancock MK, Hansen M, Harada H, Harada M, Hardt SE, Harper JW, Harris AL, Harris J, Harris SD, Hashimoto M, Haspel JA, Hayashi S, Hazelhurst LA, He C, He YW, Hebert MJ, Heidenreich KA, Helfrich MH, Helgason GV, Henske EP, Herman B, Herman PK, Hetz C, Hilfiker S, Hill JA, Hocking LJ, Hofman P, Hofmann TG, Hohfeld J, Holyoake TL, Hong MH, Hood DA, Hotamisligil GS, Houwerzijl EJ, Hoyer-Hansen M, Hu B, Hu CA, Hu HM, Hua Y, Huang C, Huang J, Huang S, Huang WP, Huber TB, Huh WK, Hung TH, Hupp TR, Hur GM, Hurley JB, Hussain SN, Hussey PJ, Hwang JJ, Hwang S, Ichihara A, Ilkhanizadeh S, Inoki K, Into T, Iovane V, Iovanna JL, Ip NY, Isaka Y, Ishida H, Isidoro C, Isobe K, Iwasaki A, Izquierdo M, Izumi Y, Jaakkola PM, Jaattela M, Jackson GR, Jackson WT, Janji B, Jendrach M, Jeon JH, Jeung EB, Jiang H, Jiang H, Jiang JX, Jiang M, Jiang Q, Jiang X, Jiang X, Jimenez A, Jin M, Jin S, Joe CO, Johansen T, Johnson DE, Johnson GV,

Jones NL, Joseph B, Joseph SK, Joubert AM, Juhasz G, Juillerat-Jeanneret L, Jung CH, Jung YK, Kaarniranta K, Kaasik A, Kabuta T, Kadowaki M, Kagedal K, Kamada Y, Kaminskyy VO, Kampinga HH, Kanamori H, Kang C, Kang KB, Kang KI, Kang R, Kang YA, Kanki T, Kanneganti TD, Kanno H, Kanthasamy AG, Kanthasamy A, Karantza V, Kaushal GP, Kaushik S, Kawazoe Y, Ke PY, Kehrl JH, Kelekar A, Kerkhoff C, Kessel DH, Khalil H, Kiel JA, Kiger AA, Kihara A, Kim DR, Kim DH, Kim DH, Kim EK, Kim HR, Kim JS, Kim JH, Kim JC, Kim JK, Kim PK, Kim SW, Kim YS, Kim Y, Kimchi A, Kimmelman AC, King JS, Kinsella TJ, Kirkin V, Kirshenbaum LA, Kitamoto K, Kitazato K, Klein L, Klimecki WT, Klucken J, Knecht E, Ko BC, Koch JC, Koga H, Koh JY, Koh YH, Koike M, Komatsu M, Kominami E, Kong HJ, Kong WJ, Korolchuk VI, Kotake Y, Koukourakis MI, Kouri Flores JB, Kovacs AL, Kraft C, Krainc D, Kramer H, Kretz-Remy C, Krichevsky AM, Kroemer G, Kruger R, Krut O, Ktistakis NT, Kuan CY, Kucharczyk R, Kumar A, Kumar R, Kumar S, Kundu M, Kung HJ, Kurz T, Kwon HJ, La Spada AR, Lafont F, Lamark T, Landry J, Lane JD, Lapaquette P, Laporte JF, Laszlo L, Lavandero S, Lavoie JN, Layfield R, Lazo PA, Le W, Le Cam L, Ledbetter DJ, Lee AJ, Lee BW, Lee GM, Lee J, Lee JH, Lee M, Lee MS, Lee SH, Leeuwenburgh C, Legembre P, Legouis R, Lehmann M, Lei HY, Lei QY, Leib DA, Leiro J, Lemasters JJ, Lemoine A, Lesniak MS, Lev D, Levenson VV, Levine B, Levy E, Li F, Li JL, Li L, Li S, Li W, Li XJ, Li YB, Li YP, Liang C, Liang Q, Liao YF, Liberski PP, Lieberman A, Lim HJ, Lim KL, Lim K, Lin CF, Lin FC, Lin J, Lin JD, Lin K, Lin WW, Lin WC, Lin YL, Linden R, Lingor P, Lippincott-Schwartz J, Lisanti MP, Liton PB, Liu B, Liu CF, Liu K, Liu L, Liu QA, Liu W, Liu YC, Liu Y, Lockshin RA, Lok CN, Lonial S, Loos B, Lopez-Berestein G, Lopez-Otin C, Lossi L, Lotze MT, Low P, Lu B, Lu B, Lu B, Lu Z, Luciano F, Lukacs NW, Lund AH, Lynch-Day MA, Ma Y, Macian F, MacKeigan JP, Macleod KF, Madeo F, Maiuri L, Maiuri MC, Malagoli D, Malicdan MC, Malorni W, Man N, Mandelkow EM, Manon S, Manov I, Mao K, Mao X, Mao Z, Marambaud P, Marazziti D, Marcel YL, Marchbank K, Marchetti P, Marciniak SJ, Marcondes M, Mardi M, Marfe G, Marino G, Markaki M, Marten MR, Martin SJ, Martinand-Mari C, Martinet W, Martinez-Vicente M, Masini M, Matarrese P, Matsuo S, Matteoni R, Mayer A, Mazure NM, McConkey DJ, McConnell MJ, McDermott C, McDonald C, McInerney GM, McKenna SL, McLaughlin B, McLean PJ, McMaster CR, McQuibban GA, Meijer AJ, Meisler MH, Melendez A, Melia TJ, Melino G, Mena MA, Menendez JA, Menna-Barreto RF, Menon MB, Menzies FM, Mercer CA, Merighi A, Merry DE, Meschini S, Meyer CG, Meyer TF, Miao CY, Miao JY, Michels PA, Michiels C, Mijaljica D, Milojkovic A, Minucci S, Miracco C, Miranti CK, Mitroulis I, Miyazawa K, Mizushima N, Mograbi B, Mohseni S, Molero X, Mollereau B, Mollinedo F, Momoi T, Monastyrska I, Monick MM, Monteiro MJ, Moore MN, Mora R, Moreau K, Moreira PI, Moriyasu Y, Moscat J, Mostowy S, Mottram JC, Motyl T, Moussa CE, Muller S, Muller S, Munger K, Munz C, Murphy LO, Murphy ME, Musaro A, Mysorekar I, Nagata E, Nagata K, Nahimana A, Nair U, Nakagawa T, Nakahira K, Nakano H, Nakatogawa H, Nanjundan M, Naqvi NI, Narendra DP, Narita M, Navarro M, Nawrocki ST, Nazarko TY, Nemchenko A, Netea MG, Neufeld TP, Ney PA, Nezis IP, Nguyen HP, Nie D, Nishino I, Nislow C, Nixon RA, Noda T, Noegel AA, Nogalska A, Noguchi S, Notterpek L, Novak I, Nozaki T, Nukina N, Nurnberger T, Nyfeler B, Obara K, Oberley TD, Oddo S, Ogawa M, Ohashi T, Okamoto K, Oleinick NL, Oliver FJ, Olsen LJ, Olsson S, Opota O, Osborne TF, Ostrander GK, Otsu K, Ou JH, Ouimet M, Overholtzer M, Ozpolat B, Paganetti P, Pagnini U, Pallet N, Palmer GE, Palumbo C, Pan T, Panaretakis T, Pandey UB, Papackova Z, Papassideri I, Paris I, Park J, Park OK, Parys JB, Parzych KR, Patschan S, Patterson C, Pattingre S, Pawelek JM, Peng J, Perlmutter DH, Perrotta I, Perry G, Pervaiz S, Peter M, Peters GJ, Petersen M, Petrovski G, Phang JM, Piacentini M, Pierre P, Pierrefite-Carle V, Pierron G, Pinkas-Kramarski R, Piras A, Piri N, Platanias LC, Poggeler S, Poirot M, Poletti A, Pous C, Pozuelo-Rubio M, Praetorius-Ibba M, Prasad A, Prescott M, Priault M, Produit-Zengaffinen N, Progulske-Fox A, Proikas-Cezanne T, Przedborski S, Przyklenk K, Puertollano R, Puyal J, Qian SB, Qin L, Qin ZH, Quaggin SE, Raben N, Rabinowich H, Rabkin SW, Rahman I, Rami A, Ramm G, Randall G, Randow F, Rao VA, Rathmell JC, Ravikumar B, Ray SK, Reed BH, Reed JC, Reggiori F, Regnier-Vigouroux A, Reichert AS, Reiners JJ, Jr., Reiter RJ, Ren J, Revuelta JL, Rhodes CJ, Ritis K, Rizzo E, Robbins J, Roberge M, Roca H, Roccheri MC, Rocchi S, Rodemann HP, Rodriguez de Cordoba S, Rohrer B, Roninson IB, Rosen K, Rost-Roszkowska MM, Rouis M, Rouschop KM, Rovetta F, Rubin BP, Rubinsztein DC, Ruckdeschel K, Rucker EB, 3rd, Rudich A, Rudolf E, Ruiz-Opazo N, Russo R, Rusten TE, Ryan KM, Ryter SW, Sabatini DM, Sadoshima J, Saha T, Saitoh T, Sakagami H, Sakai Y, Salekdeh GH, Salomoni P, Salvaterra

PM, Salvesen G, Salvioli R, Sanchez AM, Sanchez-Alcazar JA, Sanchez-Prieto R, Sandri M, Sankar U, Sansanwal P, Santambrogio L, Saran S, Sarkar S, Sarwal M, Sasakawa C, Sasnauskiene A, Sass M, Sato K, Sato M, Schapira AH, Scharl M, Schatzl HM, Scheper W, Schiaffino S, Schneider C, Schneider ME, Schneider-Stock R, Schoenlein PV, Schorderet DF, Schuller C, Schwartz GK, Scorrano L, Sealy L, Seglen PO, Segura-Aguilar J, Seiliez I, Seleverstov O, Sell C, Seo JB, Separovic D, Setaluri V, Setoguchi T, Settembre C, Shacka JJ, Shanmugam M, Shapiro IM, Shaulian E, Shaw RJ, Shelhamer JH, Shen HM, Shen WC, Sheng ZH, Shi Y, Shibuya K, Shidoji Y, Shieh JJ, Shih CM, Shimada Y, Shimizu S, Shintani T, Shirihai OS, Shore GC, Sibirny AA, Sidhu SB, Sikorska B, Silva-Zacarin EC, Simmons A, Simon AK, Simon HU, Simone C, Simonsen A, Sinclair DA, Singh R, Sinha D, Sinicrope FA, Sirko A, Siu PM, Sivridis E, Skop V, Skulachev VP, Slack RS, Smaili SS, Smith DR, Soengas MS, Soldati T, Song X, Sood AK, Soong TW, Sotgia F, Spector SA, Spies CD, Springer W, Srinivasula SM, Stefanis L, Steffan JS, Stendel R, Stenmark H, Stephanou A, Stern ST, Sternberg C, Stork B, Stralfors P, Subauste CS, Sui X, Sulzer D, Sun J, Sun SY, Sun ZJ, Sung JJ, Suzuki K, Suzuki T, Swanson MS, Swanton C, Sweeney ST, Sy LK, Szabadkai G, Tabas I, Taegtmeyer H, Tafani M, Takacs-Vellai K, Takano Y, Takegawa K, Takemura G, Takeshita F, Talbot NJ, Tan KS, Tanaka K, Tanaka K, Tang D, Tang D, Tanida I, Tannous BA, Tavernarakis N, Taylor GS, Taylor GA, Taylor JP, Terada LS, Terman A, Tettamanti G, Thevissen K, Thompson CB, Thorburn A, Thumm M, Tian F, Tian Y, Tocchini-Valentini G, Tolkovsky AM, Tomino Y, Tonges L, Tooze SA, Tournier C, Tower J, Towns R, Trajkovic V, Travassos LH, Tsai TF, Tschan MP, Tsubata T, Tsung A, Turk B, Turner LS, Tyagi SC, Uchiyama Y, Ueno T, Umekawa M, Umemiya-Shirafuji R, Unni VK, Vaccaro MI, Valente EM, Van den Berghe G, van der Klei IJ, van Doorn W, van Dyk LF, van Egmond M, van Grunsven LA, Vandenabeele P, Vandenberghe WP, Vanhorebeek I, Vaquero EC, Velasco G, Vellai T, Vicencio JM, Vierstra RD, Vila M, Vindis C, Viola G, Viscomi MT, Voitsekhovskaja OV, von Haefen C, Votruba M, Wada K, Wade-Martins R, Walker CL, Walsh CM, Walter J, Wan XB, Wang A, Wang C, Wang D, Wang F, Wang F, Wang G, Wang H, Wang HG, Wang HD, Wang J, Wang K, Wang M, Wang RC, Wang X, Wang X, Wang YJ, Wang Y, Wang Z, Wang ZC, Wang Z, Wansink DG, Ward DM, Watada H, Waters SL, Webster P, Wei L, Weihl CC, Weiss WA, Welford SM, Wen LP, Whitehouse CA, Whitton JL, Whitworth AJ, Wileman T, Wiley JW, Wilkinson S, Willbold D, Williams RL, Williamson PR, Wouters BG, Wu C, Wu DC, Wu WK, Wyttenbach A, Xavier RJ, Xi Z, Xia P, Xiao G, Xie Z, Xie Z, Xu DZ, Xu J, Xu L, Xu X, Yamamoto A, Yamamoto A, Yamashina S, Yamashita M, Yan X, Yanagida M, Yang DS, Yang E, Yang JM, Yang SY, Yang W, Yang WY, Yang Z, Yao MC, Yao TP, Yeganeh B, Yen WL, Yin JJ, Yin XM, Yoo OJ, Yoon G, Yoon SY, Yorimitsu T, Yoshikawa Y, Yoshimori T, Yoshimoto K, You HJ, Youle RJ, Younes A, Yu L, Yu L, Yu SW, Yu WH, Yuan ZM, Yue Z, Yun CH, Yuzaki M, Zabirnyk O, Silva-Zacarin E, Zacks D, Zacksenhaus E, Zaffaroni N, Zakeri Z, Zeh HJ, 3rd, Zeitlin SO, Zhang H, Zhang HL, Zhang J, Zhang JP, Zhang L, Zhang L, Zhang MY, Zhang XD, Zhao M, Zhao YF, Zhao Y, Zhao ZJ, Zheng X, Zhivotovsky B, Zhong Q, Zhou CZ, Zhu C, Zhu WG, Zhu XF, Zhu X, Zhu Y, Zoladek T, Zong WX, Zorzano A, Zschocke J, Zuckerbraun B (2012) Guidelines for the use and interpretation of assays for monitoring autophagy. Autophagy 8(4):445–544. doi: 10.4161/auto.19496

16. Thompson AR, Doelling JH, Suttangkakul A, Vierstra RD (2005) Autophagic nutrient recycling in Arabidopsis directed by the ATG8 and ATG12 conjugation pathways. Plant Physiol 138(4): 2097–2110. doi:10.1104/pp.105.060673

Part III

Proteomic Analysis and Other Post-Translational Modification Studies

Chapter 14

Protocols for Studying Protein Stability in an Arabidopsis Protoplast Transient Expression System

Séverine Planchais, Laurent Camborde, and Isabelle Jupin

Abstract

Protein stability influences many aspects of biology, and measuring their stability in vivo can provide important insights into biological systems.

This chapter describes in details two methods to assess the stability of a specific protein based on its transient expression in Arabidopsis protoplasts. First, a pulse-chase assay based on radioactive metabolic labeling of cellular proteins, followed by immunoprecipitation of the protein of interest. The decrease in radioactive signal is monitored over time and can be used to determine the protein's half-life.

Alternatively, we also present a nonradioactive assay based on the use of reporter proteins, whose ratio can be quantified. This assay can be used to determine the relative stability of a protein of interest under specific conditions.

Key words Protein stability, Arabidopsis thaliana, Protoplasts, Transient expression, Pulse-chase, UPR assay

1 Introduction

Transient expression of proteins in Arabidopsis protoplasts provides an important and versatile tool for conducting cell-based experiments to analyze the function of signaling pathways and cellular machineries [1, 2], including the ubiquitin proteasome degradation system [3, 4].

Transient expression allows a relatively large number of samples to be analyzed in a short period of time, and gene expression is not biased by position effects, as observed in stably transformed plants. Such method also allows for treatment of the cells with various pharmaceutical drugs, or for the co-expression of proteins without requiring time-consuming plant crosses. However, a major requirement is the preparation of viable protoplasts by enzymatic removal of the cell wall, and subsequent transfection of plasmid expression vectors encoding the proteins of interest. Here, we describe the obtention of Arabidopsis protoplasts from suspension

L. Maria Lois and Rune Matthiesen (eds.), *Plant Proteostasis: Methods and Protocols*, Methods in Molecular Biology, vol. 1450, DOI 10.1007/978-1-4939-3759-2_14, © Springer Science+Business Media New York 2016

cultured cells and their transfection using polyethylene glycol (PEG), using a technique routinely used in our laboratory for many years [5, 6].

Measuring the stability of a protein in vivo is a critical step in assessing whether its function may or not be regulated by proteolysis under specific physiological conditions. Two different methods to analyze protein stability in vivo are reported therein, one based on metabolic labeling and pulse-chase experiments, and the other one based on the expression of reporter proteins. Both methods have limitations which have been comprehensively reviewed [7].

Pulse-chase analysis is a method for examining how degradation of a specific protein occurs over time by successively exposing the cells to a labeled compound (the pulse), and then to an excess of the same compound in an unlabeled form (the chase period) (Fig. 1). To follow protein stability, the labeled compound used consists in radioactively labeled [^{35}S]-methionine and -cysteine amino acids that will be taken up by the cell and incorporated into all proteins synthesized during the pulse period. During the chase period, an excess of unlabeled methionine and cysteine is added, so that all proteins synthesized afterward will not be visible using radioactive detection methods. However, the amount of radioactive proteins synthesized during the pulse period can still be detected, and their remaining amount determined over time. In order to follow disappearance of the sole protein of interest, immunoprecipitation experiments using a specific antibody are required, immediately after the pulse experiment ($t=0$), and at regular intervals during the chase period. After normalizing the amount of radioactive protein to the total amount of immunoprecipitated proteins, the kinetics of protein degradation can be followed, and the half-life of the protein determined. This classical method is the most direct approach to study protein degradation, but as it relies on the incorporation of radioactive amino acid isotopes, strict compliance with safety measures and local regulation procedures for handling and waste-disposal is required.

Alternatively, reporter-dependent approaches have also been described, in which the open reading frame of the protein of interest is expressed as a fusion protein with a reporter protein, and its stability is assessed by measurement of the reporter activity. To allow normalization, a second reporter protein is encoded in the same construct and serves as a reference protein. Determining the steady-state molar ratios of the test and reference reporter proteins in cell extracts allows a direct ranking of their metabolic stability. While more simple and allowing multiple samples to be processed simultaneously, the tagging process may however interfere with the folding or proper subcellular targeting of the protein of interest, which may exert unpredictable effects on the stability of particular proteins. The protocol we describe is based on the method initially referred to as the "ubiquitin/protein/reference" (UPR)

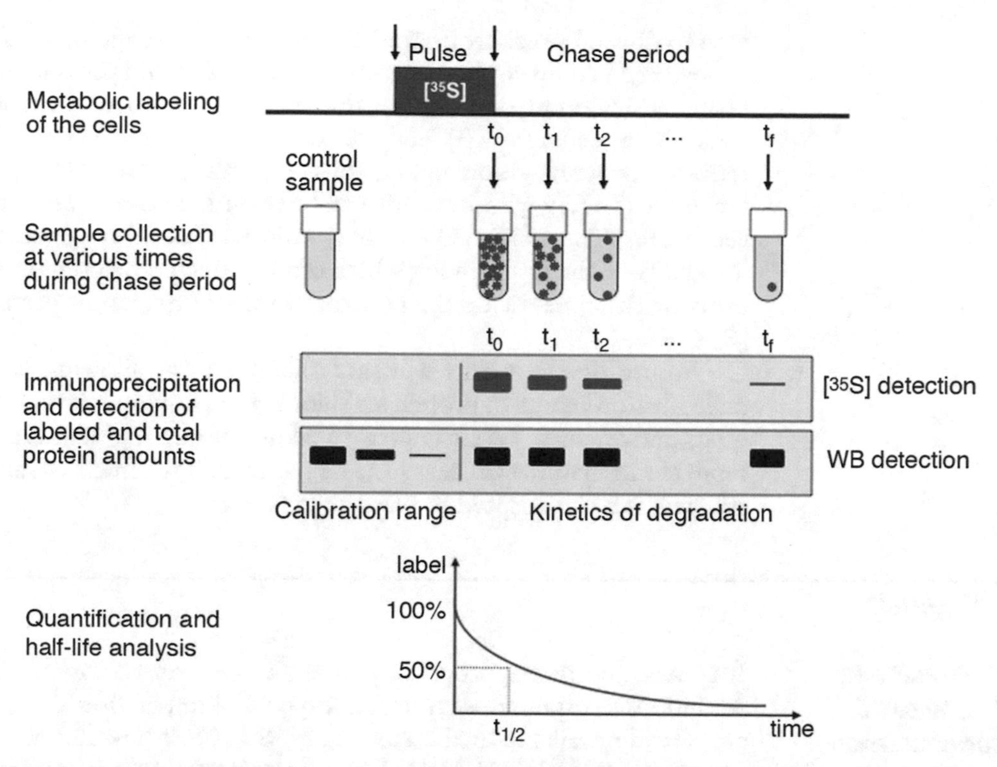

Fig. 1 Schematic representation of the pulse-chase experiment. Radioactively labeled [^{35}S]-methionine and -cysteine amino acids are taken up by the cells during the pulse period and incorporated into all proteins synthesized. An excess of unlabeled methionine and cysteine is then added during the chase period, so that all proteins synthesized afterward are not visible using radioactive detection methods. The remaining amount of radioactive protein of interest is determined over time by performing immunoprecipitation experiments, immediately after the pulse experiment (t_0), and at regular intervals during the chase period (t_1, t_2, …, tf). After normalizing the amount of radioactive protein to the total amount of immunoprecipitated proteins detected by western-blotting (WB), the kinetics of protein degradation can be followed, and the half-life ($t_{1/2}$) of the protein determined

Fig. 2 Schematic representation of the chimeric protein used in the ubiquitin/protein/reference (UPR) assay. Reference and test proteins are separated by a ubiquitin moiety (UbK48R) that is cleaved by cellular ubiquitin-specific processing proteases (UBP). Chloramphenicol acetyl transferase (CAT) serves as the internal control, and luciferase (LUC), N-terminally fused to the protein of interest, serves as the test protein. The LUC/CAT activity ratio reflects the instability of the test protein

technique [8] (Fig. 2). In this system, the test protein is produced as a translational fusion to the reference protein, separated by an ubiquitin (Ub) monomer. This Ub monomer contains a K48R substitution to prevent the conjugation of further Ub moieties which would lead to protein degradation. Such translational fusions are rapidly and precisely cleaved by cellular Ub-specific processing proteases, yielding equimolar amounts of the test and the reference proteins. Different variations of this method have been used successfully in plants [9, 10], and we adapted it using the two stable reporter proteins chloramphenicol acetyl transferase (CAT) and luciferase (LUC), whose activity can be quantified directly in crude cell lysates. The LUC/CAT activity ratio was found to reflect the instability of the test protein, which can be affected by point mutations or deletions or by the co-expression of interacting partners [3, 4].

Future developments will most likely aim at allowing large-scale measurements of protein stability at the proteome level using quantitative mass spectrometry or flow cytometry analyses, as reported in mammalian cells [11, 12], but such technical advances still awaits to be adapted to plant cell systems.

2 Materials

2.1 Maintenance of Arabidopsis Cell Suspension Culture, Preparation of Arabidopsis Protoplasts and Transfections

It is expected that appropriate plasmid expression vectors and the Arabidopsis suspension-cultured ecotype Columbia, line T87 [13] are already available in the laboratory. Such cell line can also be obtained from RIKEN BRC [14] but optimization of the subculture method may be necessary to adapt to each laboratory conditions. All experiments have to be performed in axenic conditions using a safety cabinet and standard in vitro culture/cell biology procedures.

1. Gamborg's B5 Basal medium (Sigma: G5893-10L): prepare Gamborg medium (5×) by diluting the powder in 2 l of ultrapure water. Aliquot by 200 ml and store at −20 °C.
2. 1-Naphthaleneacetic acid (NAA) (Sigma N0640): 5 mM solution in 85% ethanol, store at 4 °C.
3. Macerozyme R-10 (Yakult). Store powder at −20 °C.
4. Cellulase "Onozuka" RS (Yakult). Store powder at −20 °C.
5. "Arabidopsis culture medium": Mix 200 ml of Gamborg medium (5×), 30 g of saccharose, 200 μl of 5 mM NAA, make up to 1 l with ultrapure water and adjust pH to 5.8 using KOH. Dispense medium per 40 ml in 250-ml culture erlenmeyers, plug with cotton wool plugs covered with aluminum foil. Sterilize by autoclaving for 20 min at 110 °C to prevent sugar alteration. Store at RT.
6. "0.34 M medium": Mix 200 ml of Gamborg medium (5×), 30.4 g of glucose monohydrate, 30.4 g of mannitol (Sigma M1902), 200 μl of 5 mM NAA, make up to 1 l with ultrapure water and adjust to pH 5.8 using KOH. Sterilize by autoclaving for 20 min at 110 °C to prevent sugar alteration. Store at RT.

7. "0.28 M medium": Mix 200 ml of Gamborg medium (5×), 96 g of saccharose, 200 μl of 5 mM NAA, make up to 1 l with ultrapure water and adjust to pH 5.8 using KOH. Sterilize by autoclaving for 20 min at 110 °C to prevent sugar alteration. Store at RT.
8. "Jussieu medium": Mix 200 ml of Gamborg medium (5×), 18 g of glucose monohydrate, 45.6 g of mannitol (Sigma M1902), 200 μl of 5 mM NAA, make up to 1 l with ultrapure water and adjust to pH 5.8 using KOH. Sterilize by autoclaving for 20 min at 110 °C to prevent sugar alteration. Store at RT.
9. PEG solution: Mix 25 g of PEG 6000 (Sigma 81253), 2.36 g of $Ca(NO_3)_2$, $4H_2O$, 8.2 g of mannitol (Sigma M1902), make up to 100 ml with ultrapure water and adjust to pH 9 using NaOH. Aliquot by 10 ml and store at −20 °C. Re-adjust to pH 9 immediately before use. Sterilize by filtration on a 0.45-μm syringe filter.
10. $Ca(NO_3)_2$ solution: 275 mM $Ca(NO_3)_2$, $4H_2O$ solution in water. Sterilize by autoclaving. Store at RT.
11. 0.22 and 0.45 μm syringe filters.
12. Sterile 1.5- and 2-ml Safelock Eppendorf tubes.
13. Sterile pipette tips with large orifice (Starlab E1011-9500 and E1011-8400).
14. Sterile 5-, 10-, and 25-ml individually wrapped plastic pipettes.
15. Sterile polystyrene 14- and 50-ml centrifugation tubes (i.e. Falcon, Corning).
16. Sterile polystyrene 35-mm dishes (Easy-Grip Falcon ref 353001).
17. Variable-speed automatic pipettor (i.e. Drummond Pipet-aid).
18. Refrigerated rotating shaker with photosynthetic illumination (i.e. New Brunswick Innova 44R with Photosynthetic Light bank).
19. Microbiology rotating shaker.
20. Microbiological safety cabinet.
21. Refrigerated microbiological incubator.
22. Refrigerated low-speed centrifuge with swinging bucket rotor (i.e. Eppendorf centrifuge 5810R).
23. Inverted Microscope for routine microscopy (i.e. Nikon Eclipse TS100).
24. Fluorescence microscope (optional).

2.2 Metabolic Labeling and Pulse-Chase Experiments

It is expected that specific antibodies raised against the protein of interest are already available, that western-blotting and immuno-precipitation conditions are already set up and that standard protein electrophoresis and western-blotting laboratory equipment is available.

1. Easytag Express ^{35}S protein labeling mix (Perkin Elmer NEG772002MC). Store at 4 °C.
2. L-Methionine: 250 mM solution in water. Sterilize by filtration on a 0.22-μm syringe filter. Store at –20 °C.
3. L-Cysteine: 250 mM solution in water. Sterilize by filtration on a 0.22-μm syringe filter. Store at –20 °C.
4. Protease inhibitors ×25 (Complete Roche tablets): dissolve 1 tablet in 2 ml of ultrapure water. Store at –20 °C.
5. Bovine Serum Albumin (BSA): 100 mg/ml solution in water. Store at –20 °C.
6. Pansorbin (Calbiochem).
7. Benzyloxycarbonyl-L-leucyl-L-leucyl-L-leucinal, Z-Leu-Leu-Leu-al (MG132) (Calbiochem): 100 mM stock solution in DMSO. Store at –20 °C.
8. Clasto-Lactacystine β-lactone (Calbiochem): 10 mM stock solution in DMSO. Store at –20 °C.
9. Phosphate Buffer Saline (PBS): 137 mM NaCl, 2.7 mM KCl, 10 mM Na_2HPO_4, 1.76 mM KH_2PO_4. Sterilize by autoclaving. Store at RT.
10. Protein loading buffer (L×3): 180 mM Tris–HCl pH 6.8, 6% SDS, 30% Glycerol, 0.03% bromophenol blue. Store at –20 °C.
11. IP buffer: 50 mM Tris–HCl pH 7.5, 150 mM NaCl, 5 mM EDTA, 0.1% SDS, 0.5% Sodium deoxycholate, 1% Triton X-100, 1× protease inhibitors. Prepare fresh before use.
12. Washing buffer: 50 mM Tris–HCl pH 7.5, 100 mM NaCl, 2 mM EDTA, 0.5%, SDS, 2% Triton X-100, 1× protease inhibitors. Prepare fresh before use.
13. Primary antibody raised against the protein of interest.
14. Secondary antibody conjugated to a reporter enzyme (i.e. alkaline phosphatase or horseradish peroxidase conjugate).
15. Western-blotting substrate: either nitro blue tetrazolium (NBT)/5-bromo-4-chloro-3-indolyl-phosphate (BCIP) for alkaline phosphatase detection, or enhanced luminol-based (ECL) chemiluminescent substrate for horseradish peroxidase detection.
16. 0.22-μm syringe filters.
17. Sterile 1.5- and 2-ml Safelock Eppendorf tubes.
18. Sterile pipette tips with large orifice (i.e. Starlab E1011-9500 and E1011-8400).
19. Nitrocellulose membrane 0.22 μm or polyvinylidene difluoride (PVDF) membrane.
20. Plastic container.

21. Portable Geiger counter.
22. Containers for disposal of solid and liquid [^{35}S] radioactive waste.
23. Protein PAGE gels.
24. Protein PAGE migration apparatus.
25. Protein transfer apparatus.
26. Refrigerated low-speed centrifuge with swinging bucket rotor (i.e. Eppendorf centrifuge 5810R).
27. Heating block.
28. Refrigerated centrifuge for Eppendorf tubes.
29. Rotating shaker.
30. Image acquisition system (i.e. GE Healthcare Imagequant LAS-3000).
31. Phosphor Imager screen and cassette, or autoradiography cassette with film.
32. Phosphor Imager (i.e. Molecular Dynamics Storm or GE Healthcare Typhoon) or autoradiography film developing device.

2.3 Stability Measurements Using Reporter-Dependent Assays

1. Phosphate Buffer Saline (PBS): NaCl, 2.7 mM KCl, 10 mM Na_2HPO_4, 1.76 mM KH_2PO_4. Sterilize by autoclaving. Store at RT.
2. Protease inhibitors ×25 (Complete Roche tablets): dissolve 1 tablet in 2 ml of ultrapure water. Store at −20 °C.
3. Luciferase Reporter 1000 Assay System (Promega Cat. # E4550). Store at −20 °C.
4. Luciferase cell culture lysis reagent (CCLR) ×5 (included in the luciferase assay): 125 mM Tris-phosphate pH 7.8, 10 mM DTT, 10 mM 1,2-diaminocyclohexane-*N*,*N*,*N′*,*N′*-tetraacetic acid, 50% glycerol, 5% Triton X-100. Store at −20 °C.
5. Liquid nitrogen.
6. CAT ELISA kit assay (Roche 11363727001).
7. Sterile 1.5- and 2-ml Safelock Eppendorf tubes.
8. Sterile pipette tips with large orifice (i.e. Starlab E1011-9500 and E1011-8400).
9. Low-speed centrifuge with swinging bucket rotor (i.e. Eppendorf centrifuge 5810R).
10. Vortex.
11. Centrifuge for Eppendorf tubes.
12. 96-well opaque microtitration plate (i.e. Corning 3696).
13. Luminometer (i.e. Berthold Centro LB960).

14. Repetitive dispensing pipette (i.e. Ripette).
15. Microplate spectrophotometer capable of reading absorbance at 405 nm (i.e. Molecular Devices SpectraMax).

3 Methods

3.1 Preparation of Arabidopsis Protoplasts and Transfection

1. Subculture weekly the Arabidopsis cell suspension culture by adding 8 ml of a 7-day-old culture to 40 ml of Arabidopsis culture medium. Grow at 20–22 °C for 7 days in an incubator with a shaking platform rotating at 130 rpm, with a 16 h/8 h photoperiod.
2. For protoplast preparation, add 20 ml of a 7-days-old culture to 40 ml of Arabidopsis culture medium. Grow at 20–22 °C for 40 h in an incubator with a shaking platform rotating at 130 rpm, with a 16 h/8 h photoperiod.
3. Dissolve 50 mg of Macerozyme R-10 and 300 mg de Cellulase "Onozuka" RS in 25 ml of "0.34 M medium". Stir well for at least 30 min with a strong agitation. Sterilize solution by filtering on a 0.45-μM syringe filter.
4. Transfer 45 ml of the cell culture (from **step 2**) in a sterile 50-ml centrifugation tube, and centrifuge at 80 × *g* for 3 min at room temperature (RT), without brake. Remove and discard supernatant (*see* **Note 1**).
5. Add the Macerozyme and Cellulase solution (from **step 3**) and gently resuspend the cell pellet by slow pipetting. Adjust the volume to 50 ml using "0.34 M medium", and transfer to a sterile erlenmeyer containing 25 ml of "0.34 M medium" (total volume = 75 ml). Incubate at 30 °C on a shaking platform rotating at 130 rpm for a duration of 50–90 min. The extent of cell wall digestion has to be followed at regular intervals by placing an aliquote (100 μl) of the digestion on a glass slide and observing the cells by light microscopy. Cell clusters should progressively dissociate into isolated near-spherical protoplasts. Care has to be taken not to over-digest the cells, as evidenced by the appearance of numerous cellular debris in the medium (*see* **Note 2**).
6. Stop digestion by transferring the erlenmeyer on ice, and transfer its content into two sterile 50-ml centrifugation tubes (*see* **Note 3**). Keep the tubes on ice during the whole procedure. Centrifuge protoplasts at 80 × *g* for 3 min at 4 °C, without brake. Carefully remove and discard supernatant.
7. Wash the protoplast pellets with 25 ml of "0.34 M medium", and carefully resuspend cells by gently swirling or inverting the tube 3–4 times. Centrifuge at 80 × *g* for 3 min at 4 °C, without brake. Carefully remove and discard supernatant. Repeat the washing step with 25 ml of "0.34 M medium".

8. Resuspend each protoplast pellet with 5 ml of "0.28 M medium", and pool them together within a 14-ml centrifugation tube. Centrifuge at 80 × *g* for 3 min at 4 °C, without brake. The intact protoplasts will float on the top of the "0.28 M medium", whereas cellular debris and undigested cell clusters will sediment at the bottom of the tube (*see* **Note 4**).
9. Collect intact protoplasts floating on the top of the "0.28 M medium" using a 1-ml tip with a large orifice and transfer them to a 14-ml centrifugation tube. Keep the tube on ice (*see* **Note 5**).
10. To estimate the number and concentration of protoplasts obtained, a 10-μl aliquot of the protoplast suspension is diluted 50× in "Jussieu medium", and counted using a hemocytometer (i.e. Malassez counting chamber). Calculate the protoplast concentration accordingly (i.e. 100 squares of the Malassez counting chamber correspond to a volume of 1 μl). Adjust the protoplast suspension to a concentration of 10^7 protoplasts/ml using "Jussieu medium". Keep the cells on ice for 1–3 h (*see* **Note 6**).
11. In the meantime, prepare 2-ml Eppendorf tubes containing the nucleic acids to be transfected (0.1–10 μg per transfection in a volume of 10–20 μl) (*see* **Note 7**).
12. Allow the PEG solution to thaw at RT, and readjust to pH 9 using NaOH. Sterilize by filtration on a 0.45-μm syringe filter.
13. After incubation of the cells on ice for at least 1 h (**step 10**), add 50 μl of the protoplast suspension (=5×10^5 protoplasts) to the 2-ml Eppendorf tubes containing the nucleic acids, and immediately add 200 μl of "Jussieu medium" and 250 μl of PEG solution (*see* **Note 8**).

Mix by gently patting the tube with fingertips, and incubate the tube for 25 min at RT in the dark by placing the tube on a rack covered with an aluminum foil. Proceed similarly for all transfections, keeping a 1-min interval between each sample (*see* **Note 9**).

14. Upon the 25-min incubation with PEG, add 500 μl of $Ca(NO_3)_2$ solution, mix gently by inverting the tube 4–5 times and incubate the tube for 5 min at RT in the dark. Proceed similarly for all transfections one after the other, respecting the 1-min interval between each sample. After the 5-min incubation, add another 500 μl of $Ca(NO_3)_2$ solution and mix gently by inverting the tube 4–5 times. Proceed similarly for all transfections, respecting the 1-min interval between each sample.
15. Centrifuge at 80 × *g* for 3 min at RT, without brake. Carefully remove and discard supernatant, taking care not to remove cells, as the protoplast pellet is rather loose.
16. Add 1 ml of "Jussieu medium" to the protoplast pellets and transfer to sterile 35-mm polystyrene dishes. Incubate in the dark at 24 °C in an incubator for the desired period of time

(i.e. 16–48 h) to allow protoplasts to express the protein of interest.

17. Optional: in case transfection with an expression vector encoding a fluorescent protein has been performed, the percentage of transfected cells can be estimated by observing an aliquot of the protoplasts with a fluorescent microscope, and relating the number of fluorescent cells to the total number of cells. The percentage of transfection may vary from less than 10–30% depending on the cell physiological status and the extent of cell wall digestion.

3.2 Pulse-Chase Experiments Using Radioactive Metabolic Labeling of Transfected Protoplasts

There can be great variation in the half-life of different proteins, so for an unfamiliar protein, it is recommended to chase until 24 h using 0, 2, 4, 8, and 24 h time points. According to the results obtained, the time frame can then be narrowed down in subsequent experiments. The general outline of the experiment is schematized in Fig. 3.

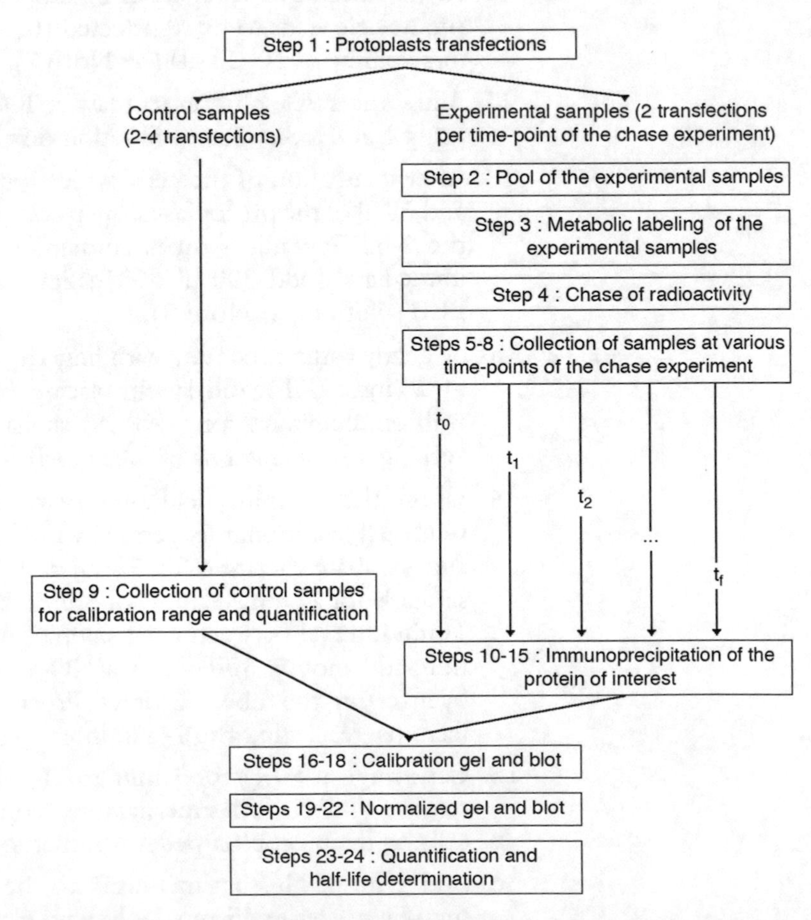

Fig. 3 Schematic outline of the various experimental steps in Subheading 3.2

1. Transfect Arabidopsis protoplasts with an appropriate expression vector encoding the protein of interest as described from **step 13** in Subheading 3.1 (*see* **Note 10**). As each time point of the chase period will be performed in duplicates, perform 2 transfections per time point of the chase experiment. Additionally, 2–4 transfections are also required to verify proper expression of the protein and to be used as internal loading controls. Those samples do not need to be metabolically labeled. Incubate transfected protoplasts in the dark in an incubator at 24 °C, to allow proper expression of the protein of interest.
2. 24 h post-transfection, pool all the transfected protoplasts dedicated to the pulse-chase experiment within a sterile 15-ml or 50-ml centrifugation tube, to ensure homogeneity of the protoplast samples and labeling reaction.
3. For metabolic labeling of total cellular proteins, add [^{35}S]-radioactively labeled Methionine and Cysteine (Easytag Express ^{35}S protein labeling mix) to the pool of transfected protoplasts using 50 μCi (1.85 MBq)/ml of protoplasts. Mix gently by inverting the tube several times and place it horizontally to maximize the contact surface with the air. Place it in a plastic container to avoid any risk of spilling and return it to the incubator (*see* **Note 11**).
4. After a labeling period of 2 h, chase radioactivity from the cells by adding nonradioactive L-Methionine and L-Cysteine, each to a 5 mM final concentration. This corresponds to the time-zero of the chase period.
5. Immediately collect 2 samples of 1 ml of protoplasts and place them into a 1.5-ml Safelock Eppendorf tube. Those samples will serve as references for the subsequent decrease in radioactive labeling of the protein of interest over time.
6. Centrifuge tubes at 80 × g for 2 min at RT (no brake), and carefully remove supernatant (*see* **Note 12**). Wash protoplasts by gently adding 400 μl of PBS containing 1× protease inhibitors.

   Centrifuge at 80 × g for 2 min (no brake), and carefully remove supernatant, taking care not to remove cells, as the cell pellet is rather loose (*see* **Note 12**).
7. Measure the volume of cells with a pipetman (*see* **Note 13**), and add half-volume of protein loading buffer (L×3). Heat tubes at 100 °C in a heating block for 10 min and centrifuge the tubes at 13,000 × g for 5 min at RT. Store tubes at −20 °C until all samples are collected.
8. Collect samples in duplicate at each other time points of the chase experiment (i.e. t = 2, 4, 8, 24 h according to the experimental design) and proceed similarly to **steps 5–7**.

9. At the end of the experiment (e.g. 48 h post-transfection), proceed similarly to collect the remaining samples which have not been metabolically labeled, and which will be used to verify proper expression of the protein and serve as internal standards for calibration of protein immunoprecipitation (see below **step 15**).
10. Once all metabolically labeled samples have been collected, proceed with the immunoprecipitation (IP) of the protein of interest. The suitable conditions for immunoprecipitating the protein of interest has to be set-up beforehand. The conditions described here are those previously described for the TYMV 66K protein [6, 15], but they may greatly vary depending on each antigen/antibody combinations, and are provided only as a guideline.
11. The protoplast samples collected at **step** 7 are allowed to thaw at RT and centrifuged at 13,000 × *g* for 5 min at RT. The supernatant (~ 50 μl) is transferred to a clean Safelock 1.5-ml Eppendorf tube (*see* **Note 12**) containing 750 μl of IP buffer supplemented with BSA (1 mg/ml final concentration) and 0.75 μl of specific antibody (*see* **Note 14**). Incubate o/n at 4 °C on a rotating shaker to allow antigen/antibody complexes to form.
12. Equilibrate Pansorbin in IP buffer according to suppliers' instructions (*see* **Note 15**). Add 20 μl of Pansorbin to each tube of protoplast cell lysates. Incubate 2 h at 4 °C on a rotating shaker to allow antigen/antibody/Pansorbin complexes to form.
13. Centrifuge samples at 6000 × *g* for 5 min at 4 °C, and carefully remove supernatant (*see* **Note 12**). Add 500 μl of washing buffer and resuspend Pansorbin by vortexing for 45 s. Incubate on ice for 5–10 min.
14. Proceed to **step 12** four more times (five washes in total), and during the last wash, transfer the Pansorbin suspension to a clean Safelock 1.5-ml Eppendorf tube. Centrifuge samples at 6000 × *g* for 5 min at 4 °C, and carefully remove supernatant (*see* **Note 12**).
15. Add 25 μl of protein loading buffer (L×3) to each tube and carefully resuspend Pansorbin by vortexing. Heat tubes at 100 °C in a heating block for 10 min and centrifuge them at 13,000 × *g* for 5 min at RT. Transfer the supernatant to a clean 1.5-ml Eppendorf tube (*see* **Note 12**). Store IP samples at –20 °C.
16. Because the efficiency of IP may vary from one tube to another, it is strongly advisable to first perform a control western blot (WB) of the material eluted at **step 15**, which will allow to quantify the amount of protein immunoprecipitated in each sample, and to subsequently normalize them.

To allow quantification, analyze 5 μl of the material eluted at **step 15** by SDS-PAGE, together with increasing amounts (i.e. 2, 4, and 8 μl) of the control samples collected at **step 9**, so as to have identical internal standards on each gel.

17. Transfer the gel to nitrocellulose or PVDF membrane, incubate the blot with a primary antibody raised specifically against the protein of interest, then with a secondary antibody conjugated to a reporter enzyme (e.g. alkaline phosphatase or horseradish peroxidase) and reveal the WB using corresponding substrates. Acquire signal with an image acquisition system, in case ECL chemiluminescent substrate is used.
18. For quantification, select regions of interest and quantify each lane of the blot using Imagequant software, or any other image analysis software such as NIH Image J (*see* **Note 16**). The internal standards can be used to normalize signals from one blot to another, and to adjust the volumes of samples of immunoprecipitated protein to be loaded in order to achieve equal loading.
19. Perform a second—normalized—SDS-PAGE of the material eluted at **step 15** based on the quantification performed at **step 18**. Also include increasing amounts (i.e. 2, 4, and 8 μl) of the control samples collected at **step 9**, so as to have identical internal standards on each gel.
20. Transfer the gel to nitrocellulose or PVDF membrane, and allow it to dry for 2 h at 37 °C (*see* **Note 17**). Place the blot in a cassette with a Phosphor Imager screen—or if not available, an autoradiography film (*see* **Note 18**) in order to detect the amount of radioactive protein present in the immunoprecipitate.
21. After 24 h exposure, scan the Phosphor Imager screen using a Phosphor Imager and analyze signals with appropriate software (i.e. Imagequant). Alternatively, develop the autoradiography film and analyze its scanned image using NIH Image J, or any other image analysis software (*see* **Note 19**).
22. Pursue revelation of the WB as described in **step 16**, by incubating the blot from **step 20** with a primary antibody raised against the protein of interest, then with a secondary antibody conjugated to a reporter enzyme, and revealing the blot using corresponding substrates. Acquire signal with an image acquisition system, in case ECL chemiluminescent substrate is used.
23. Quantify signals as described in **step 18**, using the internal standards to normalize western-blotting signals.
24. For each time point of the chase experiment, calculate the ratio between the radioactive signal detected by Phosphor Imaging and the total amount of protein detected by western-blotting (in arbitrary units). Express as a percentage of the ratio calculated at the time zero of the chase period, and plot the

corresponding data. Estimate the half-life of the protein by extrapolating the time when 50% of the radioactive protein has disappeared.

25. To determine whether protein instability is due to proteasome degradation, such experiments may be performed in the presence of proteasome inhibitors such as MG132 or clastolactacystine β-lactone. Dissolve inhibitors in DMSO and use at a final concentration of 100 μM and 25 μM, respectively. Perform control samples containing DMSO at the same final concentration (*see* **Note 20**).

3.3 Stability Measurements Using Reporter-Dependent Assays

Different reporter proteins have been used in plant cells to measure protein stability [9, 10, 16], and we made use of CAT and LUC as both essays could be performed using the same cell lysate.

For the CAT assay, we favored the use of a colorimetric enzyme immunoassay (CAT ELISA) as compared to an acetyl group transfer activity-based assay [17], as it is safer (no radioisotopes are used), more accurate as it measures the amount of CAT protein synthesized, and not just CAT activity, and allows the assay to be performed using the same cell lysate as the LUC assay (no inhibition by the detergent present in cell lysis reagent). Both assays are easily performed using commercially available kits.

1. Construct an appropriate expression vector encoding the protein of interest fused in-frame to test and reference reporters (*see* **Note 21**).
2. Transfect Arabidopsis protoplasts with the expression vector as described in Subheading 3.1.

As a control, transfect an expression vector encoding only test and reference proteins, i.e. pΩ-CAT-Ub:LUC [3]. To assess the reproducibility of the measurements and allow their subsequent statistical analyses, perform 6–12 transfections with each construct.

3. 48 h post-transfection, collect protoplasts samples and place them into a 1.5-ml Safelock Eppendorf tube using pipette tips with large orifice. Centrifuge tubes at 80 × *g* for 2 min at RT (no brake), and carefully remove supernatant. Wash protoplasts by gently adding 400 μl of PBS containing 1× protease inhibitors. Centrifuge at 80 × *g* for 2 min (no brake), and carefully remove supernatant, taking care not to remove cells, as the cell pellet is rather loose. Keep tubes on ice.
4. Prepare the lysis buffer by diluting 5× CCLR in water (provided in the Luciferase Reporter Assay System), and add 500 μl of 1× CCLR to each sample. Vortex vigorously each tube for 2 × 45 s, keep tubes on ice. Centrifuge at 13,000 × *g* for 1 min at 4 °C to remove cell debris. Transfer 50 μl of supernatant to a 1.5-ml Safelock Eppendorf tube to be used for luciferase assay, and

400 μl to a second 1.5-ml Safelock Eppendorf tube to be used for CAT assay. Freeze samples in liquid nitrogen, and store at −80 °C (*see* **Note 22**).

5. Prior to the luciferase assay, prepare the Luciferase Assay Reagent by resuspending the Luciferase Assay substrate (lyophilized) in the Luciferase Assay buffer according to the supplier's instructions. Aliquot by 1 ml and store frozen at −80 °C.
6. For luciferase assay, equilibrate the necessary amount of Luciferase Assay Reagent to room temperature in the dark. Each reaction requires 100 μl of luciferase assay reagent, plus the void volume of the injection system of the microplate luminometer (~1 ml for a Berthold Centro LB960 luminometer) (*see* **Note 23**).
7. Program the luminometer for appropriate delay and measurement times. Typical delay time is 2 s, followed by a 1- to 10-s measurement read for luciferase activity. Injection volume of Luciferase Assay Reagent is 100 μl per well. The time required for measurement has to be determined empirically as it depends on the expression level of the protein of interest, and sensitivity of the luminometer (*see* **Note 24**).
8. Thaw the 50-μl aliquot of protoplast cell lysate at RT, and dispense 20 μl of each sample to the wells of a 96-well opaque microtitration plate.
9. Place the microplate in the luminometer and initiate reading by injecting 100 μl of Luciferase Assay Reagent into each well. Record the values which are expressed in "Relative Light Units" (RLU) (*see* **Note 25**).
10. The CAT assay is based on a colorimetric enzyme immunoassay (CAT ELISA), and is performed as described by the supplier. Prior to the assay, dissolve the CAT enzyme that will be used to obtain a standard curve in ultrapure water, aliquot by 40 μl and store at −20 °C. Reconstitute the anti-CAT-DIG antibody in 500 μl of ultrapure water, aliquote by 50 μl, and store at −20 °C. Reconstitute the anti-DIG-peroxidase antibody in 500 μl of ultrapure water and store at 4 °C.
11. Prepare the samples for the CAT standard curve, by making serial dilutions of the CAT enzyme as recommended by the supplier (i.e. from 0.125 to 1 pg/μl of enzyme), and dispense 200 μl of each dilution into the wells of the CAT ELISA microplate. A standard curve, preferably in duplicate, must be established for each experiment.
12. Thaw the 400-μl aliquot of protoplast cell lysate on ice, and dispense 200 μl of cell extract into each well of the CAT ELISA microplate (*see* **Note 26**). Cover the wells with the adhesive foil and incubate at 37 °C for 1 h (*see* **Note 27**). This step

allows the binding of the CAT enzyme present in the lysate to the anti-CAT antibody that is pre-coated to the wells of the CAT ELISA microplate.

13. In the meantime, prepare wash buffer by diluting the 10× wash buffer stock solution in ultrapure water as recommended by the supplier, and dilute the anti-CAT-DIG antibody 1/100e as recommended by the supplier.
14. Remove the foil, empty the plate, and blot dry by tapping the inverted plate on absorbent paper. Pipette 250 μl of wash buffer into the wells, incubate for 30 s with gentle agitation, empty the plate and blot dry by tapping the inverted plate on absorbent paper. Repeat the washing step four more times (*see* **Note 28**).
15. Dispense 200 μl of anti-CAT-DIG antibody to each well, cover with the adhesive foil and incubate at 37 °C for 1 h.
16. In the meantime, dilute the anti-DIG-peroxidase antibody 1/133e as recommended by the supplier.
17. Perform five washes as described in **step 14**. Dispense 200 μl of anti-DIG-peroxidase antibody to each well, cover with the adhesive foil and incubate at 37 °C for 1 h. In the meantime, equilibrate at RT the ABTS solution (ready-to-use peroxidase substrate).
18. Perform five washes as described in **step 14**. Dispense 200 μl of the ABTS solution and incubate at RT. Read the absorbance of the solution in the wells within 10–30 min, using a microplate reader set to 405 nm. Record the values.
19. Plot the standard curve (absorbance readings against the concentration of the CAT enzyme), and use a linear regression for curve fitting. The concentration of CAT enzyme in the protoplast cell lysate samples can thus be determined from the standard curve.
20. Calculate the ratio of LUC/CAT activities per μl of cell lysate (expressed in RLU/pg of CAT enzyme). Normalization is done by expressing the results as a percentage of the control samples. For statistical significance, we advise the use of a Mann–Whitney rank test, a nonparametric test that allows two groups of samples to be compared without making the assumption that values are normally distributed.
21. Such experiments may be performed in the presence of proteasome inhibitors such as MG132 or clastolactacystine β-lactone (*see* **Note 29**). Dissolve inhibitors in DMSO and use at a final concentration of 100 μM and 25 μM, respectively (*see* **Note 30**). Perform control samples containing DMSO at the same final concentration.

4 Notes

1. The cell pellet should correspond approximately to a volume of 15 ml.
2. Digestion time depends on the cell culture physiology and should be determined empirically. This is the most critical step of the experiment.
3. Protoplasts are fragile and should be handled with care, using either plastic pipettes (with an automatic pipetor set up at slow speed) or pipeting tips with large orifice.
4. The volume of intact protoplasts obtained may vary from 1 to 5 ml depending on the extent of cell digestion.
5. Remove all the 0.28 M medium that may have been inadvertently pipetted, and that will spontaneously decant at the bottom of the tube.
6. If the concentration of protoplasts is between 2×10^6 and 10^7 protoplasts/ml, no dilution is required at this step. Below 2×10^6 protoplasts/ml, transfection will not be possible, and the digestion conditions will have to be optimized first.
7. It is also possible to use a mixture of expression vectors when co-expression of several proteins is required. As the efficiency of transfection depends on the amount of DNA to be transfected, we advise that each sample contains the same amount of total nucleic acids (i.e. 10 μg). For that purpose, empty expression vector, or any unrelated plasmid (e.g. pUC vector) may be added to the nucleic acids of interest to reach the same amount of DNA in every sample. We also strongly advise to perform one transfection with an expression vector encoding a fluorescent protein (i.e. GFP), as it will allow to estimate the efficiency of transfection (**step 16**).
8. In case the concentration of protoplasts is between 2×10^6 and 10^7 protoplasts/ml, add 5×10^5 protoplasts to the nucleic acids and adjust the volume to 250 μl using "Jussieu medium".
9. Due to constraints in incubation times, a maximum of 25 transfections can be performed at once. However, two to three series can be performed one after the other, provided the pH of the PEG solution is readjusted to pH 9 immediately before use.
10. Expression of the protein of interest has to be verified by western-blotting beforehand.
11. Radioactive material has to be handled and disposed with appropriate safety measures and according to local regulation. Regularly check gloves, pipettes, and materials for potential contamination using a portable Geiger counter.

12. Discard supernatant, tubes, and tips according to radioactive safety procedures.
13. The typical volume measured for 5×10^5 transfected cells is about 40 μl.
14. Centrifuge the antiserum for 10 min at $15,000 \times g$ at 4 °C prior to use, to avoid pipetting aggregates.
15. Protein A coupled to agarose beads can be used as an alternative to Pansorbin cells. Protein A agarose pellets are softer and easier to resuspend than Pansorbin pellets, but the risk of aspirating beads and loosing material during the washing steps is higher as well.
16. An image analysis guide should be referred to for a first-time user.
17. Make sure the blot is completely dry before exposure, as humidity will cause significant deterioration of the Phosphor Imager screens.
18. Phosphor Imaging is a more efficient and quantitative method than traditional film autoradiography.
19. Longer exposure time (up to 1 week) may be required if autoradiography film is used.
20. Because MG132 is a reversible proteasome inhibitor, its inhibitory effect is transient, and we advise to add the inhibitor at regular intervals (i.e. every 6–8 h) to maintain proteasome inhibition throughout the chase period.
21. We advise the use of PCR-based techniques [18], as it allows in-frame gene fusions to be obtained conveniently without the need for restriction enzymes.
22. As protein activities may decrease over time, we advise to measure CAT and LUC activities shortly (within 1 week) after sample collection.
23. The most convenient method for performing a large number of luciferase assays is to use a luminometer capable of processing a 96-well microplate with an injector, so that each well is measured right after injection of the Luciferase assay reagent. However, if not available, light intensity can also be measured using manual luminometers, in which single tubes are processed manually one after the other.
24. Because luminometers can experience signal saturation at high light intensities, it is essential to verify in the first experiment that the values obtained fall within the linear range of light detection. For that purpose, prepare serial dilutions of protoplast cell lysate and plot the values obtained to produce a standard curve.
25. As the light intensity of the reaction is stable for about 1 min, and then slowly decays with a half-life of approximately 10 min,

we do not advise to make a second read of the same plate, as the variations in time interval between the injection of each sample and their subsequent readings may cause inaccuracies.

26. Use only the number of wells modules required for the experiment. Unused modules must be stored at 4 °C in the foil pouch.
27. Keep the plate on a clean paper towel to avoid damaging the bottom surface of the wells, which may cause inaccuracies in sample reading.
28. It is most convenient to use a repetitive dispensing pipette to dispense the wash buffer.
29. We and others reported that both inhibitors exert a strong inhibitory effect on reporter protein production [3, 9, 19]. It is therefore essential to correct this effect using a control plasmid (i.e. pΩ-CAT-Ub:LUC) treated with the same inhibitors to normalize the results.
30. Because MG132 is a reversible proteasome inhibitor, its inhibitory effect is transient, and we advise to add the inhibitor at regular intervals (i.e. every 6–8 h) to maintain proteasome inhibition throughout the period of protein expression.

References

1. Doelling JH, Pikaard CS (1993) Transient expression in Arabidopsis thaliana protoplasts derived from rapidly established cell suspension cultures. Plant Cell Rep 12:241–244
2. Yoo SD, Cho YH, Sheen J (2007) Arabidopsis mesophyll protoplasts: a versatile cell system for transient gene expression analysis. Nat Protoc 2:1565–1572
3. Camborde L, Planchais S, Tournier V, Jakubiec A, Drugeon G, Lacassagne E, Pflieger S, Chenon M, Jupin I (2010) The ubiquitin-proteasome system regulates the accumulation of Turnip yellow mosaic virus RNA-dependent RNA polymerase during viral infection. Plant Cell 22:3142–3152
4. Chenon M, Camborde L, Cheminant S, Jupin I (2012) A viral deubiquitylating enzyme targets viral RNA-dependent RNA polymerase and affects viral infectivity. EMBO J 31:741–753
5. Schirawski J, Planchais S, Haenni AL (2000) An improved protocol for the preparation of protoplasts from an established Arabidopsis thaliana cell suspension culture and infection with RNA of turnip yellow mosaic tymovirus: a simple and reliable method. J Virol Methods 86:85–94
6. Prod'homme D, Le Panse S, Drugeon G, Jupin I (2001) Detection and subcellular localization of the turnip yellow mosaic virus 66K replication protein in infected cells. Virology 281:88–101
7. Alvarez-Castelao B, Ruiz-Rivas C, Castaño J (2012) A critical appraisal of quantitative studies of protein degradation in the framework of cellular proteostasis. Biochem Res Int, Article ID 823597, 11 p. doi: 10.1155/2012/823597
8. Levy F, Johnson N, Rümenapf T, Varshavsky A (1996) Using ubiquitin to follow the metabolic fate of a protein. Proc Natl Acad Sci 93:4907–4912
9. Karsies A, Hohn T, Leclerc D (2001) Degradation signals within both terminal domains of the cauliflower mosaic virus capsid protein precursor. Plant J 27:335–343
10. Stary S, Yin XJ, Potuschak T, Schlögelhofer P, Nizhynska V, Bachmair A (2003) PRT1 of Arabidopsis is a ubiquitin protein ligase of the plant N-end rule pathway with specificity for aromatic amino-terminal residues. Plant Physiol 133:1360–1366
11. Yen HC, Xu Q, Chou DM, Zhao Z, Elledge SJ (2008) Global protein stability profiling in mammalian cells. Science 322:918–923
12. Doherty MK, Hammond DE, Clague MJ, Gaskell SJ, Beynon RJ (2009) Turnover of the human proteome: determination of protein intracellular stability by dynamic SILAC. J Proteome Res 8:104–112

13. Axelos M, Curie C, Mazzolini L, Bardet C, Lescure B (1992) A protocol for transient gene expression in Arabidopsis thaliana protoplasts isolated from cell suspension culture. Plant Physiol Biochem 30:123–128
14. http://epd.brc.riken.jp/en/pcellc
15. Jakubiec A, Notaise J, Tournier V, Héricourt F, Block MA, Drugeon G, van Aelst L, Jupin I (2004) Assembly of turnip yellow mosaic virus replication complexes: interaction between the proteinase and polymerase domains of the replication proteins. J Virol 78:7945–7957
16. Worley CK, Ling R, Callis J (1998) Engineering in vivo instability of firefly luciferase and Escherichia coli beta-glucuronidase in higher plants using recognition elements from the ubiquitin pathway. Plant Mol Biol 37:337–347
17. Gorman CM, Moffat LF, Howard BH (1982) Recombinant genomes which express chloramphenicol acetyltransferase in mammalian cells. Mol Cell Biol 2:1044–1051
18. Gibson DG, Young L, Chuang RY, Venter JC, Hutchison CA, Smith HO (2009) Enzymatic assembly of DNA molecules up to several hundred kilobases. Nat Methods 6:343–345
19. Deroo BJ, Archer TK (2002) Proteasome inhibitors reduce luciferase and beta-galactosidase activity in tissue culture cells. J Biol Chem 277:20120–20123

Chapter 15

Detection and Quantification of Protein Aggregates in Plants

Marc Planas-Marquès, Saul Lema A., and Núria S. Coll

Abstract

Plants are constantly exposed to a complex and changing environment that challenges their cellular homeostasis. Stress responses triggered as a consequence of unfavorable conditions result in increased protein aggregate formation at the cellular level. When the formation of misfolded proteins surpasses the capacity of the cell to remove them, insoluble protein aggregates accumulate. In the animal field, an enormous effort is being placed to uncover the mechanisms regulating aggregate formation because of its implications in many important human diseases. Because of its importance for cellular functionality and fitness, it is equally important to expand plant research in this field. Here, we describe a cell fractionation-based method to obtain very pure insoluble protein aggregate fractions that can be subsequently semi-quantified using image analysis. This method can be used as a first step to evaluate whether a particular condition results in an alteration of protein aggregate formation levels.

Key words Protein fractionation, Protein aggregates, Immunoblot, Silver stain, Proteostasis, Plants, Protocols

1 Introduction

Protein aggregation occurs as a result of protein misfolding. Cells are equipped with protein quality control systems that help refolding misfolded proteins or dispose of them when their repair is not possible to prevent the formation of protein aggregates [1]. Insoluble protein aggregates have to be eliminated to prevent sustained damage that can lead to defects in growth, a decrease in yield, accelerated aging, and even death.

In the animal field, the study of the processes leading to aggregation of misfolded proteins is the focus of extensive research as it is associated with various important human diseases such as Alzheimer's, Huntington's, and Parkinson's, among others [2, 3]. Previously thought as an uncontrolled and unspecific process, protein aggregation and disaggregation is emerging as a very complex, tightly regulated process conserved across all kingdoms [1].

L. Maria Lois and Rune Matthiesen (eds.), *Plant Proteostasis: Methods and Protocols*, Methods in Molecular Biology, vol. 1450, DOI 10.1007/978-1-4939-3759-2_15, © Springer Science+Business Media New York 2016

In plants, the study of protein aggregate formation has mostly focused on chloroplastic processes [4–6]. In contrast, the cytoplasmic regulation of protein aggregation remains poorly understood. Recent work from our laboratory has shown that autophagic components and the death protease metacaspase 1 (AtMC1) are required for clearance of protein aggregates [7]. Autophagy and AtMC1 are emerging as central players in the proteostasis network, conserved across kingdoms [8–10]. We present here protocols developed in our laboratory to isolate and quantify protein aggregates. We use these to analyze differences in the formation of protein aggregates between different plant lines and physiological conditions.

2 Materials

2.1 Plant Growth

1. Arabidopsis thaliana seeds Col-0 ecotype.
2. Soil mix: 4.5 parts peat + 2 parts sand + 1 part vermiculite.
3. Controlled growth chamber: Aralab chamber D1200PLH with controlled temperature, humidity, and photoperiod (Aralab, Albarraque, Portugal).

2.2 Protein Extraction and Aggregate Isolation

1. Fractionation buffer: 20 mM Tris–HCl (pH 8), 1 mM EDTA (pH 8), and 0.33 M sucrose. Prepare the buffer from autoclaved 1 M stock solutions (*see* **Note 1**) and store it at 4 °C.
2. Fractionation buffer 0.3% Triton X-100: same buffer composition as in Fractionation buffer but adding Triton X-100 to reach a final concentration of 0.3%. Triton X-100 is added from a 10% stock solution.
3. Prior to the use of fractionation buffers, add 1 tablet of COMPLETE protease inhibitor (Roche, Basel, Switzerland) cocktail for every 10 ml of buffer and keep them on ice.
4. Protein Assay (Bio-Rad, Hercules, CA, USA).
5. SDS-loading buffer (5×): 250 mM Tris–HCl (pH 6.8), 50% glycerol, 0.5% bromophenol blue (BPB), 10% SDS, and 500 mM DTT (*see* **Note 2**).
6. Miracloth (Millipore, Billerica, MA, USA).
7. Beckman Coulter Optima™ L-100 XP Ultracentrifuge, SW60Ti rotor and 4 ml ultracentrifuge tubes.
8. Bioruptor® (Diagenode, Denville, NJ, USA).

2.3 Immunoblot

1. Anti-HA monoclonal antibody (3F10, Roche, Basel, Switzerland) at 1:5000 dilution.
2. Anti-cAPX antibody (Agrisera, Vännäs, Sweden) at 1:10,000 dilution.
3. Anti-PM ATPase antibody (Agrisera, Vännäs, Sweden) at 1:10,000 dilution.

2.4 Silver Stain

Prepare all solutions fresh with ultrapure water under a fume hood and keep at room temperature (*see* **Note 3**).

1. Fixing solution: 50% methanol, 37% formaldehyde, and 12% acetic acid. To prepare 100 ml, mix 50 ml of methanol, 37 ml of formaldehyde, and 12 ml of acetic acid.
2. 50% ethanol.
3. Pretreatment solution: 0.02% sodium thiosulfate ($Na_2S_2O_4$). To prepare, dissolve 20 mg in 100 ml.
4. Staining solution: 0.2% silver nitrate ($AgNO_3$), 0.03% formaldehyde. To prepare, dissolve 200 mg of silver nitrate in 100 ml of water and add 30 μl of formaldehyde.
5. Revealing solution: 6% sodium carbonate (Na_2CO_3), 0.02% formaldehyde, 0.0005% sodium thiosulfate. To prepare, dissolve 6 g of sodium carbonate in 90 ml of water, add 20 μl of formaldehyde and 2.5 ml of a 0.02% solution of sodium thiosulfate. Bring up to 100 ml with water.
6. Stop solution: 50% methanol and 12% acetic acid. To prepare mix 50 ml of methanol and 12 ml of acetic acid and bring up to 100 ml with water.
7. Gel image analysis: Scanned gel images can be analyzed using the Multi Gauge V3.0 software (Fujifilm, Minato, Japan).

3 Methods

3.1 Plant Growth and Senescence Induction

1. Sow around 50 Arabidopsis seeds on pots filled with soil mix watered to field capacity and vernalize at 4 °C for 3 days. Keep them on a tray and covered with a plastic film to maintain high humidity.
2. Transfer plants to a controlled growth chamber and grow under 9 h light at 21 °C and 15 h dark at 18 °C for 3 weeks (*see* **Note 4**).
3. To induce senescence transfer plants to long-day conditions: 16 h light at 21 °C and 8 h dark at 18 °C. Grow plants for 4 weeks under these conditions (*see* **Note 5**).

3.2 Cell Fractionation (Scheme of the Process Shown in Fig. 1)

1. Freeze 200 mg of rosette leaves (leaves are harvested regardless of their observed senescence degree) from different plants on liquid nitrogen.
2. Grind frozen samples using a mortar and pestle on ice with liquid nitrogen.
3. Transfer ground powder to 15 ml tubes placed on ice (*see* **Note 6**).
4. Add 4 ml of Fractionation buffer to the 15 ml tubes containing the powdered samples (scale up for larger sample amounts).
5. Vortex vigorously to ensure a good mixture.

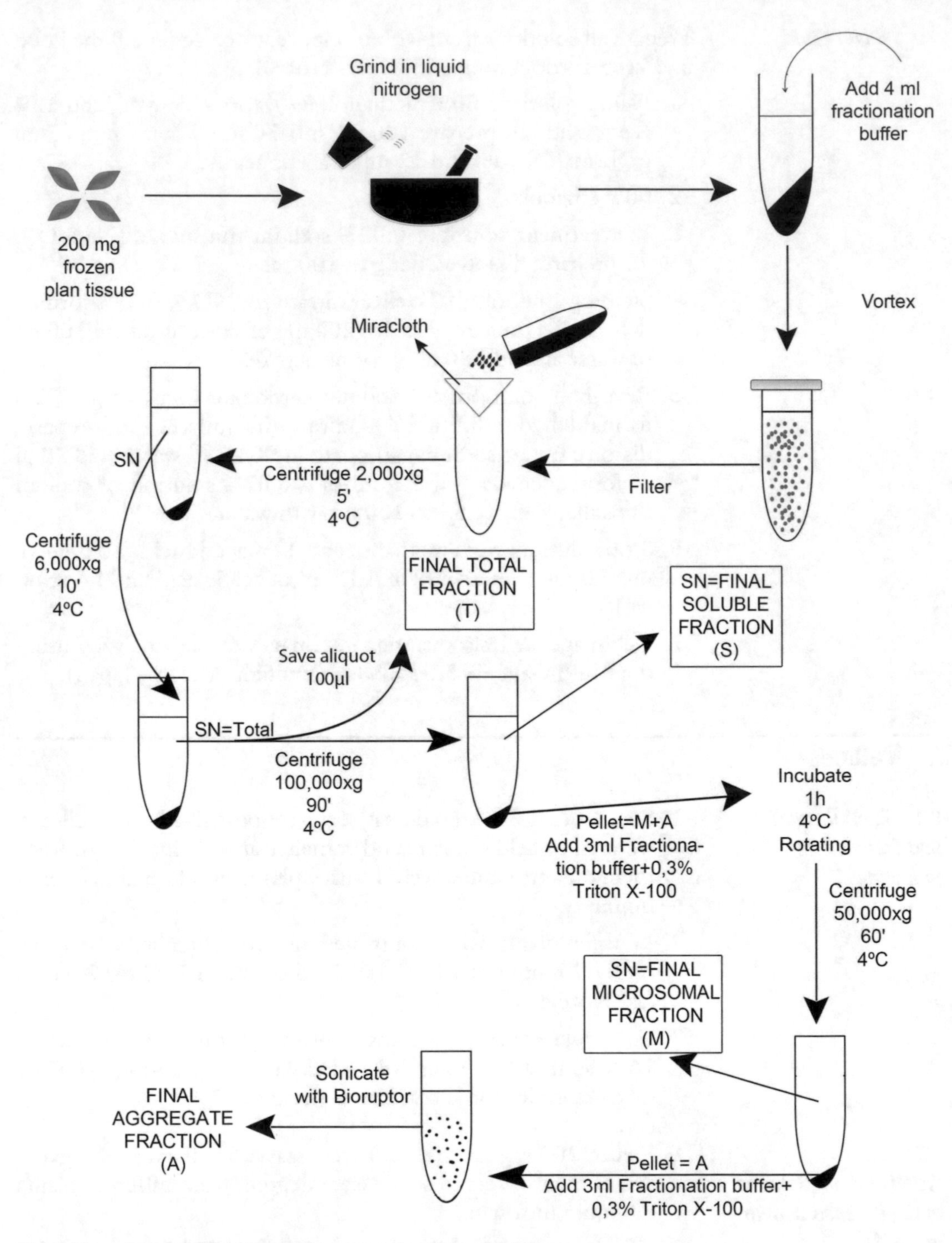

Fig. 1 Flow chart to obtain insoluble aggregates from plant tissue

6. Pass through a Miracloth filter to new falcon tubes in order to eliminate cell debris.
7. Centrifuge the suspensions at 2000 × *g* for 5 min at 4 °C to remove large particles.
8. Collect the supernatants and subsequently centrifuge them at 6000 × *g* for 10 min at 4 °C.
9. The resulting supernatants are the total (T) protein fractions. Transfer these supernatants to new falcon tubes (*see* **Note 7**).
10. Measure protein concentration of the different samples using the Protein Assay (Bio-Rad) following manufacturer's instructions.
11. Equal protein concentration among all samples, adjusting them at 3 ml (*see* **Note 8**).
12. Save an aliquot (200 μl) of each sample on new 1.5 ml tubes labeled T and keep them on ice.
13. Transfer the supernatants to Beckman Coulter 4 ml ultracentrifuge tubes and centrifuge them at 100,000 × *g* for 90 min at 4 °C (*see* **Note 9**) to separate the soluble fraction (S) (supernatant) from the fraction containing microsomal membranes and insoluble aggregates (M + A) (pellet).
14. Collect the soluble fractions and transfer them to new 15 ml tubes. Save an aliquot of 200 μl of each soluble fraction and keep them on ice.

3.3 Aggregate Isolation

1. To separate microsomal proteins from insoluble protein aggregates in the pellet, add 3 ml of fractionation buffer supplemented with 0.3% Triton X-100. Resuspend the pellet by pipetting, transfer to 15 ml tubes, and incubate in a rotating shaker at 4 °C for 1 h (*see* **Note 10**).
2. Transfer the Triton X-100-treated M + A fractions back to 4 ml ultracentrifuge tubes, equilibrate them and centrifuge at 50,000 × *g* for 60 min at 4 °C. The supernatant of this centrifugation step corresponds to the microsomal fraction (M), whereas the pellet contains the insoluble protein aggregates (A).
3. Collect the microsomal fractions and transfer them to fresh 15 ml tubes. Be careful not to disturb the pellet but ensure that no M fraction remains on the tube (*see* **Note 11**).
4. Add 3 ml of fractionation buffer supplemented with 0.3% Triton X-100 to the pellet, containing the insoluble protein aggregates and pipet up and down repeatedly to mix.
5. To solubilize the proteins, sonicate the samples using a Bioruptor® at 4 °C set on "High" with the following parameters: 3 cycles, 30 s "ON", 30 s "OFF" (*see* **Note 12**).
6. Add the corresponding amount of 5× SDS-loading buffer to every fraction collected (*see* **Note 13**). Boil them for 5 min and store at −20 °C.

3.4 Localization of Particular Proteins in the Aggregate Fraction

1. Run equal volumes of each fraction on SDS-PAGE gels (*see* **Note 14**).
2. Presence of a particular protein in one of the fractions can be analyzed by immunoblot (Fig. 2) using an antibody against the protein of interest.
3. Available control antibodies to show purity of the different fractions should be used in parallel (*see* **Note 15** and Fig. 2).
4. Coomassie staining of the membranes once the immunoblot is completed is highly recommended as an additional method to test purity of the fractions.

3.5 Relative Quantification of Aggregates

Use silver staining to compare the amount of proteins present in each fraction.

1. Run equal volumes of each fraction on SDS-PAGE gels.
2. Incubate the gels 1 h in the Fixing solution. Perform all steps at room temperature and using a rocking shaker.
3. Wash three times 20 s with 50% ethanol.
4. Incubate the gels 1 min in the Pre-treatment solution.
5. Wash three times 20 s with ultrapure water.
6. Incubate 20 min with the Staining solution in the dark (*see* **Note 16**).
7. Wash three times 20 s with ultrapure water.
8. Incubate with the Revealing solution until the bands become visible (*see* **Note 17**).

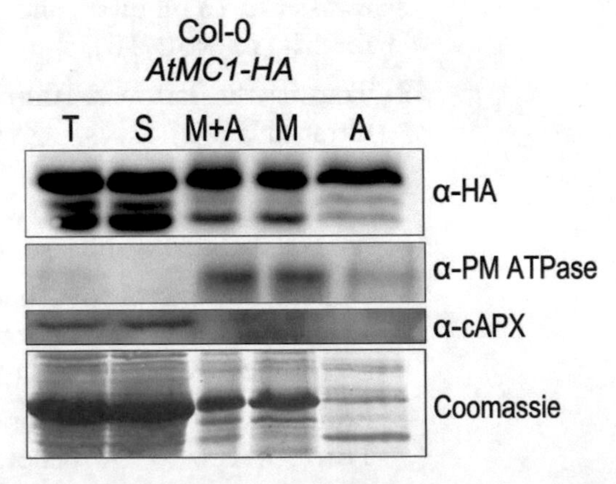

Fig. 2 Immunoblot of the protein fractions. Equal volumes of fractionated protein extracts of Arabidopsis Col-0 plants expressing the protein of interest (in this case, AtMC1-HA) were run on SDS-PAGE gels. After separation, the gels were either stained with Coomassie or analyzed by immunoblot using anti-HA antisera (α-HA), anti-cytosolic ascorbate peroxidase (α-cAPX) and anti plasma membrane H+ ATPase (α-PM ATPase). Reproduced with modifications from [7] with permission of Nature Publishing Group

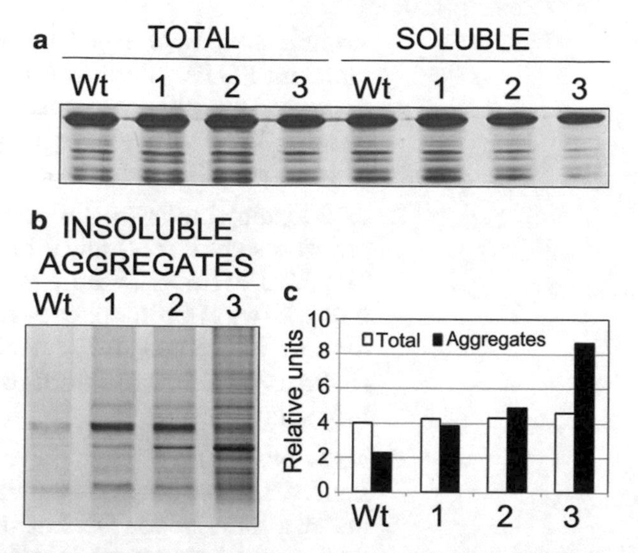

Fig. 3 Relative quantification of protein aggregates. Equal volumes of fractionated protein extracts of Arabidopsis wild-type (Wt) and mutant (1, 2, and 3) Col-0 plants were run on SDS-PAGE polyacrylamide gels and silver-stained. (**a**) Total and soluble fractions. (**b**) Insoluble protein aggregates. (**c**) Relative quantification of Total versus Aggregate fractions using the Multi Gauge V3.0 software. Silver stain images are reproduced with modifications from [7] with permission of Nature Publishing Group

9. Wash 5 s with water and add the Stop solution.
10. Gels can be stored at 4 °C in water for some weeks.
11. Since the intensity of the stain correlates with protein content, a representation of the quantity of proteins present in the aggregate fraction can be estimated by comparing the A and T fractions (Fig. 3a, b).
12. To calculate the intensity of the different lanes, analyze the scanned image of the gel scan using the image J software.
13. Analyze the bitmap (.bmp) images of the stained gels using an image analysis software to obtain numeric values for signal intensity and plot values on a graph (Fig. 3c).
14. Finally, divide the aggregate lane value by the total lane value and plot it.

4 Notes

1. Tris–HCl 1 M stock solution pH 8: weigh 121.14 g Tris base and transfer it to a 1 l bottle or glass beaker containing 900 ml water. Mix and adjust pH with HCl. Make up to 1 l with water. Ethylenediaminetetraacetic acid (EDTA) 1 M stock solution pH 8 is prepared dissolving 186.12 g EDTA in 1 l water.

Around 20 g NaOH pellet is required to adjust pH 8. Tris–HCl and EDTA solutions can be stored at room temperature for several months. To prepare the sucrose 1 M stock solution, dissolve 342.3 g sucrose in 1 l water. Store at 4 °C. Autoclave all these solutions before using them.

2. SDS-loading buffer (5×) is prepared from stock solutions as in previous step. For 50 ml of buffer, add 12.5 ml of 1 M stock Tris–HCl pH 6.8, 25 ml glycerol, 3.8 g DTT, 5 g SDS and 0.002% (w/v) bromophenol blue. Leave one aliquot at 4 °C for current use and store remaining aliquots at −20 °C. Prior to the use of a frozen aliquot, thaw and heat briefly at 37 °C to help solubilize.
3. Most buffers used for silver staining contain hazardous substances. To avoid exposure, always prepare and handle solutions under a fume hood. Do not discard down the drain and use dedicated waste containers. Containers used to prepare the silver stain solution must be rinsed immediately, as silver-ammonia solutions become explosive when dry.
4. Under these conditions wild-type plants establish a rosette of true leaves, all green.
5. Under these conditions wild-type plants start showing signs of senescence, with yellow patches appearing on the rosette leaves.
6. Add liquid nitrogen every now and then to help grinding the tissue sample and avoid hydration. When transferring the powdered sample to the 15 ml tube use a spatula that has been previously frozen in liquid nitrogen to recover as much sample as possible.
7. After the initial filtering/centrifugation steps, approximately 500 μl of the initial suspension is lost and the remaining volume is approximately 3.5 ml.
8. Since protein concentration may differ from sample to sample, you may need to prepare dilutions in order to equal them. Adjust all samples to a volume of 3 ml of the sample with the lowest protein concentration.
9. Place tubes on the adaptors and equilibrate on a precision scale by adding, if necessary, sucrose buffer to the samples.
10. Triton X-100 will solubilize membranes, which will allow for a proper separation of membranes from insoluble aggregates during the next centrifugation step.
11. Glass Pasteur pipettes with a very thin end are useful to clean the remaining solution without disturbing the pellet.
12. Sonication is needed to help dissolve protein aggregates in order to be able to properly run them on SDS-PAGE gels. The water in the sonication bath must be kept at 4 °C to ensure

preservation of the samples and to prevent damage to the instrument. Precooling the bath must be done at least 15 min before the sonication process. Water level on the bath must reach the red line on the tank. Use only distilled water to fill the tank (do not use deionized water).

13. It is sufficient to keep 200 μl per fraction.
14. We normally run 40 μl of each fraction per lane.
15. There are several antibodies that work well as soluble and microsomal fraction purity controls. In this case we used anti-cytosolic ascorbate peroxidase antisera (α-cAPX) as a soluble fraction marker and anti-plasma membrane ATPase antisera (α-PM ATPase) as microsomal fraction marker.
16. To keep the solution with the gels in the dark, wrap the incubation boxes with aluminum foil.
17. Usually takes very short. In less than 1 min bands start appearing. Stop once the desired intensity is achieved.

References

1. Tyedmers J, Mogk A, Bukau B (2013) Cellular strategies for controlling protein aggregation. Nat Rev 11:777–788
2. Shrestha A, Puente LG, Brunette S et al (2013) The role of Yca1 in proteostasis. Yca1 regulates the composition of the insoluble proteome. J Proteomics 81:24–30
3. Kim YE, Hipp MS, Bracher A et al (2013) Molecular chaperone functions in protein folding and proteostasis. Annu Rev Biochem 82:323–355
4. Lee U, Rioflorido I, Hong SW et al (2006) The Arabidopsis ChlpB/Hsp100 family of proteins: chaperones for stress and chloroplast development. Plant J 49:115–127
5. Howell HH (2013) Endoplasmic reticulum stress responses in plants. Annu Rev Plant Biol 64:477–499
6. Rajan VBV, D'Silva P (2009) Arabidopsis thaliana J-class heat shock proteins: cellular stress sensors. Funct Integr Genomics 9: 433–446
7. Coll NS, Smidler A, Puigvert M et al (2014) The plant metacaspase AtMC1 in pathogen-triggered programmed cell death and aging: functional linkage with autophagy. Cell Death Differ 21:1399–1408
8. Munch D, Rodriguez E, Bressendorff S et al (2014) Autophagy deficiency leads to accumulation of ubiquitinated proteins, ER stress, and cell death in Arabidopsis. Autophagy 10:1579–1587
9. Lee RE, Brunette S, Puente LG et al (2010) Metacaspase Yca1 is required for clearance of insoluble protein aggregates. Proc Natl Acad Sci U S A 107:13348–13353
10. Hill SM, Hao X, Liu B et al (2014) Life-span extension by a metacaspase in the yeast Saccharomyces cerevisiae. Science 344: 1389–1392

Chapter 16

Determination of Protein Carbonylation and Proteasome Activity in Seeds

Qiong Xia, Hayat El-Maarouf-Bouteau, Christophe Bailly, and Patrice Meimoun

Abstract

Reactive oxygen species (ROS) have been shown to be toxic but also function as signaling molecules in a process called redox signaling. In seeds, ROS are produced at different developmental stages including dormancy release and germination. Main targets of oxidation events by ROS in cell are lipids, nucleic acids, and proteins. Protein oxidation has various effects on their function, stability, location, and degradation. Carbonylation represents an irreversible and unrepairable modification that can lead to protein degradation through the action of the 20S proteasome. Here, we present techniques which allow the quantification of protein carbonyls in complex protein samples after derivatization by 2,4-dinitrophenylhydrazine (DNPH) and the determination proteasome activity by an activity-based protein profiling (ABPP) using the probe MV151. These techniques, routinely easy to handle, allow the rapid assessment of protein carbonyls and proteasome activity in seeds in various physiological conditions where ROS may act as signaling or toxic elements.

Key words Protein oxidation, Carbonylation, Degradation, Proteasome activity, Seeds

1 Introduction

Reactive oxygen species (ROS), such as hydrogen peroxide or superoxide anion, are known to be toxic but can also act as signaling molecules in cells. In seeds, ROS, which can be produced during both dry storage and imbibition, have been shown to be involved in seed dormancy alleviation, germination, and aging [1–3]. ROS can act by oxidizing macromolecules in cell as lipids, nucleic acids, and proteins. The oxidation of specific proteins or mRNAs, thus preventing their translation, has been proposed as being a key mechanism regulating seed dormancy alleviation and germination [4–6]. ROS are considered as being the signal between environmental factors, such as temperature or oxygen, and internal determinants of dormancy and germination, such as hormones [7]. Proteins are major determinants of seed germination since seeds

L. Maria Lois and Rune Matthiesen (eds.), *Plant Proteostasis: Methods and Protocols,* Methods in Molecular Biology, vol. 1450, DOI 10.1007/978-1-4939-3759-2_16, © Springer Science+Business Media New York 2016

contain all RNA species needed for germination [8] and that de novo transcription is not necessary for germination [9]. Protein abundance, location, degradation, and modulation of their activity are regulated at different levels, during transcription, translation, and by post-translational modifications (PTMs). It has been shown that PTMs, including S-nitrosylation, carbonylation, and glycosylation, are also involved in the regulation of seed germination [4, 10–13]. Protein oxidation is often associated with change in function, stability, localization, or degradation [14], and it also interacts with other PTMs such as phosphorylation [15]. Oxidative modification of enzymes has been shown to inhibit a wide array of enzyme activities thus leading to either mild or severe effects on cellular or systemic metabolism [16]. Among oxidative modifications, the involvement of protein carbonylation has been demonstrated in seed, from dormancy breaking to aging, in different species [4, 6]. Oxidative attack of exposed residues, such as Lys, Arg, Pro, and Thr, induces the formation of carbonyl groups [17]. Carbonylation of proteins is irreversible and not repairable [18]. To avoid their accumulation that could lead to aggregation of oxidized proteins and to a toxic effect in cell, carbonylated proteins are degraded through the action of the 20S proteasome in the cytosol [17]. In seeds, the role of proteasome in dormancy and germination is not completely understood. The involvement of the 20S proteasome was suggested during germination of spinach, Arabidopsis, and wheat seeds [11, 19, 20].

The study of protein carbonyls relies on many tests developed for several decades. They are often based on derivatization of the carbonyl group by 2,4-dinitrophenylhydrazine (DNPH), which leads to the formation of a stable 2,4-dinitrophenyl (DNP) hydrazone product [21], thus allowing the detection of protein carbonyls. One of the techniques presented here is the quantification of protein carbonyls in a complex protein sample. This technique, routinely used in the laboratory, is based on the spectral property of DNP at 370 nm. The second technique, complementary to the first one, is an *in gel* activity-based protein profiling (ABPP) using MV151, a proteasome inhibitor, containing a Bodipy fluorescent group for fluorescent imaging that allow the proteasome activity investigation [22]. These two techniques allow to obtain robust, reliable, and rapid results about protein oxidative modifications and proteolytic events that can be involved in any seed developmental process.

2 Materials

2.1 Determination of Protein Carbonylation by Spectrophotometry

1. Protein extraction buffer: 10 mM Hepes-NaOH buffer pH 7.5 containing 0.1% protease inhibitor cocktail (Sigma-Aldrich) and 0.07% (v/v) β-mercaptoethanol.
2. Protein Assay Dye Reagent Concentrate (Bio-Rad).

3. Derivatization solution: 0.2% DNPH in 2 M HCl (*see* **Note 1**).
4. Derivatization blank solution: 2 M HCl.
5. 100% (w/v) trichloroacetic acid.
6. Absolute ethanol-ethyl acetate (1:1).
7. 6 M guanidine-HCl (pH 2.3) (*see* **Note 2**).
8. 2-D Quant Kit (GE Healthcare).
9. 1.5 ml quartz cuvette.

2.2 Proteasome Activity Profiling

1. Protein extraction buffer: 50 mM Tris–HCl, pH 7.4.
2. Protein Assay Dye Reagent Concentrate (Bio-Rad).
3. MV151 probe (*see* **Note 3**).
4. 12% Mini-PROTEAN TGX Precast Gels (Bio-Rad).

3 Methods

Protein carbonylation is a type of protein oxidation that can be promoted by reactive oxygen species [23]. Detection and quantification of carbonylated proteins is accomplished after derivatization of the carbonyl groups. It usually refers to a process that forms reactive ketones or aldehydes that can be reacted by 2,4-dinitrophenylhydrazine (DNPH) to form hydrazones (Fig. 1). Carbonyl concentration in mol carbonyl/l is the sample absorbance at 370 nm divided by the molar absorptivity of the hydrazone (22,000 mol^{-1} cm^{-1}).

3.1 Determination of Protein Carbonylation by Spectrophotometry

3.1.1 Protein Extraction and Quantification

1. Ground approximately 0.1 g dry weight seeds or tissue in an ice-cold mortar and homogenize with 1.2 ml Hepes extraction buffer.
2. Incubate the solution on ice for 20 min with occasional mixing.
3. Centrifuge for 20 min at 16,000 × *g* at 4 °C.
4. Determine protein concentration in supernatant by Bradford protein assay according to the manufacturer's recommendations (*see* **Note 4**).

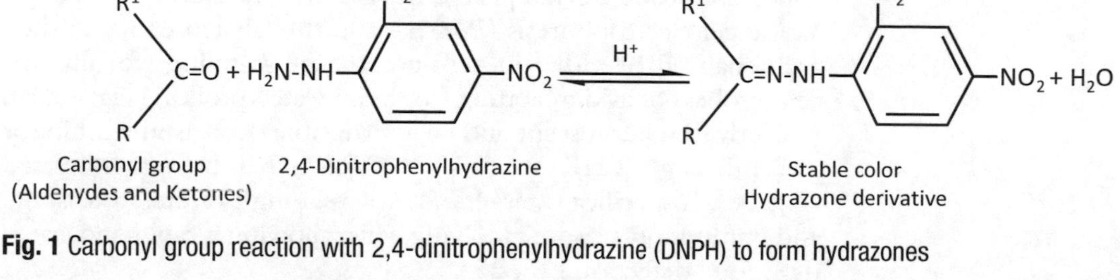

R^1RC=O + H_2N-NH-$C_6H_3(NO_2)_2$ $\overset{H^+}{\rightleftharpoons}$ R^1RC=N-NH-$C_6H_3(NO_2)_2$ + H_2O

Carbonyl group (Aldehydes and Ketones) 2,4-Dinitrophenylhydrazine Stable color Hydrazone derivative

Fig. 1 Carbonyl group reaction with 2,4-dinitrophenylhydrazine (DNPH) to form hydrazones

3.1.2 Determination of Carbonyl Groups

1. Prepare two 1.5 ml polypropylene tubes containing 1 mg of protein for each sample, one tube being the derivatization blank (control), another being the derivatization treatment.
2. Add 0.5 ml of 2 M HCl in the control and 0.5 ml of 0.2% DNPH in the derivatization tube respectively, vortex to suspend the sample.
3. Incubate tubes 15 min at room temperature in the dark, vortex occasionally.
4. Precipitate proteins in 10% (v/v) TCA and incubate 15 min on ice.
5. Concentrate the extract using a slow-speed centrifugation for 2 min at 3,000 × *g* which is adequate to form a loose, easily dispersed pellet.
6. Discard the supernatant carefully, wash the pellet with 1 ml ethanol-ethyl acetate 1:1 three times, and every time shake smoothly to disperse the pellet and centrifuge for 2 min at no more than 5,000 × *g* to avoid compacting the pellet (*see* **Note 5**).
7. Dry the pellet thoroughly, and dissolve it in 1.2 ml 6 M guanidine pH 2.3 (at 37 °C for 1 h), and then centrifuge at 10,000 × *g* for 5 min to remove insoluble material (*see* **Note 6**).
8. Take 1 ml supernatant to measure carbonyl at the absorbance of 370 nm, and read spectra against a blank of the related control.
9. Take 4 μl supernatant to measure protein concentration by 2-D Quant Kit (*see* **Note 7**).

   Calculate the results by the following equations:

$$\text{Carbonyl}\left(\text{nmol}/\text{ml}\right) = \frac{\left(\text{A370}\right)}{22{,}000/\text{mol}/\text{cm}} = \text{A370} \times 45.45$$

$$\text{Carbonyl}\left(\text{nmol carbonyl}/\text{mg protein}\right) = \frac{\text{A370} \times 45.45\left(\text{nmol}/\text{ml}\right)}{\text{Protein concentration}\left(\text{mg}/\text{ml}\right)}$$

10. An example of the results obtained is presented in Fig. 2.

3.2 Proteasome Activity Profiling

Activity-based protein profiling (ABPP) has emerged as a powerful chemical proteomic strategy to characterize enzyme function directly in native biological systems at a global scale [24]. In this approach, probe-treated proteomes are first resolved by polyacrylamide gel electrophoresis (PAGE), and the labeled enzymes then are visualized by either in-gel fluorescence scanning (for fluorescent probes) or avidin blotting (for biotinylated probes) (Fig. 3a, b). Recently, the fluorescent and cell-permeable proteasome inhibitor probe Bodipy TMR-Ahx_3L_3VS (MV151) has been synthesized (Fig. 3c). It specifically targets all active subunits of the proteasome and immunoproteasome in living cells, allowing a rapid and sensitive in-gel detection [23, 25].

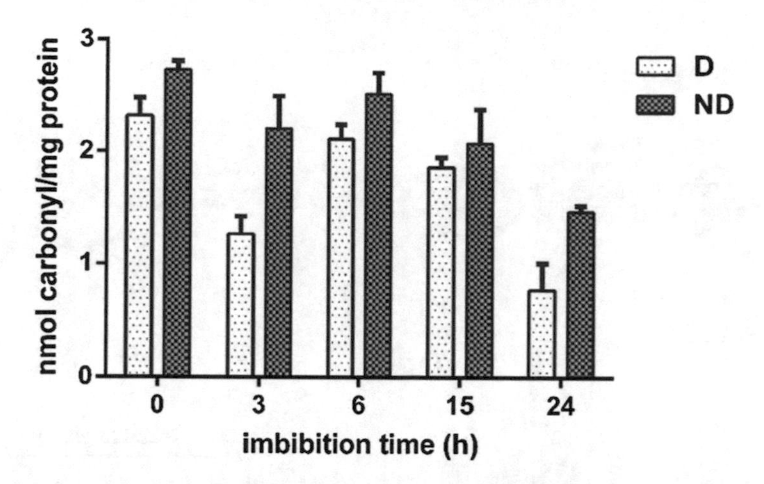

Fig. 2 Changes in protein oxidation in dormant and non-dormant sunflower seeds during their imbibition at 10 °C. Dormant (D) seeds: freshly harvested, unable to germinate under favorable conditions; non-dormant (ND) seeds: able to germinate after a dry storage period called after ripening. As showed in the graph, ND seeds exhibited higher protein carbonylation level compared to dormant seeds at all tested imbibition times. Data are mean ± SD of three independent biological replicates

3.2.1 Protein Extraction and Quantification

1. Grind approximately 0.1 g dry weight seeds or tissue in an ice-cold mortar and homogenize with 1 ml 50 mM Tris–HCl pH 7.4 extraction buffer.
2. Centrifuge for 20 min at 16,000 × *g*, 4 °C.
3. Filter the supernatant with Miracloth to remove remaining cell debris.
4. Determine protein concentration by Bradford protein assay.

3.2.2 In-gel Proteasome Activity-Based Protein Profiling

1. Incubate 15 μg protein in 1 μM MV151 (*see* **Note 8**) for 3–6 h at related temperature (*see* **Note 9**) in the dark.
2. Load the labeled proteins on 12% precast gels (Bio-Rad). Electrophoresis can be performed in a Mini Protean III (Bio-Rad) by using a Tris-Glycine-SDS running buffer (Bio-Rad), at 200 V and 40 mA for 45 min.
3. Visualize activity by revealing fluorescence using for example a Typhoon 8600 scanner (GE Healthcare Life Sciences) with excitation and emission wavelengths at 532 nm and 580 nm respectively (Fig. 4).

4 Notes

1. When preparing 0.2% DNPH in 2 M HCl, please take into account the actual content of DNPH in the reagent as most manufactures supply DNPH with at least 30% of water

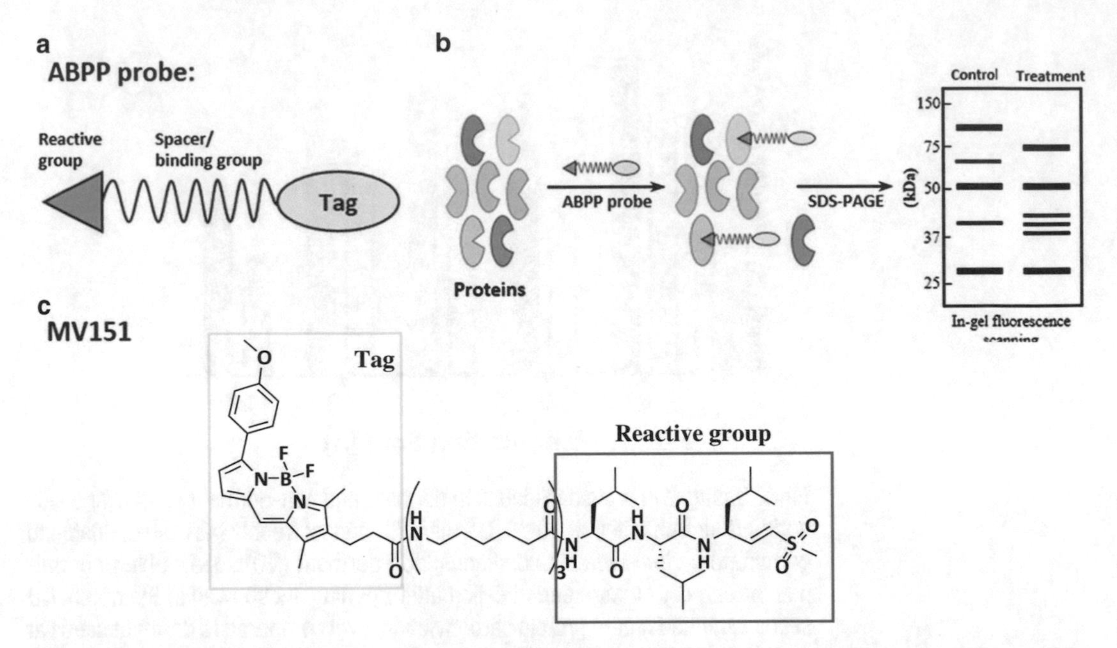

Fig. 3 Gel-based activity-based protein profiling (ABPP) and MV151 probe structures. (**a**) Representative structure of an ABPP probe, which contains a reactive group (*blue*), a spacer or binding group (*black*), and a reporter tag (*yellow*). A variety of reporter tags can be used for enzyme visualization and enrichment, including fluorophores and biotin; (**b**) Probe-labeled enzymes are visualized and quantified across proteomes by in-gel fluorescence scanning, adapted from [24]; (**c**) Molecular structures of MV151 probe used in this experiment. MV151 carries a VS reactive group, a Bodipy fluorescent reporter tag, a leucine tripeptide binding group, and a long linker region, adapted from [25]

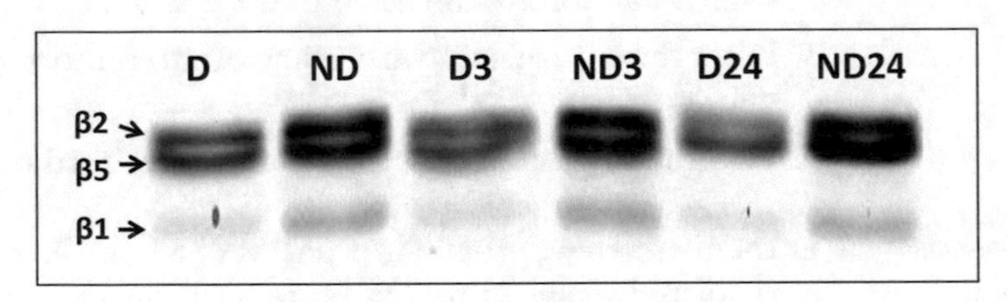

Fig. 4 Proteasome activity assessed using Activity-Based Protein Profiling (ABPP, a functional proteomic technology method) in which the fluorochrome (MV151) reacts with the active site of catalytic subunits β2, β5, and β1 of the proteasome in an activity-dependent manner. D: dormant sunflower seeds; ND: non-dormant sunflower seeds; D3/ND3: dormant/non-dormant seeds after 3 h of imbibition at 10 °C; D24/ND24: dormant/non-dormant seeds after 24 h of imbibition at 10 °C. As showed on the gel, ND seeds exhibited higher proteasome activity for each catalytic subunits compared to dormant seeds at all tested imbibition times

content, and be sure that the dry weight/volume ratio reached 0.2 %. After dissolving DNPH, stir the solution several hours in the dark and filter before use. Store protected from light at 4 °C for a maximum of 7 days.

2. Initial pH of 6 M guanidine solution is almost 5 and it decreases very rapidly.
3. In plants, three of the seven β subunits are responsible for the proteolytic activities of the proteasome. The vinyl sulfone

(VS)-based probe MV151 can react with the active site of these three catalytic subunits [22].

4. Nucleic acids are carbonyl positive, thus nucleic acid contamination of extracts in the spectrophotometric carbonyl assay can cause artifactual elevation in protein carbonyl measurements. In order to remove nucleic acids after protein extraction, add streptomycin sulfate to a final concentration of 1 % (v/v), then incubate 30 min at room temperature and centrifuge at 6,000 × *g* for 10 min at 4 °C.
5. If the pellet still look yellow after three times washing, more washes are needed to remove all free DNPH.
6. The time of pellet solubilization can be shortened, but not less than 15 min, and centrifugation of the samples is recommended to avoid light scatter.
7. A part of proteins may be lost during the washing process, so an accurate protein concentration must be obtained. Considering the low amount of proteins finally dissolved in guanidine pH 2.3, the 2-D Quant Kit is proposed for quantification.
8. A fluorochrome (MV151) reacts with the active site of catalytic subunits of the proteasome in an activity-dependent manner [22].
9. Concerning the MV151 incubation temperature, room temperature is normally used, but the proteasome activity is very sensitive to temperature, so if the samples are treated under different temperatures, the incubation also should be made at the same temperature. In order to obtain a good profiling, the incubation time should be extended when temperature decreases.

References

1. Bailly C, Leymarie J, Lehner A, Rousseau S, Come D, Corbineau F (2004) Catalase activity and expression in developing sunflower seeds as related to drying. J Exp Bot 55(396):475–483. doi:10.1093/jxb/erh050
2. Bailly C, El-Maarouf-Bouteau H, Corbineau F (2008) Seed dormancy alleviation and oxidative signaling. J Soc Biol 202(3):241–248. doi:10.1051/jbio:2008025
3. Bailly C, El-Maarouf-Bouteau H, Corbineau F (2008) From intracellular signaling networks to cell death: the dual role of reactive oxygen species in seed physiology. C R Biol 331(10):806–814. doi:10.1016/j.crvi.2008.07.022
4. Oracz K, El-Maarouf Bouteau H, Farrant JM, Cooper K, Belghazi M, Job C, Job D, Corbineau F, Bailly C (2007) ROS production and protein oxidation as a novel mechanism for seed dormancy alleviation. Plant J 50(3):452–465. doi:10.1111/j.1365-313X.2007.03063.x
5. Bazin J, Langlade N, Vincourt P, Arribat S, Balzergue S, El-Maarouf-Bouteau H, Bailly C (2011) Targeted mRNA oxidation regulates sunflower seed dormancy alleviation during dry after-ripening. Plant Cell 23(6):2196–2208. doi:10.1105/tpc.111.086694
6. El-Maarouf-Bouteau H, Meimoun P, Job C, Job D, Bailly C (2013) Role of protein and mRNA oxidation in seed dormancy and germination. Front Plant Sci 4:77. doi:10.3389/fpls.2013.00077
7. El-Maarouf-Bouteau H, Sajjad Y, Bazin J, Langlade N, Cristescu SM, Balzergue S, Baudouin E, Bailly C (2015) Reactive oxygen species, abscisic acid and ethylene interact to regulate sunflower seed germination. Plant Cell Environ 38(2):364–374. doi:10.1111/pce.12371
8. Bewley JD (1997) Seed germination and dormancy. Plant Cell 9(7):1055–1066. doi:10.1105/tpc.9.7.1055

9. Rajjou L, Gallardo K, Debeaujon I, Vandekerckhove J, Job C, Job D (2004) The effect of alpha-amanitin on the Arabidopsis seed proteome highlights the distinct roles of stored and neosynthesized mRNAs during germination. Plant Physiol 134(4):1598–1613. doi:10.1104/pp.103.036293
10. Berger S, Menudier A, Julien R, Karamanos Y (1996) Regulation of De-N-glycosylation enzymes in germinating radish seeds. Plant Physiol 112(1):259–264
11. Job C, Rajjou L, Lovigny Y, Belghazi M, Job D (2005) Patterns of protein oxidation in Arabidopsis seeds and during germination. Plant Physiol 138(2):790–802. doi:10.1104/pp.105.062778
12. Bethke PC, Libourel IG, Jones RL (2006) Nitric oxide reduces seed dormancy in Arabidopsis. J Exp Bot 57(3):517–526. doi:10.1093/jxb/erj060
13. Zhao MG, Tian QY, Zhang WH (2007) Nitric oxide synthase-dependent nitric oxide production is associated with salt tolerance in Arabidopsis. Plant Physiol 144(1):206–217. doi:10.1104/pp.107.096842
14. Davies MJ (2005) The oxidative environment and protein damage. Biochim Biophys Acta 1703(2):93–109. doi:10.1016/j.bbapap.2004.08.007
15. Hardin SC, Larue CT, Oh MH, Jain V, Huber SC (2009) Coupling oxidative signals to protein phosphorylation via methionine oxidation in Arabidopsis. Biochem J 422(2):305–312. doi:10.1042/BJ20090764
16. Shacter E (2000) Quantification and significance of protein oxidation in biological samples. Drug Metab Rev 32(3-4):307–326. doi:10.1081/DMR-100102336
17. Nystrom T (2005) Role of oxidative carbonylation in protein quality control and senescence. EMBO J 24(7):1311–1317. doi:10.1038/sj.emboj.7600599
18. Dalle-Donne I, Giustarini D, Colombo R, Rossi R, Milzani A (2003) Protein carbonylation in human diseases. Trends Mol Med 9(4):169–176
19. Miyawaki M, Aito M, Ito N, Yanagawa Y, Kendrick RE, Tanaka K, Sato T, Nakagawa H (1997) Changes in proteasome levels in spinach (Spinacia oleracea) seeds during imbibition and germination. Biosci Biotechnol Biochem 61(6):998–1001. doi:10.1271/bbb.61.998
20. Shi C, Rui Q, Xu LL (2009) Enzymatic properties of the 20S proteasome in wheat endosperm and its biochemical characteristics after seed imbibition. Plant Biol 11(6):849–858. doi:10.1111/j.1438-8677.2009.00193.x
21. Levine RL, Garland D, Oliver CN, Amici A, Climent I, Lenz AG, Ahn BW, Shaltiel S, Stadtman ER (1990) Determination of carbonyl content in oxidatively modified proteins. Methods Enzymol 186:464–478
22. Verdoes M, Florea BI, Menendez-Benito V, Maynard CJ, Witte MD, van der Linden WA, van den Nieuwendijk AM, Hofmann T, Berkers CR, van Leeuwen FW, Groothuis TA, Leeuwenburgh MA, Ovaa H, Neefjes JJ, Filippov DV, van der Marel GA, Dantuma NP, Overkleeft HS (2006) A fluorescent broad-spectrum proteasome inhibitor for labeling proteasomes in vitro and in vivo. Chem Biol 13(11):1217–1226. doi:10.1016/j.chembiol.2006.09.013
23. Wong CM, Bansal G, Marcocci L, Suzuki YJ (2012) Proposed role of primary protein carbonylation in cell signaling. Redox Rep 17(2):90–94. doi:10.1179/1351000212Y.0000000007
24. Cravatt BF, Wright AT, Kozarich JW (2008) Activity-based protein profiling: from enzyme chemistry to proteomic chemistry. Annu Rev Biochem 77:383–414. doi:10.1146/annurev.biochem.75.101304.124125
25. Gu C, Kolodziejek I, Misas-Villamil J, Shindo T, Colby T, Verdoes M, Richau KH, Schmidt J, Overkleeft HS, van der Hoorn RA (2010) Proteasome activity profiling: a simple, robust and versatile method revealing subunit-selective inhibitors and cytoplasmic, defense-induced proteasome activities. Plant J 62(1):160–170. doi:10.1111/j.1365-313X.2009.04122.x

Chapter 17

Isobaric Tag for Relative and Absolute Quantitation (iTRAQ)-Based Protein Profiling in Plants

Isabel Cristina Vélez-Bermúdez, Tuan-Nan Wen, Ping Lan, and Wolfgang Schmidt

Abstract

Isobaric tags for relative and absolute quantitation (iTRAQ) is a technology that utilizes isobaric reagents to label the primary amines of peptides and proteins and is used in proteomics to study quantitative changes in the proteome by tandem mass spectrometry. Here, we present an adaptation of the iTRAQ experimental protocol for plants that allows the identification and quantitation of more than 12,000 plant proteins in *Arabidopsis* with a false discovery rate of less than 5%.

Key words Isobaric tags, Primary amines, Absolute quantitation, Proteomics, Mass spectrometry

1 Introduction

Protein quantitation through incorporation of stable isotopes and mass spectrometric analysis provides a powerful tool to systematically and quantitatively assess the differences in protein profiles in modern proteomics research [1]. Specifically, iTRAQ-based quantitation (Fig. 1) facilitates the comparative analysis of peptides and proteins in a variety of settings including comparisons of normal or treated states, by using isobaric reagents to label peptides that can be identified and quantified through analysis of reporter groups that are generated upon fragmentation in the mass spectrometer. Currently, the iTRAQ methodology is one of the major quantitation tools used in differential plant proteomic research [2–6].

iTRAQ is based on the covalent labeling of the N-terminus and side chain amines of peptides resulting from protein digestions with tags of varying mass that contain three regions: a peptide reactive region, a reporter region, and a balance region [7]. The samples for quantitation are separately isolated, digested, and chemically labeled with one of the iTRAQ reagents. Subsequently, protein samples are pooled, fractionated by one- or two-dimensional

L. Maria Lois and Rune Matthiesen (eds.), *Plant Proteostasis: Methods and Protocols*, Methods in Molecular Biology, vol. 1450, DOI 10.1007/978-1-4939-3759-2_17,

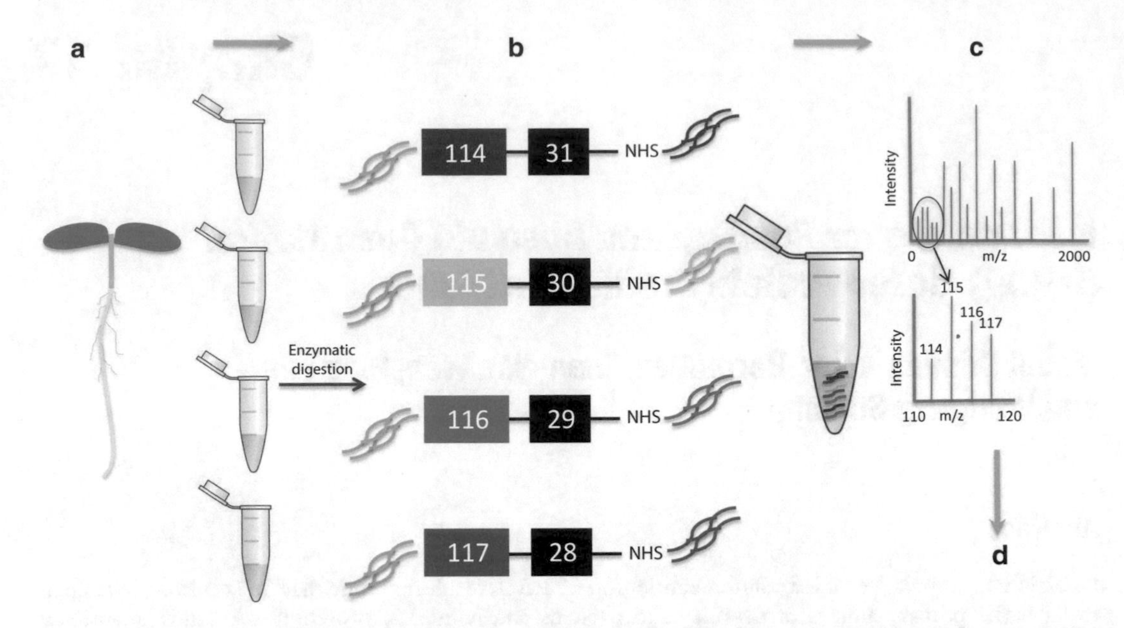

Fig. 1 Keys steps in the iTRAQ workflow. (**a**) Sample preparation. (**b**) In-solution trypsin digestion and iTRAQ labeling. Peptides from multiple samples are labeled with iTRAQ tags. (**c**) Nano-HPLC-MS/MS analysis and spectrum. A general scheme and example data for a 4-plex iTRAQ experiment are shown. (**d**) Data processing and statistical analysis. Mascot and/or SEQUEST can be used to identify and quantitate proteins

HPLC and analyzed by tandem mass spectrometry (MS/MS). During tandem mass spectrometry, the fragmentation of the isolated precursor peptide ions takes place, allowing for qualitative analysis of intense reporter ions in the tandem mass spectrum. At the peptide level, the signals of the reporter ions of each MS/MS spectrum allow for calculating the relative abundance of each peptide identified by this spectrum [8]. Then a database search is performed using the obtained data to identify the labeled peptides and the corresponding proteins. The iTRAQ data can be processed in user-friendly environments with emphasis on quality control [9]. Here, we describe an iTRAQ protocol for plants with high reproducibility that is partly complementary to conventional gel-based methods.

2 Materials

All reagents used should be of molecular biology grade. Sterile techniques and ultrapure water should always be used for the preparation of reagents to prevent contamination, modification, and degradation of proteins.

2.1 Solutions

1. Solution A: 10% of trichloroacetic acid (TCA) (Merck Millipore, Cat. No. 76-03-9) in acetone (Merck Millipore, Cat. No. 67-64-1) (*see* **Note 1**). Add 0.07% β-mercaptoethanol

(Merck Millipore, Cat. No. 60-24-2) (*see* **Note 2**), store the solution at −20 °C.

2. Solution B: 100% acetone, 0.07% β-mercaptoethanol, and 1 mM phenylmethanesulfonyl fluoride (PMSF) (Merck Millipore, Cat. No. 329-98-6) (*see* **Note 3**). Store at −20 °C.
3. Dissolving solution A: 50 mM Tris–HCl, pH 8.5 (Bio Basic, Cat. No. SD8141), and 8 M urea (Merck Millipore, Cat. No. 57-13-6) (*see* **Note 4**).

2.2 Preparation of Protein Extracts

1. *Arabidopsis thaliana* samples (*see* **Note 5**).
2. Mortars and pestles.
3. 50 ml centrifuge tubes (Nalgene) (Sigma/Aldrich, Cat. No. T1418).
4. Liquid nitrogen.
5. Spoon and spatula.
6. Centrifuge.
7. Vacuum.
8. Eppendorf tubes (Life Technologies, Cat. No. AM12400).
9. Pierce™ 660 nm protein assay (Thermo Scientific, Cat. No. 22660).
10. NuPAGE® Novex® 12% Bis-Tris protein gels, 1.0 mm, 12 well (Life Technologies, Cat. No. NP0342BOX).

2.3 Trypsin Digestion and iTRAQ Labeling

1. 100 mM DL-dithiothreitol solution (DTT) (Merck Millipore, Cat. No. 3483-12-3).
2. 500 mM iodoacetamide (Merck Millipore, Cat. No. 144-48-9).
3. 50 mM Tris–HCl, pH 8.5 (Bio Basic, Cat. No. SD8141).
4. 0.5 μg lysyl endopeptidase®, mass spectrometry grade (Wako, Cat. No. 125-05061).
5. 2 μg trypsin (Promega, Cat. No. V5111).
6. 50 mM Tris–HCl, pH 8.0 (Life Technologies, Cat. No. 15568-025).
7. 10% trifluoroacetic acid (Life Technologies, Cat. No. 28904).
8. C18 solid-phase extraction cartridge (Thermo Scientific, Cat. No. 60108-305).
9. iTRAQ™ Reagents Methods Development Kit (Applied Biosystems).
10. PolySulfoethyl A, 5 μm, 200-Å bead (PolyLC Inc., Cat. No. 202SE0502).
11. Vacuum centrifuge.

2.4 MS/MS Analysis

1. Dionex UltiMate™ 3000 RSLCnano System (Thermo Scientific).
2. Q Exactive hybrid quadrupole-Orbitrap mass spectrometer (Thermo Scientific).
3. Acclaim PepMap RSLC column (Thermo Scientific, Cat. No. 164536).
4. Solvent A: 0.1% formic acid in water (J.T. Baker, 9834-03).
5. Solvent B: acetonitrile with 0.1% formic acid (J.T. Baker, 9832-03).

2.5 Protein Identification

1. Mascot software (Matrix Science) version 2.4 and SEQUEST (integrated in the Proteome Discoverer software version 1.4, Thermo Scientific).
2. Arabidopsis protein database (TAIR10 20110103, 27416 sequences; ftp://ftp.arabidopsis.org/home/tair/Sequences/blast_datasets/TAIR10_blastsets/ TAIR10 pep 20110103 representative gene model).

3 Methods

3.1 Sample Preparation

1. Collect *Arabidopsis* samples and store at −80 °C.
2. Store sterile mortars and pestles at −20 °C.
3. Grind the samples into a fine powder with liquid nitrogen.
4. Suspend 1 g of sample in 10× volume of solution A.
5. Place the samples at −20 °C for 2 h.
6. Centrifuge the samples at 35,000 × *g* at 4 °C for 30 min and discard the supernatant (*see* **Note 6**).
7. Resuspend the pellet in 10× volume of solution B (*see* **Note 7**).
8. Vortex the tubes and use a spoon or spatulas to break the pellet. Mix well to obtain a homogeneous suspension.
9. Incubate at −20 °C for 1 h and centrifuge at 35,000 × *g* at 4 °C for 30 min and remove the supernatant (*see* **Note 8**).
10. Resuspend the pellet in 20× volume of solution B (*see* **Note 9**) and centrifuge the samples at 35,000 × *g* at 4 °C for 30 min and discard the supernatant. Repeat this step three times.
11. Dry the pellet overnight under vacuum (*see* **Note 10**).
12. Transfer the dry pellet to new Eppendorf tube and weigh the pellet (*see* **Note 11**).
13. Suspend the pellet in 25× volume of dissolving solution A and transfer the dissolved sample solution to a new Eppendorf tube (*see* **Note 12**).
14. Place the tubes on shaker for 2 h at 4 °C.
15. Centrifuge at 19,000 × *g* at 4 °C for 10 min.

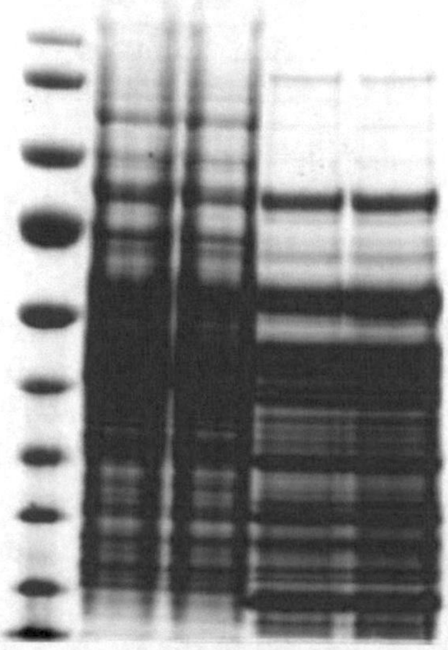

Fig. 2 SDS-PAGE gel containing iTRAQ *Arabidopsis* samples. (**a**) Prestained SDS-PAGE molecular weight standard. (**b**) Roots from control (Fe-sufficient) plants. (**c**) Root from Fe-deficient plants. (**d**) Shoot from control (Fe-sufficient) plants. (**e**) Shoot from Fe-deficient plants. 10 μg of total proteins were loaded per lane

16. Transfer the supernatant to new Eppendorf tube.
17. Quantitate the protein concentration with Pierce™ 660 nm protein assay following the manufacturer's instructions (*see* **Note 13**).
18. Normalize the protein concentration to 5 μg/μl and run the samples in a 12% Bis–Tris protein gel (Fig. 2).
19. The samples can be stored at −20 °C or −80 °C.

3.2 In-Solution Trypsin Digestion and iTRAQ Labeling

1. Add dithiothreitol to a final concentration of 10 mM per 100 μg of total protein.
2. Incubate for 1 h at room temperature.
3. Add iodoacetamide to a final concentration of 50 mM and incubate for 30 min at room temperature in the dark.
4. Add 30 mM dithiothreitol to the mixture to consume any free iodoacetamide and incubate for 1 h at room temperature in the dark (note that the total concentration in the final mixture would be 35 mM approximately).
5. Dilute the proteins with 50 mM Tris–HCl, pH 8.5, to reduce the urea concentration to 4 M or less.
6. Digest with 0.5 μg lysyl endopeptidase (Lys-C) for 4 h at room temperature.
7. Dilute the solution with 50 mM Tris–HCl, pH 8.0, to reduce the urea concentration to less than 1 M.

8. Digest the Lys-C digested proteins by adding 2 μg of modified trypsin (Promega) at room temperature overnight.
9. Add trifluoroacetic acid to a final concentration of 10% to acidify the solution.
10. Desalt the samples on a C18 solid-phase extraction cartridge.
11. Label the samples with iTRAQ reagents (Applied Biosystems) according to the manufacturer's instructions for 1 h at room temperature.
12. Combine and fractionate the peptides offline using high-resolution strong cation-exchange chromatography coupled to a HPLC system (PolySulfoethyl A, 5 μm, 200-Å bead) on HPLC. Collect and combine the fractions according to the peak area (*see* **Note 14**).
13. Lyophilize the fractions in a centrifugal speed vacuum concentrator.
14. Store the samples at −80 °C.

3.3 Nano-HPLC-MS/MS Analysis

1. Perform the liquid chromatography on a Dionex UltiMate 3000 RSLCnano system coupled to a Q Exactive hybrid quadrupole-Orbitrap mass spectrometer equipped with a Nanospray Flex Ion Source.
2. Redissolve the peptides of each fraction in solvent A and centrifuge at 20,000 × *g* for 10 min.
3. Load the supernatant (peptide mixtures) onto the LC-MS/MS.
4. Separate the samples using a segmented gradient in 120 min from 5 to 40% solvent B at a flow rate of 300 nl/min.
5. Maintain the samples at 8 °C in the autosampler.
6. Operate the Q Exactive hybrid quadrupole-Orbitrap mass spectrometer in positive ionization mode (*see* **Note 15**).

3.4 Database Search and Quantitation

1. Use Mascot and/or SEQUEST to identify and quantitate proteins.
2. Make the searches against the Arabidopsis protein database (TAIR10) and concatenate with a decoy database containing the randomized sequences of the original database.
3. For each technical repeat, combine the spectra from all the fractions into one MGF (Mascot generic format) file after loading the raw data, and use the MGF files to query protein databases.
4. For each biological repeat, spectra from the three technical repeats should be combined into one file and searched. The search parameters should be as follows: trypsin is chosen as the enzyme with two missed cleavages allowed; fixed modifications of carbamidomethylation at Cys, iTRAQ at N-terminus and

Lys, variable modifications of oxidation at Met and iTRAQ at Tyr; peptide tolerance is set at 10 ppm, and MS/MS tolerance is set at 0.05 Da. Peptide charge is set M_r, and monoisotopic mass is chosen. iTRAQ is chosen for quantitation during the search simultaneously.

5. Pass the search results through additional filters before exporting the data. For protein identification, the filters are set as follows: significance threshold $p < 0.05$ (with 95% confidence) and ion score or expected cutoff less than 0.05 (with 95% confidence).
6. For protein quantitation, the filters are set as follows: "weighted" is chosen for protein ratio type (http://mascot-pc/mascot/help/quant_config_help.html); minimum precursor charge is set to 1 and minimum peptides is set to 2; only unique peptides should be used to quantitate proteins.
7. Summed intensities are set as normalization, and outliers are removed automatically. The peptide threshold is set as above for homology.

3.5 Data Processing and Statistical Analysis

For data analysis, we suggest to use the method described by Cox and Mann [10].

1. Calculate the log2 ratios for the quantified proteins detected in at least two biological repeats and analyze for normal distribution.
2. Calculate the mean and SD and use the 95% confidence (Z score = 1.96) to select the proteins with a distribution far from the main distribution.
3. For downregulated proteins, calculate a confidence interval (mean ratio − 1.96 × SD), corresponding to a protein ratio of 0.83.
4. For upregulated proteins, calculate a confidence interval (mean ratio + 1.96 × SD), corresponding to a protein ratio of 1.29 (*see* **Note 16**).

4 Notes

1. For example, 10 g of TCA in 100 ml of acetone.
2. Solution A can be prepared 1 day before and stored in −20 °C overnight, however β-mercaptoethanol should be added freshly just before use.
3. The acetone should be stored at −20 °C overnight. β-mercaptoethanol and phenylmethanesulfonyl fluoride (PMSF) should be added freshly just before use. PMSF should be prepared freshly, immediately before to be used in a stock solution of 100 mM or 200 mM in ethanol.

4. The dissolving solution should be prepared freshly just before use. Optional for small quantities of pellet, the dissolving solution B can be prepared as: 0.1 M Tris–HCl, pH 7.6 (Sigma/Aldrich, T2788-1L), 4% Sodium dodecyl sulfate (SDS) (Merck Millipore, Cat. No. 151-21-3), and 0.1 M DL-Dithiothreitol solution (DTT) (Merck Millipore, Cat. No. 3483-12-3).
5. Reproducible extract preparations depend mainly on a standardized sampling procedure.
6. Weigh the Nalgene centrifuge tubes containing the samples and balance them using the solution A before centrifugation.
7. Usually, in our hands, we get 200 mg of pellet from 1 g of fresh Arabidopsis material (roots or shoots). Use 20× volume of solution B for large pellets (400 mg).
8. Weigh the Nalgene centrifuge tubes containing the samples and balance them with solution B before centrifugation.
9. Use 30× volume of solution B for large pellets (400 mg).
10. The pellet should be totally dry (you can test this using a spatula, it should be a fine powder). The pellet can be broken with a spatula. Vacuum drying may be continued when necessary.
11. Weigh the Eppendorf tube empty and then transfer the dry pellet with a spatula and weigh the Eppendorf tube containing the pellet again. Calculate the weight of the pellet subtracting the weight of the Eppendorf tube. Be careful because the dry pellet can be easily lost.
12. To get better efficiency in dissolving proteins, the Eppendorf tube should not contain more than 0.02 g of dry pellet. An optional procedure can be performed for small pellets: resuspend the pellet in 5× volume of dissolving solution B and transfer the dissolved sample solution to a new Eppendorf tube. Boil the samples at 95 °C for 5 min. Sonicate in a water bath sonicator ten times for 15 s–1 min at room temperature. Continue to mix for 1 h at room temperature. Centrifuge the samples to 14,100 × *g* for 30 min at room temperature. Transfer the supernatant to new Eppendorf tube. Quantitate the protein with RC DC™ Protein Assay (BIO-RAD, Cat. No. 500-0121) and run 5 μg of protein sample in a 12% Bis–Tris protein gel.
13. A 660 nm protein assay should be used to quantitate the proteins (suitable for samples containing 8 M urea).
14. For example, 40 fractions can be collected and combined into 20 final fractions.
15. Acquisition cycle: acquire a full scan (*m*/*z* 350–1600) in the Orbitrap analyzer at resolution 70,000, then perform the MS/MS of the ten most intense peptide ions with HCD acquisition of the same precursor ion. Make the HCD with collision

energy of 30% and detect HCD-generated fragment ions in the Orbitrap at resolution 17,500.

16. Protein ratios outside this range can be defined as being significantly different at a threshold of $p = 0.05$.

Acknowledgement

Work in the Schmidt lab is supported by MoST and Academia Sinica.

References

1. Aebersold R, Mann M (2003) Mass spectrometry-based proteomics. Nature 422: 198–207
2. Lan P, Li W, Schmidt W (2012) Complementary proteome and transcriptome profiling in phosphate-deficient *Arabidopsis* roots reveals multiple levels of gene regulation. Mol Cell Proteomics 11:1156–1166
3. Qin J, Gu F, Liu D, Yin C, Zhao S, Chen H, Zhang J, Yang C, Zhan X, Zhang M (2013) Proteomic analysis of elite soybean Jidou17 and its parents using iTRAQ-based quantitative approaches. Proteome Sci 11:12
4. Martínez-Esteso MJ, Vilella-Antón MT, Pedreño MA, Valero ML, Bru-Martínez R (2013) iTRAQ-based protein profiling provides insights into the central metabolism changes driving grape berry development and ripening. BMC Plant Biol 13:167
5. Zhou XX, Yang LT, Qi YP, Guo P, Chen LS (2015) Mechanisms on boron-induced alleviation of aluminum-toxicity in *Citrus grandis* seedlings at a transcriptional level revealed by cDNA-AFLP analysis. PLoS One 10:e0115485
6. Chu P, Yan GX, Yang Q, Zhai LN, Zhang C, Zhang FQ, Guan RZ (2015) iTRAQ-based quantitative proteomics analysis of *Brassica napus* leaves reveals pathways associated with chlorophyll deficiency. J Proteomics 113: 244–259
7. Ross PL, Huang YN, Marchese JN, Williamson B, Parker K, Hattan S, Khainovski N, Pillai S, Dey S, Daniels S, Purkayastha S, Juhasz P, Martin S, Bartlet-Jones M, He F, Jacobson A, Pappin DJ (2004) Multiplexed protein quantitation in saccharomyces cerevisiae using amine-reactive isobaric tagging reagents. Mol Cell Proteomics 3:1154–1169
8. Gafken PR, Lampe PD (2006) Methodologies for characterizing phosphoproteins by mass spectrometry. Cell Commun Adhes 13:249–262
9. Vaudel M, Burkhart JM, Zahedi RP, Martens L, Sickmann A (2012) iTRAQ data interpretation. In: Marcus K (ed) Quantitative methods in proteomics, vol 893, Methods in molecular biology. Springer Science+Business Media, LLC, New York, NY, pp 501–509
10. Cox J, Mann M (2008) MaxQuant enables high peptide identification rates, individualized p.p.b.-range mass accuracies and proteome-wide protein quantification. Nat Biotechnol 26:1367–1372

Chapter 18

Use of a Phosphatidylinositol Phosphate Affinity Chromatography (PIP Chromatography) for the Isolation of Proteins Involved in Protein Quality Control and Proteostasis Mechanisms in Plants

T. Farmaki

Abstract

Protein functionality depends directly on its accurately defined three-dimensional organization, correct and efficient posttranslational modification, and transport. However, proteins are continuously under a hostile environment threatening with folding aberrations, aggregation, and mistargeting. Therefore, proteins must be constantly "followed up" by a tightly regulated homeostatic mechanism specifically known as proteostasis. To this end other proteins ensure this close surveillance including chaperones as well as structural and functional members of the proteolytic mechanisms, mainly the autophagy and the proteasome related. They accomplish their action via interactions not only with other proteins but also with lipids as well as cytoskeletal components. We describe a protocol based on an affinity chromatographic approach aiming at the isolation of phosphatidyl inositol phosphate binding proteins, a procedure which results into the enrichment and purification of several members of the proteostasis mechanism, e.g. autophagy and proteasome, among other components of the cell signaling pathways.

Key words Phosphatidylinositol phosphate (PIP), Autophagy, Chaperones, Sumoylation, Proteasome, Ubiquitin

1 Introduction

The phosphatidylinositol phosphate affinity chromatography (PIP chromatography), (Fig. 1) has been used for the isolation, purification and study of proteins interacting with lipids bearing the phosphorylated inositol head group [1–3]. These polyphosphoinositides (PPIs) are membrane-localized phospholipids having their polar inositol headgroups exposed to the cytoplasmic environment and accessible to a number of modifying enzymes as well as important signaling intermediates. These interacting partners may bear a specific PIP recognizing domain enabling docking to the lipid. The enrichment of different PIP isoforms in specific membrane

L. Maria Lois and Rune Matthiesen (eds.), *Plant Proteostasis: Methods and Protocols*, Methods in Molecular Biology, vol. 1450, DOI 10.1007/978-1-4939-3759-2_18, © Springer Science+Business Media New York 2016

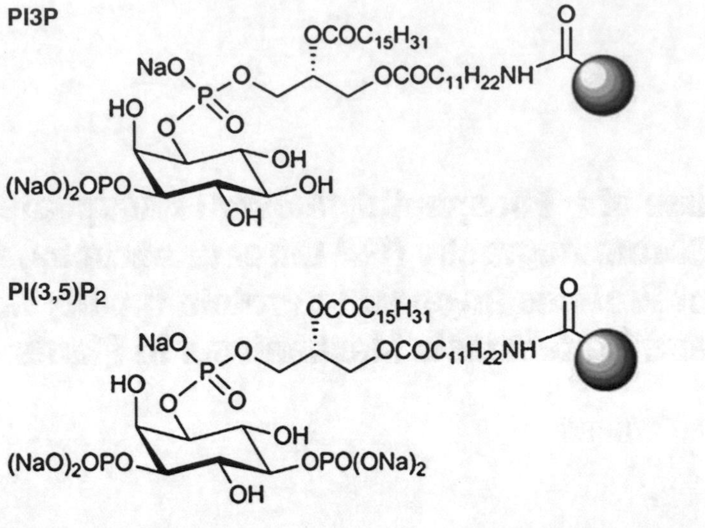

PI3P

PI(3,5)P_2

Fig. 1 PI3P and PI(3,5)P_2 polyphosphoinositols (PPIs) covalently linked to agarose beads

compartments determines the functional role of their effectors as well as their involvement in signaling pathways alongside with the suggested identity determination of the organelles [4]. Their main role is to achieve a targeted and dynamic—in space and time—translocation of cytoplasmic proteins to the membrane, becoming, therefore key regulators of trafficking and signaling. Proteins enriched in the PIP chromatography include cytoplasmic enzymes containing PI-recognizing domains as well as other proteins which directly or indirectly, e.g. via the PIP domain containing effectors, interact with the lipids.

PIPs phosphorylated at the 3′ or the 3,5′ site of the inositol ring have been proven partners of the autophagy effectors in both animal and plant species [5–7]. In addition, previous studies have shown that members of a subset of five putative AtPI4Ks contain N-terminal UBL (ubiquitin-like) domains interacting and phosphorylating UFD1 and RPN10, a mechanism by which their function can be regulated [8].

We present a detailed protocol of a PI3P and a PI(3,5)P_2 affinity chromatography designed for the isolation of proteins involved in osmotic/salinity stress responses from a suspension culture of *A. thaliana*. Synthesis of different stereoisomers of phosphatidyl inositol phosphate lipids coupled to agarose beads and their storage conditions have been described [9–11]. The protocol describes a differential approach, i.e. more than one PIP species have been involved in a comparative study (Fig. 2). In addition, treated versus untreated cultures have been used. The PIP stereoisomers have been shown to be involved in osmotic/salinity stress responses [6, 12, 13] as well as in the autophagosome function [14–16]. The differential study

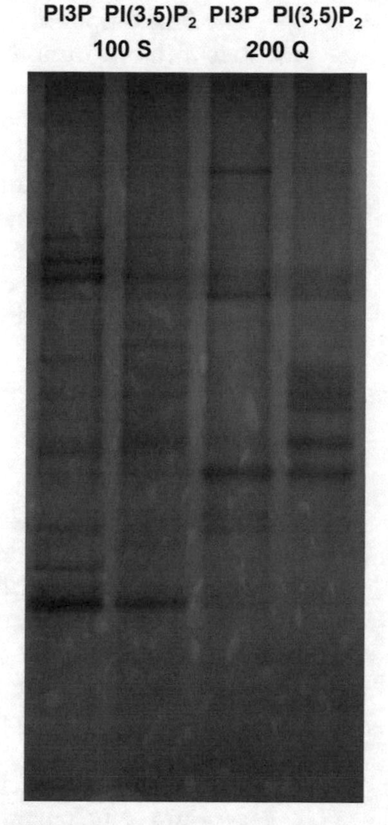

Fig. 2 Example of a polyacrylamide protein profile following PI3P and PI(3,5)P_2 affinity chromatography of the 100S and 200Q fractions from samples treated with 0.4 M NaCl in a *A. thaliana* liquid culture. Different bands appear in the same Q or S fraction following incubation with PI3P or PI(3,5)P_2 beads

approach provides the means for an in depth understanding of the mechanisms involved from the first steps of the proteomics analysis while at the same time preferential binding may be used as an indication of specificity since PIP promiscuity has always been an issue [17]. In addition, the method aims at the identification of bonafide targets of PI3P and PI(3,5)P_2. Binding specificity may be further established by the use of soluble lipid as a competitor in the assay. The nature of the starting material (e.g. suspension culture, leaf tissue, entire plants, callus culture, etc.) depends on the aims as well as the technical issues of the project. A readily extracted high cytoplasmic content is a desirable characteristic that can be provided by callus cultures; however, when uniformity of the applied stress factor is required in order to achieve reproducibility of the comparative data, a suspension culture would be preferable. Since our preliminary studies using whole extracts resulted in highly complex protein profiles, a pre-fractionation strategy based on the intrinsic charge of the proteins in the extract was followed (Q and S Sepharose purification).

The final output of the method is a list of proteins appearing to interact with one or both lipids. However, the interacting partners are not necessarily directly associated with the lipid. The direct association may be further confirmed by a lipid overlay assay [6] or other approaches, e.g. liposome assays.

Finally, bioinformatics tools (www.megabionet.org/atpid/webfile/; http://string-db.org/; www.genemania.org/; www.cytoscape.org/) may be utilized in order to show that a purified protein may be an indirect interactor [7] as well as investigate other putative interacting partners in the pathway.

2 Materials

All buffers are autoclaved or filter-sterilized and kept at 4 °C. PBS (Phosphate Buffered Saline in tablets) was provided by Sigma (UK) and dissolved in ultrapure water according to the manufacturer's instructions. Ultracentrifugations were performed using a floor ultracentrifuge (Beckman), in ultraclean sterile tubes suitable for the rotors SW28Ti and SW41Ti. Low-speed centrifugations were performed either inside a bench low-speed centrifuge in 15 or 50 mL falcon tubes or a microfuge in 1.5 mL eppendorf tubes.

2.1 Cell Culture Medium

1. Grow *A. thaliana* cell line T87 in liquid Murashige and Skoog medium supplemented with vitamins, MES (Duchefa), 3% sucrose, 1 mg/L kinetin, and 0.5 mg/L 1-naphthalene acetic acid.
2. All solutions are prepared in ultrapure water (18.2 MΩ cm at 25 °C water).
3. Add hormones to the growth medium following sterilization by autoclaving.

2.2 Extraction (TNEE) Buffer

1. Tris–HCl (1 M stock, final concentration 50 mM, pH 8.0).
2. NaCl (5 M stock, final concentration 80 mM).
3. EGTA (100 mM stock, final concentration 2 mM).
4. EDTA (500 mM stock, final concentration 1 mM).
5. Sucrose (weigh directly, final concentration 300 mM).
6. Filter sterilize using a 45 μM filter.
7. DTT (1 M stock, final concentration 2 mM, added after filtration).
8. Protease inhibitors (use a complete protease inhibitor cocktail or commercial tablets, dissolve after TNEE filtration and before extraction).
9. PVPP (1%, 10 g/L, added after filtering, at the time of extraction).
10. Store at 4 °C.

2.3 Q and S Sepharose

1. Ion exchange Q and S chromatography (Amersham Biosciences, Uppsala, Sweden).
2. Q and S equilibration buffer is prepared using half-strength TNEE buffer/50 mM Tris–HCl, pH 8.0.
3. Q Sepharose buffer is prepared using 30 mM HEPES-NaOH pH 7.2.
4. S Sepharose buffer is prepared using 30 mM HEPES-NaOH pH 6.0. Different concentrations of NaCl are added (0.1, 0.2, 0.4, 0.6, and 1 M NaCl).

2.4 IPP Buffer

50 mM Tris–HCl, pH 7.5, 150 mM NaCl, 5 mM EDTA. Add 0.1% Tween-20 after filtering or autoclaving and 0.02% sodium azide.

2.5 Laemmli Sample Buffer

20 mM Tris–HCl pH 6.8, 2% SDS, 10% glycerol, 0.1 M DTT, and 0.1% Bromophenol blue.

2.6 Bead Washing Buffer

2% SDS, 0.1 M DTT, 50 mM Tris-HCl pH 7.4.

3 Methods

3.1 PIP-Affinity Chromatography

PIPs coupled to the beads are stored in 1.5 mL Eppendorf tubes at 4 °C in IPP buffer. All steps unless otherwise stated are performed at 4 °C.

3.2 Cell Culture (*See* Note 1)

1. Shake flasks with cell cultures at 100 rpm, in a growth chamber at 22 °C, and with a photoperiod of 16 h light/8 h darkness.
2. Transfer once a week one-tenth of the culture to fresh medium.
3. Grow cultures for use in subsequent experiments in 1000 mL flasks containing 200 mL growth medium. Split them depending on the treatments (*see* **Note 2**).

3.3 Cell Extraction

1. Collect the cell culture inside a double miracloth placed inside a funnel.
2. Wash the cells three times with 30 mL of PBS discarding the buffer that passes through the miracloth (*see* **Note 3**).
3. Release the collected cells inside a conical tube using 20 mL ice-cold TNEE half strength buffer (diluted 1:1 with 50 mM Tris buffer), containing protease inhibitors and DTT. Rinse the miracloth with another 20 mL of TNEE half-strength buffer in order to collect all cells that may have remained adherent on the miracloth.
4. Add 1% PVPP and extract five times with 5 s pulses using a Polytron with a suitable probe. Leave on ice for about 15 s in between extractions (*see* **Note 4**).

5. Filter the extract through a miracloth and spin the flow-through at 10,000 × *g* in a SW28Ti rotor for 15–30 min at 4 °C. Discard the pellet.
6. Collect the resultant supernatant through a single miracloth and spin at 100,000 × *g*) for 1 h in a SW41Ti rotor at 4 °C. Discard the pellet.
7. Prepare several aliquots and store them at −80 °C, or use them directly in the subsequent steps (*see* **Note 5**).

3.4 Sample Fractionation Using Q and S Sepharose

1. Equilibrate Q and S Sepharose by washing inside a 50 mL falcon tube with half-strength TNEE buffer four times. Spin at low speed (700 × *g*) to collect the beads.
2. Take 2.5 mL bead compact volume of each sepharose bead slurry and incubate each one with 20–30 mL extract of 1–2 mg/mL protein concentration inside the 50 mL falcon tubes (*see* **Note 6**).
3. Incubate by rotating for 1 h at 4 °C.
4. Spin at 40 × *g* for 2 min.
5. Collect the supernatant (flow through, FT) and filter it through a miracloth or/and a filter to get rid of Sepharose beads. Make aliquots (*see* **Note 7**).
6. Resuspend the beads in 10 mL 0.2 M NaCl Q and S buffers (pH 7.2 for Q and pH 6.0 for S Sepharose).
7. Incubate for 1 h at 4 °C and collect as before.
8. Resuspend the beads in 0.4 M Q and S buffer.
9. Continue as before until final washing performed using 1 M Q and S buffer.
10. Collect aliquots, filter to get rid of beads, snap freeze and store at −80 °C.

3.5 PIP Incubation

1. Extracts may be used directly or following purification through Q and S Sepharose.
2. Beads are washed with IPP buffer (*see* **Note 8**).

3.5.1 For Small-Scale Trials

1. Use 1 mL aliquots. Defrost quickly at 37 °C. Shake without making bubbles. Spin at 14,000 rpm for 20 s to get rid of the Sepharose beads (only in case of pre-fractionation).
2. Split between Eppendorf tubes and bring all samples to 150 mM NaCl, Tris–HCl, pH 8.0, 0.5 % IGEPAL.
3. Add 30 μL compact volume of beads.
4. Incubate extracts with the PIP beads for 2 h.
5. Perform subsequent steps (washing, bead collection, and elution) as described for large-scale trials (Subheading 3.4, **step 2**), using 1.2 mL volumes of IPP buffer.

3.5.2 For Large-Scale Trials Followed by Mass Spectrometry

1. For large-scale experiments, use a starting total volume of 20 mL plant extract of 1–2 mg/mL concentration. If necessary, concentrate samples using the protein concentrating devices in order to achieve a final protein concentration of 1–2 mg/mL.
2. Take 10 mL samples (samples stored at −80 °C may also be used). Defrost quickly by shaking without making bubbles at 37 °C. Split them in 4 mL aliquots in 15 mL falcon tubes.
3. Spin at 700 × *g* for 2 min to remove the Sepharose beads.
4. Add 0.5 % IGEPAL (final concentration).
5. Take 250 μL beads compact volume and resuspend them in 1 mL of IPP buffer.
6. Add 60 μL of PIP beads to each extract.
7. Incubate at low rev/min in a rotator at 4 °C for 2 h and 30 min.
8. After incubation spin at 750 rpm for 3 min in a microfuge.
9. Throw the supernatant away and resuspend in 4 mL IPP buffer.
10. Wash twice in 4 mL IPP buffer.
11. Transfer to an eppendorf tube and wash twice in 1 mL IPP buffer.
12. In the last spin, leave some buffer behind and collect buffer from Eppendorf walls by spinning at 750 rpm for 30 s in a microfuge.
13. Aspirate buffer by putting in the aspirator a tip and a syringe.
14. Add 70 μL of 2× sample buffer (*see* **Note 9**).
15. Boil for 1 min and 30 s.
16. Load on a gel 60 μL sample volume.
17. Run a gel (*see* **Note 10**).
18. Bands are excised and proteins are identified by mass spectrometry.

4 Notes

1. Different cell cultures may be used, e.g. *A. thaliana* T87 (this example), *N. tabacum* Bright Yellow-2, or any other starting plant material. In particular, plant tissues such as meristems may also be used, especially when developmental studies are performed. Source material must be selected so that an adequately high cytosol concentration is attained; this is one of the reasons why cell cultures are an attractive starting material. In addition, the need for a reliable identification of the PIP interacting proteins following mass spectrometry must be taken

into account. Therefore, in cases where the identification of novel PIP-interacting partners is investigated, it is preferable to use a standardized system with a sequenced genome already available, e.g. *A. thaliana*.

2. A wide range of treatments can be applied such as salinity (NaCl), osmotic (mannitol), oxidative stress, heat, cold, toxic compounds, light, starvation (growth in the absence of light in combination with nutrient depletion), addition of inhibitors in combination with the treatments, etc.
3. This method of collecting and washing cells is preferable compared to the collection by centrifugation. Serial centrifugations and washings may result in signal attenuation due to time lapse and it is also difficult to standardize. In addition, cell clumps produced during serial centrifugations may result in nonreproducible data. In the case of plant tissues, collected material is rinsed with PBS and directly subjected to the extraction step.
4. Number and length of the extraction pulses can be variable depending on the plant material. Keep samples on ice as long as possible so that they are not heated at any point during the extraction procedure.
5. Prepare aliquots for small-scale (1 mL in Eppendorf tubes) and large-scale experiments (up to 10 mL in 15 mL falcon tubes), snap freeze them in liquid nitrogen, and store at –80 °C for few weeks up to 1 month. Since protein profiles of plant extracts are very complex, a pre-fractionation using ion exchange chromatography (Q and S Sepharose) is applied before the PIP chromatography. In this case, it is preferable to use the extract directly in the Q and S chromatography in order to avoid clamping and protein aggregation.
6. Since it is difficult to estimate protein concentration of the extract from the beginning due to imponderable factors (cytosol release from starting material, nature of the treatment, etc.), a desirable protein concentration prior and after Q and S fractionation may be attained using protein concentrating devices (Centricon, Millipore, UK).
7. Prepare 1 mL aliquots for small-scale trials. Samples used as an input for subsequent PIP incubations may need to be concentrated first using protein concentrators.
8. Wash by resuspension and avoid vortexing beads. In the last wash, keep some buffer so that beads do not dry out, wash remaining beads on the Eppendorf walls and spin for a last 30 s at 750 rpm in a microfuge. Different concentrations of IGEPAL (0.01–0.1%) may also be added to increase specificity.
9. The beads can be reused up to five times. At the end of an experiment, collect beads using PIP buffer and centrifuge.

Add to the pellet 4× bead volume of 50 mM Tris–HCl pH 7.4, 2% SDS, 100 mM DTT for 10 min. Wash beads six times with IPP buffer and store at 4 °C as a 10% suspension in the presence of sodium azide.

10. Gels are stained with silver nitrate (BDH, UK). For sequencing purposes, gels are stained with colloidal Coomassie (Bio-Rad, UK). PI(3)P- and PI(3,5) P_2-enriched proteins are differentially identified (Fig. 2).

Acknowledgments

The work was supported by a research project (PENED) co-financed by E.U.-European Social Fund (75%) and the Greek Ministry of Development-GSRT (25%) and the BBSRC.

References

1. Catimel B, Yin MX, Schieber C, Condron M, Patsiouras H, Catimel J, Robinson DE, Wong LS, Nice EC, Holmes AB (2009) PI(3,4,5)P3 Interactome. J Proteome Res 8(7):3712–3726
2. Catimel B, Schieber C, Condron M, Patsiouras H, Connolly L, Catimel J, Nice EC, Burgess AW, Holmes AB (2008) The PI(3,5)P2 and PI(4,5)P2 interactomes. J Proteome Res 7(12):5295–5313
3. Catimel B, Kapp E, Yin MX, Gregory M, Wong LS, Condron M, Church N, Kershaw N, Holmes AB, Burgess AW (2013) The PI(3)P interactome from a colon cancer cell. J Proteomics 82:35–51
4. Munro S (2004) Organelle identity and the organization of membrane traffic. Nat Cell Biol 6(6):469–472
5. Burman C, Ktistakis NT (2010) Regulation of autophagy by phosphatidylinositol 3-phosphate. FEBS Lett 584(7):1302–1312
6. Karali D, Oxley D, Runions J, Ktistakis N, Farmaki T (2012) The Arabidopsis thaliana immunophilin ROF1 directly interacts with PI(3)P and PI(3,5)P2 and affects germination under osmotic stress. PLoS One 7(11):e48241
7. Oxley D, Ktistakis N, Farmaki T (2013) Differential isolation and identification of PI(3)P and PI(3,5)P2 binding proteins from Arabidopsis thaliana using an agarose-phosphatidylinositol-phosphate affinity chromatography. J Proteomics 91:580–594
8. Galvao RM, Kota U, Soderblom EJ, Goshe MB, Boss WF (2008) Characterization of a new family of protein kinases from Arabidopsis containing phosphoinositide 3/4-kinase and ubiquitin-like domains. Biochem J 409(1):117–127
9. Manifava M, Thuring JW, Lim ZY, Packman L, Holmes AB, Ktistakis NT (2001) Differential binding of traffic-related proteins to phosphatidic acid- or phosphatidylinositol (4,5)-bisphosphate-coupled affinity reagents. J Biol Chem 276(12):8987–8994
10. Lim ZY, Thuring JW, Holmes AB, Manifava M, Ktistakis NT (2002) Synthesis and biological evaluation of a PtdIns(4,5)P-2 and a phosphatidic acid affinity matrix. J Chem Soc Perkin 1 1(8):1067–1075
11. Conway SJ, Gardiner J, Grove SJ, Johns MK, Lim ZY, Painter GF, Robinson DE, Schieber C, Thuring JW, Wong LS (2009) Synthesis and biological evaluation of phosphatidylinositol phosphate affinity probes. Org Biomol Chem 8(1):66–76
12. Dove SK, Cooke FT, Douglas MR, Sayers LG, Parker PJ, Michell RH (1997) Osmotic stress activates phosphatidylinositol-3,5-bisphosphate synthesis. Nature 390(6656):187–192
13. Zonia L, Munnik T (2004) Osmotically induced cell swelling versus cell shrinking elicits specific changes in phospholipid signals in tobacco pollen tubes. Plant Physiol 134(2):813–823
14. Baskaran S, Ragusa MJ, Boura E, Hurley JH (2012) Two-site recognition of phosphatidylinositol 3-phosphate by PROPPINs in autophagy. Mol Cell 47(3):339–348
15. Lu KY, Tao SC, Yang TC, Ho YH, Lee CH, Lin CC, Juan HF, Huang HC, Yang CY, Chen MS (2012) Profiling lipid-protein interactions

using nonquenched fluorescent liposomal nanovesicles and proteome microarrays. Mol Cell Proteomics 11(11):1177–1190

16. Polson HE, de Lartigue J, Rigden DJ, Reedijk M, Urbe S, Clague MJ, Tooze SA (2010) Mammalian Atg18 (WIPI2) localizes to omegasome-anchored phagophores and positively regulates LC3 lipidation. Autophagy 6(4):506–522

17. Lietzke SE, Bose S, Cronin T, Klarlund J, Chawla A, Czech MP, Lambright DG (2000) Structural basis of 3-phosphoinositide recognition by pleckstrin homology domains. Mol Cell 6(2):385–394

Chapter 19

In Vivo Radiolabeling of *Arabidopsis* Chloroplast Proteins and Separation of Thylakoid Membrane Complexes by Blue Native PAGE

Catharina Nickel, Thomas Brylok, and Serena Schwenkert

Abstract

The investigation of membrane protein complex assembly and degradation is essential to understand cellular protein dynamics. Blue native PAGE provides a powerful tool to analyze the composition and formation of protein complexes. Combined with in vivo radiolabeling, the synthesis and decay of protein complexes can be monitored on a timescale ranging from minutes to several hours. Here, we describe a protocol to analyze thylakoid membrane complexes starting either with ^{35}S-methionine labeling of intact *Arabidopsis* leaves to investigate protein complex dynamics or with unlabeled leaf material to monitor steady-state complex composition.

Key words Radiolabeling, BN-PAGE, Photosynthetic protein complexes, Thylakoid membrane, *Arabidopsis thaliana*

1 Introduction

Thylakoids are internal membrane systems found inside chloroplasts and cyanobacteria and provide the platform for the light-dependent reactions of photosynthesis. Due to the endosymbiotic origin of chloroplasts, the thylakoid membrane complexes are built up of a mosaic of nuclear as well as plastid encoded subunits. The main complexes are photosystem I and II (PS I and PS II), light harvesting complex II (LHC II), cytochrome b_6f complex (Cyt b_6f), and ATP synthase. All of these complexes consist of multiple single subunits and are themselves able to form higher-molecular-weight complexes with each other [1]. Therefore, the thylakoid membrane system is an excellent target system to be studied by blue native polyacrylamide gel electrophoresis (BN-PAGE) [2]. Furthermore, pulse labeling with ^{35}S-methionine or with other beta emitting radioisotopes like ^{35}S-cysteine is a widely used method to investigate metabolism, biosynthesis, maturation, and degradation of proteins within the thylakoid membrane [3]. Combined

L. Maria Lois and Rune Matthiesen (eds.), *Plant Proteostasis: Methods and Protocols*, Methods in Molecular Biology, vol. 1450, DOI 10.1007/978-1-4939-3759-2_19,

with BN-PAGE and 2D SDS-PAGE, it is a powerful tool to evaluate protein synthesis and assembly of multi-subunit protein complexes.

In general, BN-PAGE can be used to isolate protein complexes from biological membranes, determine the oligomerization state of proteins, or analyze physiological protein–protein interactions. Investigating thylakoid membranes, in vivo radiolabeling studies and native gel experiments can be applied for different types of mutant analyses. For example, they were used to elucidate PSII biogenesis and repair, to establish the role of several assembly factors or to characterize the functions of individual components of thylakoid membrane complexes [4–8].

The crucial step for a native gel electrophoresis is the solubilization of the membranes without denaturating the proteins and by this destroying protein–protein interactions. Therefore, nonionic detergents are used. For the solubilization of thylakoid membranes, dodecyl-maltoside has proven to be an appropriate detergent. It is a mild neutral detergent leaving photosynthetic complexes intact, even up to high-molecular-weight complexes consisting of PSII and LHCII. However, labile hydrophobic interactions are dissociated. One of the mildest detergents is digitonin which can be used for the identification of protein–protein interaction without applying chemical crosslinkers. To be able to separate the proteins via gel electrophoresis, they have to be charged. For that the anionic dye Coomassie blue G-250 is used. Although water-soluble, due to hydrophobic properties, the dye binds to membrane proteins causing them to migrate to the anode. In an acrylamide gradient gel with its decreasing pore size, the protein complexes are separated according to their size.

In a following second dimension, these complexes are then split into their single components. Firstly, the BN gel strip is incubated in SDS solution, which denaturizes the proteins in the gel. Subsequently, one lane of the BN-PAGE is applied on a SDS-PAGE which separates all proteins according to their molecular weight. By a highly sensitive silver staining of this gel, the proteins are made visible as individual spots [9]. In a vertical line, all proteins belonging initially to one complex can be found; while in a horizontal line, a single protein can be tracked and assigned to all different complexes it was associated with. An overview of the workflow is provided in Fig. 1.

2 Materials

2.1 Plant Growth

Arabidopsis thaliana (Col-0) is grown on soil or on agar plates with half-strength MS medium: 1% (w/v) sucrose, 0.05% (w/v) MES, 0.237% (w/v) MS salts, 1.2% (w/v) plant agar, pH 5.7. Standard growth conditions for both are: 22 °C and a 16 h/8 h light/dark cycle of 120 µE/m^2 s (*see* **Note 1**).

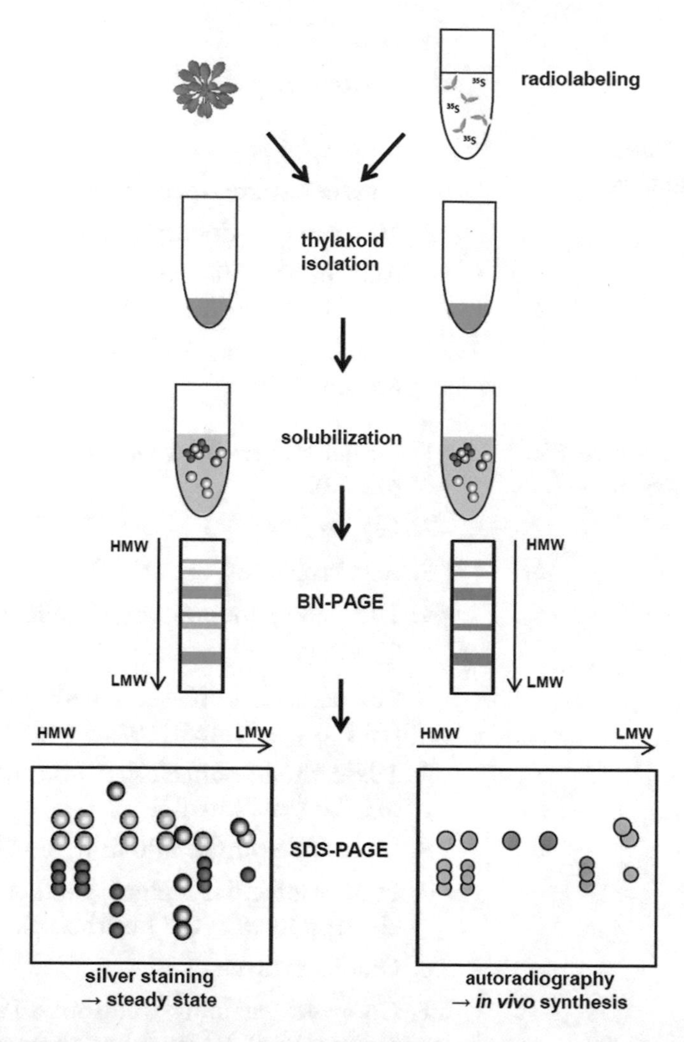

Fig. 1 Schematic workflow overview. Thylakoids are prepared from *Arabidopsis* leaves (*left panel*) or *Arabidopsis* seedlings incubated with ^{35}S-methionine prior to thylakoid preparation to investigate in vivo protein synthesis (*right panel*). Isolated thylakoids are solubilized with detergent to release the protein complexes from the membrane. Protein complexes are separated by BN-PAGE, where chlorophyll-containing complexes can be immediately monitored. Unlabeled protein complexes are further resolved into their components by 2D SDS-PAGE and silver staining (*left*). ^{35}S-labeled proteins are resolved by 2D SDS-PAGE and detected by autoradiography, preferably after blotting on PVDF membrane (*right*)

2.2 Thylakoid Membrane Isolation

1. Isolation buffer: 50 mM HEPES/KOH pH 7.5, 330 mM sorbitol, 2 mM EDTA-Na_2, 1 mM $MgCl_2$, 5 mM ascorbic acid (*see* **Note 2**).
2. Wash buffer: 50 mM HEPES/KOH pH 7.5, 5 mM sorbitol.
3. TMK buffer: 50 mM HEPES/KOH pH 7.5, 100 mM sorbitol, 5 mM $MgCl_2$.

4. Gauze.
5. Polytron homogenizer.

2.3 Sample Preparation

1. 80% (v/v) acetone.
2. Quartz cuvette, spectrophotometer.
3. 10% (w/v) *n*-dodecyl β-d-maltoside (β-DM).
4. ACA buffer: 750 mM ε-aminocapronic acid, 50 mM Bis–Tris pH 7.0, 0.5 mM EDTA-Na_2.
5. Loading buffer: 750 mM ε-aminocapronic acid, 5% (w/v) Serva-G 250.

2.4 BN-PAGE (First Dimension)

1. 6× gel buffer: 3 M ε-aminocapronic acid, 300 mM Bis–Tris pH 7.0.
2. Glycerol.
3. Acrylamide (37.5:1).
4. 10% (w/v) ammonium persulfate (APS).
5. TEMED.
6. 10× cathode buffer blue: 500 mM tricine, 150 mM Bis–Tris pH 7.0 (*see* **Note 3**), 0.2% (w/v) Serva G-250.
7. 10× cathode buffer clear: 500 mM tricine, 150 mM Bis–Tris pH 7.0 (*see* **Note 3**).
8. 10× anode buffer: 500 mM Bis–Tris pH 7.0 (*see* **Note 3**).
9. High-molecular-weight marker calibration kit for native electrophoresis (GE healthcare).
10. Gradient mixer.
11. Coomassie staining solution: 45% (v/v) methanol, 9% (v/v) acetic acid, 0.2% (w/v) coomassie brilliant blue R-250.
12. Coomassie destaining solution: 45% (v/v) methanol, 9% (v/v) acetic acid.

2.5 Denaturating SDS-PAGE (Second Dimension)

1. 3 M Tris–HCl pH 8.8.
2. 1 M Tris–HCl pH 6.8.
3. Acrylamide (30:1).
4. Urea.
5. 10% (w/v) Sodium dodecyl sulfate (SDS).
6. 10% (w/v) APS.
7. TEMED.
8. SDS solution: 67 mM SDS, 67 mM Na_2CO_3.
9. 10× SDS running buffer: 250 mM Tris base, 1.92 M glycine, 1% (w/v) SDS.

2.6 Silver Staining

1. Fixation solution: 50% (v/v) ethanol, 12% (v/v) acetic acid, 0.05% (v/v) formaldehyde.
2. 50% (v/v) ethanol (denaturated).
3. Pre-impregnation solution: 0.02% (w/v) sodium thiosulfate.
4. Impregnation solution: 0.2% (w/v) silver nitrate, 0.075% (v/v) formaldehyde (keep dark).
5. Development solution: 6% (w/v) Na_2CO_3, 0.05% (v/v) formaldehyde, 0.0004% (w/v) sodium thiosulfate.
6. Stop solution: 50% (v/v) ethanol, 12% (v/v) acetic acid.

2.7 Radiolabeling

1. Reaction buffer: 1 mM KH_2PO_4 pH 6.3, 0.1% (w/v) Tween 20.
2. ^{35}S-methionine/cysteine mix (specific activity 0.1000 Ci/mmol).
3. Optional: chloramphenicol (100 μg/ml) and cycloheximide (80 μg/ml) (*see* **Note 4**).
4. 20 mM Na_2CO_3.
5. For subsequent BN-PAGE: isolation buffer, wash buffer, and TMK buffer (*see* Subheading 2.2).
6. Stainless-steel micro-pestle for 1.5 ml reaction tubes.

2.8 Western Blot Using Semi-dry Electroblotting

1. Anode buffer 1: 300 mM Tris base, 20% (v/v) methanol.
2. Anode buffer 2: 25 mM Tris base, 20% (v/v) methanol.
3. Cathode buffer: 40 mM ε-aminocapronic acid, 20% (v/v) methanol.
4. Methanol.
5. PVDF membrane.
6. Blotting (Whatman) paper.

3 Methods

3.1 Thylakoid Membrane Isolation of Unlabeled Leaf Material

1. In our experiences, the best results are obtained with 3-week-old plants. Harvest the plants after a dark period to reduce the amount of starch. Either start early in the morning or place the plants in the dark for at least 1 h directly before you start the isolation.
2. All procedures are carried out at 4 °C. 1 g leaves are homogenized with a polytron homogenizer in 25 ml isolation medium and filtered through two layers of gauze. The homogenate is centrifuged for 4 min at 760×*g* and the supernatant is discarded.
3. The pellet is carefully resuspended in 3 ml wash medium and centrifuged again for 4 min at 760×g.

4. The thylakoid/chloroplast (*see* **Note 5**) pellet is resuspended in 1 ml TMK buffer, transferred to 1.5 ml reaction tube, and incubated 10 min on ice in the dark. Then the homogenate is centrifuged for 3 min at 760 × *g* and the pellet is resuspended in 500 μl TMK buffer.

3.2 Sample Preparation

1. To determine the chlorophyll concentration, 1 μl thylakoid membranes in TMK buffer are mixed with 1 ml of 80% acetone and the optical density is measured at 645, 663, and 750 nm against the solvent. The chlorophyll concentration can be calculated with this formula: μg chlorophyll/μl = 8.02 × ($E_{663} - E_{750}$) + 20.2 × ($E_{645} - E_{750}$).
2. For each lane of the BN gel, a total amount of thylakoids corresponding to 30 μg of chlorophyll is used and pelleted for 3 min at 3300 × *g*. The supernatant is discarded.
3. To solubilize membrane protein complexes, the pellet is resuspended in 70 μl ACA buffer and 8 μl of 10% β-DM. Digitonin can also be used for solubilization, if a milder solubilization is required [8] (*see* **Note 6**). After 10 min incubation on ice, the samples are centrifuged for 10 min at 18,000 × *g*. Take care not to disturb the pellet. The supernatant is added to 5 μl loading buffer.

3.3 BN-PAGE (First Dimension)

1. It is recommended to use a separate BN gel unit which does not come into contact with denaturating detergents. We use the PROTEAN II xi system (Bio-Rad, USA). Clean the plates with 70% ethanol and assemble the apparatus using 0.75 mm spacers. Also clean the gradient mixer with ultrapure H_2O and make sure you have a constant flow before you pour your gel solutions inside. The flow rate should be 2–3 ml/min. The gradient mixer is placed on a magnetic stirrer. A syringe allows the solution being cleanly poured in at the upper center between the plates (*see* **Note 7**).
2. For the separation of thylakoid membrane complexes, a 6–15% gradient gel is suitable. Prepare the following solutions on ice:

   6% *solution*: 1.2 ml of acrylamide, 1.26 ml of 6× gel buffer, 1.72 ml of ultrapure H_2O, 3.57 μl of TEMED, and 14.31 μl of 10% APS.

   15% *solution*: 1.5 g of glycerin, 3 ml of acrylamide, 1.26 ml of 6× gel buffer, 5.02 ml of ultrapure H_2O, 3.57 μl of TEMED, and 14.31 μl of 10% APS.

   While the connection between the two chambers is still closed, the 15% solution and a magnetic stirring bar are placed in the first chamber of the gradient mixer. The 6% solution is poured in the back chamber. Start the gradient mixer and let the 15% solution migrate a few centimeters in the tube before opening the connection to the back chamber with the 6% solution. Overlay the casted gel with isopropanol and let it polymerize for at least 1 h.

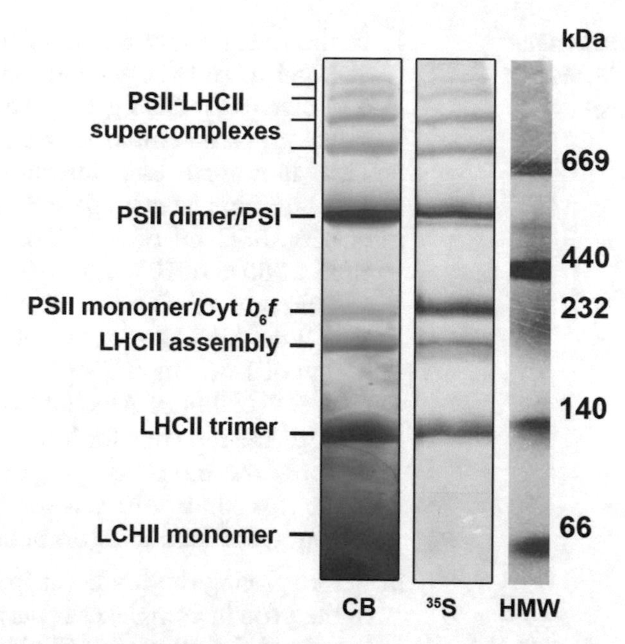

Fig. 2 Separation of thylakoid membrane complexes by BN-PAGE. Unlabeled protein complexes are visualized as green bands due to the bound chlorophyll (PSII, PSI, and LHC) or are slightly stained by the coomassie (CB), which is present during the run (Cyt b_6f) (*left panel*). ^{35}S-labeled and separated protein complexes were detected by autoradiography. Pulse labeling was performed for 20 min (*middle panel*). HMW, high-molecular-weight marker (*right panel*)

3. Remove isopropanol and insert the comb. Prepare the stacking gel and pipet it on top of the separating gel: 0.56 ml of acrylamide, 0.58 ml of 6× gel buffer, 2.36 ml of ultrapure H_2O, 4.2 μl of TEMED, and 21.2 μl of 10% APS. Let polymerize for approx. 30 min.
4. Assemble the gel unit. Fill blue cathode buffer (approx. 350 ml) in the upper part and anode buffer (approx. 1900 ml) in the lower tank of the running chamber. Load your samples slowly with a 100 μl Hamilton syringe (also one for BN-PAGE only). Use 50 μl of HMW marker.
5. The gels are run overnight at 30 V and 4 °C. After approx. 17 h the dye front should have migrated to the middle of the gel. At this point the blue cathode buffer is exchanged for clear cathode buffer to better visualize the protein complexes. If the voltage is increased to 200, the run is completed in approx. 3 h.
6. Remove the stacking gel and cut out the individual lanes of the BN gel. The molecular marker together with one lane is stained with coomassie solution. An example of the stained marker as well as the visible separated thylakoid membrane complexes is presented in Fig. 2. The gel strips can either be transferred to the second dimension immediately or they can also be wrapped in aluminum foil and stored at −20 °C for later usage.

3.4 Denaturating SDS-PAGE (Second Dimension)

1. In the next step, one lane of the BN gel is laid on top of an SDS gel to run the second denaturating dimension. We add 4 M urea to the SDS gel to increase the resolution of the individual spots. Assemble your glass plates using 1.5 mm spacer. This allows an easy insertion of the BN gel strip. For 16×17.5 cm plates prepare 30 ml separating gel using 7.21 g of urea, 3.75 ml of 3 M Tris/HCl pH 8.8, 12 ml of acrylamide, 300 μl of 10% SDS, 300 μl of 10% APS, 12 μl TEMED, and ultrapure H_2O up to 30 ml. The stacking gel should be at least 2 cm high. Prepare 6 ml with 4.1 ml of ultrapure H_2O, 750 μl of 1 M Tris/HCl pH 6.8, 1 ml of acrylamide, 60 μl of 10% SDS, 60 μl of APS, and 6 μl of TEMED. Leave 1–2 cm space to the top edge for insertion of the BN gel strip. In order to apply the molecular weight marker, we use a homemade single slot comb. Alternatively, there are also available combs leaving space for a gel strip besides a regular pocket.
2. Before placing the BN gel strip on the SDS gel, a denaturation of the protein complexes is performed. For this the gel strip is incubated for 20 min in SDS solution on a shaker at 60 rpm (*see* **Note 8**).
3. Insert the BN gel strip between the glass plates on top of the SDS gel. The easiest way is to start with high percentage part of the strip (lower, blue part of the gel). Make sure there are no air bubbles between the strip and the gel (*see* **Note 9**). Load molecular weight marker. We run the second dimension overnight at 8 mA or at 35 mA for 4–5 h.

3.5 Silver Staining

1. To analyze all photosynthetic protein complexes, a silver staining of the second dimension is suitable. Looking at silver staining protocols, there exist different approaches. We use silver nitrate for impregnation which is easier in handling and more compatible with different electrophoretic systems than the alternative silver-ammonia complex. Keep in mind that basic proteins are stained less efficiently than acidic ones (*see* **Note 10**). Even though there exist fast staining protocols, we recommend a procedure with extensive washing and a thiosulfate treatment to increase sensitivity and reduce background binding. Metal lightproof trays are recommended. All steps are performed shaking the gels at 60 rpm at room temperature.
2. First the gel is incubated for 1 h (or overnight) in fixation solution, then it is washed three times for 30 min in 50% ethanol.
3. After a pre-impregnation for 1.5 min, the gel is washed three times for 30 s in ultrapure H_2O.
4. The following impregnation is performed for 30 min. Be aware that the impregnation solution must be disposed of separately as heavy metal waste.

5. The gel is washed three times for 30 s in ultrapure H_2O and then the development solution is applied. During development, keep shaking the tray manually and constantly survey the gel. Usually it takes about 1–2 min to get a suitable staining (*see* **Note 11**).
6. Exchange the development solution for stop solution. After 10 min, wash the gel in ultrapure H_2O. A stained gel is shown in Fig. 3.

3.6 Radiolabeling

1. Highest labeling efficiency is obtained with 12- to 14-day-old plants. Harvest roughly 20 plants and transfer them into 1.5 ml tubes with 50 μl reaction buffer. Keep at 4 °C.
2. If inhibition of the synthesis of either nuclear or plastid encoded proteins is desired, add chloramphenicol or cycloheximide to the reaction buffer, respectively. Incubate 15 min in darkness at 4 °C.
3. Add ^{35}S-methionine/cysteine (30 μCi) to each sample.
4. To infiltrate the plants, centrifuge samples in a vacuum centrifuge for 1 min keeping the reaction tubes open.
5. Immediately move the infiltrated plants into a water bath exposed to a high light source to stimulate D1 turnover (800–1000 $\mu E/m^2$ s). Take care to maintain the temperature of the water bath at 25 °C. The incubation period can range from 5 min to 1 h. Initial assembly intermediates may already be visible after 5–10 min incubation. Also, be aware that prolonging the incubation will result in visualizing not only protein synthesis, but also degradation (*see* **Note 12**).
6. After the incubation, the reaction buffer is removed and the samples are washed with 20 mM Na_2CO_3 to remove residual ^{35}S-methionine/cysteine.
7. For isolation of thylakoid membranes after the labeling process and subsequent analysis by BN-PAGE, the plants are transferred to fresh reaction tubes containing 100 μl isolation buffer and homogenization is performed with a stainless-steel micropestle. To prevent damage of the protein complexes, the samples are homogenized briefly for intervals of 5 s. Take care that the temperature of the samples is maintained at 4 °C.
8. Make sure that there is as little residual root and plant material remaining as possible. Samples are centrifuged for 10 min at $760 \times g$ and 4 °C to pellet the thylakoids.
9. The pellet is resuspended in 100–150 μl of wash buffer. The samples are centrifuged again as above.
10. The supernatant is discarded and the thylakoid pellet is resuspended in 100–150 μl TMK Puffer. Samples are further treated as described in Subheading 3.2 and samples are analyzed by

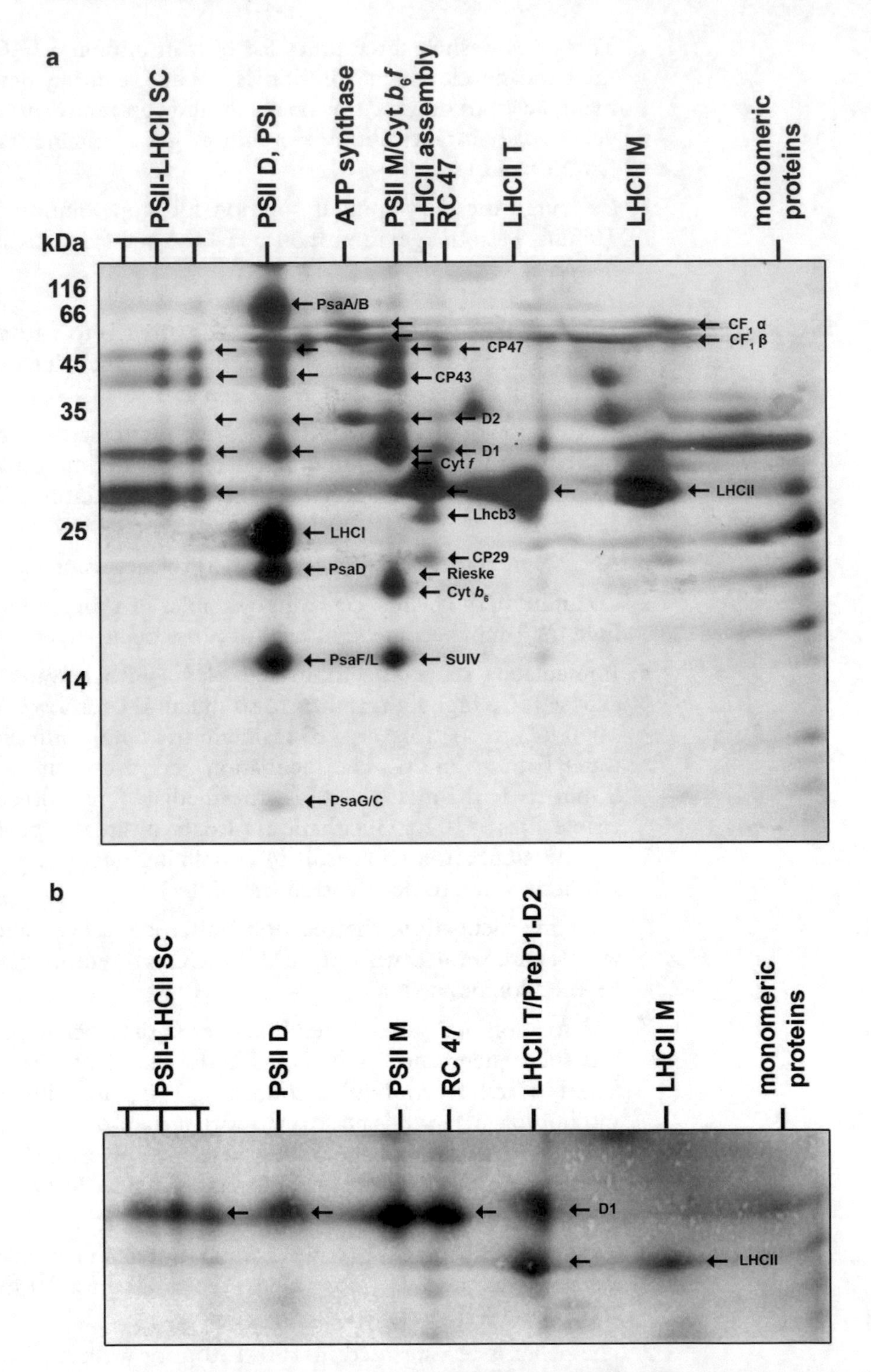

Fig. 3 Separation of thylakoid membrane complexes by 2D SDS-PAGE. (**a**) BN-PAGE lanes were resolved in a 2D SDS-PAGE and proteins were detected by silver staining. Individual proteins were assigned according to the gelmap database (https://gelmap.de/arabidopsis-chloro) as indicated. Consecutive arrows indicate their appearance in increasingly higher-molecular-weight complexes. (**b**) BN-PAGE lanes run with ^{35}S-labeled proteins were resolved in a 2D SDS-PAGE, blotted on PVDF membrane and detected by autoradiography. The most strongly labeled proteins (D1 and LHCII) and their successive assembly into higher-molecular-weight complexes is indicated by *arrows*. Pulse labeling was performed for 20 min

BN-PAGE (*see* **Note 13**). The first dimension can be dried on a vacuum gel dryer and exposed to an X-ray film or phosphorimaging plate, or used to run a second dimension SDS-PAGE as described in Subheading 3.4.

11. 1.5 mm thick gels are prone to break while drying. Therefore, we recommend to blot the gels on PVDF or nitrocellulose membrane before detection of radiolabeled proteins. We use a semi-dry blot system.
12. Assemble the blot as follows (from bottom to top): 3 blotting papers soaked in anode buffer 1, 2 blotting papers soaked in anode buffer 2, membrane (pre-activated with 100% methanol), gel, 3 blotting papers soaked in cathode buffer.
13. The blot is run at 0.8 mA/cm^2 for 2 h.
14. Dry the membrane and expose overnight or longer to an X-ray film or phosphorimaging plate. Labeled proteins separated by BN-PAGE as well as by SDS-PAGE are shown in Figs. 2 and 3, respectively.

4 Notes

1. Performing a radiolabeling experiment which is analyzed by BN-PAGE, plants grown on MS agar plates work best in our hands. Plants can be grown on MS agar plates without sucrose, if preferred.
2. Ascorbic acid has to be added freshly to the isolation medium. All other buffers for the thylakoid preparation can be prepared in advance and stored at 4 °C. If investigating phosphorylation of proteins 10 mM NaF, an unspecific phosphatase inhibitor, can be added to isolation, wash and TMK buffer.
3. Adjust pH with concentrated HCl.
4. Chloramphenicol and cycloheximide have to be added freshly to the reaction medium.
5. Instead of this crude thylakoid extract you can use intact chloroplasts, if stromal protein complexes are of interest.
6. Pellet 100 μg of chlorophyll and resuspend in 60 μl ACA buffer. Add 30 μl of 5% digitonin and rotate for 1 h at 4 °C at an overhead shaker. Centrifuge for 1 h at 4 °C (pellet is quite loose!) at 25,000 × *g* and load 70 μl of the supernatant on the gel.
7. It is the safest to fix the syringe with a tape at the plates. If air bubbles or irregularities of the surface appear during pouring the gel, tap at the plates. Do not shake the casting apparatus, this would destroy the gradient. When having problems with leaky assembly of the glass plates, we recommend to fill in a quickly polymerizing plug solution first. We have a mixture out

of 40 ml of acrylamide, 60 ml ultrapure H_2O, and 450 µl TEMED which we store at 4 °C. To 1 ml of this plug solution, 30 µl 10% APS are given and immediately poured between the plates. If differently sized gel plates are used, make sure to adjust the volumes of the gel solutions to still covert the whole percentage from 6 to 15%. Gel can be stored overnight at 4 °C in wet paper towels and a plastic bag.

8. When using frozen gel strips, start the denaturation after thawing.
9. You might encounter the problem that the gel strip is flushed out of the glass plates when filling the running chamber with running buffer. This can be prevented by putting two small pieces of folded blotting paper above the gel strip between the glass plates which block the way out. Nevertheless, be careful filling in the running buffer.
10. Silver-ammonia, in contrast to silver nitrate, stains basic proteins more efficiently than acidic ones. It offers more flexibility in the control of staining but only works with glycine and taurine electrophoresis systems.
11. Using 0.75 mm thick gels, the development of the silver staining is very quick. Since we take 1.5 mm thick gels for the second dimension it can take longer than 2 min.
12. To perform pulse as well as chase experiments replace the reaction buffer with ^{35}S-methionine/cysteine with reaction buffer with unlabeled methionine/cysteine after the pulse and incubate for 1–6 h to monitor D1 decay.
13. In addition to BN-PAGE, the samples can also be analyzed by denaturating 1D SDS-PAGE to determine the overall labeling efficiency.

Acknowledgements

Financial support from the German Research Council (DFG, SFB1035, project A4, to SS) is acknowledged. We would further like to thank Manuela Urbischek for providing figures as well as Peter Hagl for critical reading of the manuscript.

References

1. Nevo R, Charuvi D, Tsabari O, Reich Z (2012) Composition, architecture and dynamics of the photosynthetic apparatus in higher plants. Plant J 70(1):157–176. doi:10.1111/j.1365-313X.2011.04876.x
2. Granvogl B, Reisinger V, Eichacker LA (2006) Mapping the proteome of thylakoid membranes by de novo sequencing of intermembrane peptide domains. Proteomics 6(12):3681–3695. doi:10.1002/pmic.200500924

3. Meurer J, Plucken H, Kowallik KV, Westhoff P (1998) A nuclear-encoded protein of prokaryotic origin is essential for the stability of photosystem II in Arabidopsis thaliana. EMBO J 17(18):5286–5297
4. Rokka A, Suorsa M, Saleem A, Battchikova N, Aro EM (2005) Synthesis and assembly of thylakoid protein complexes: multiple assembly steps of photosystem II. Biochem J 388(Pt 1): 159–168. doi:10.1042/BJ20042098, BJ20042098 [pii]
5. Aro EM, Suorsa M, Rokka A, Allahverdiyeva Y, Paakkarinen V, Saleem A, Battchikova N, Rintamaki E (2005) Dynamics of photosystem II: a proteomic approach to thylakoid protein complexes. J Exp Bot 56(411):347–356. doi:10.1093/jxb/eri041, eri041 [pii]
6. Ploscher M, Reisinger V, Eichacker LA (2011) Proteomic comparison of etioplast and chloroplast protein complexes. J Proteomics 74(8):1256–1265. doi:10.1016/j.jprot.2011.03.020, S1874-3919(11)00111-4 [pii]
7. Armbruster U, Zuhlke J, Rengstl B, Kreller R, Makarenko E, Ruhle T, Schunemann D, Jahns P, Weisshaar B, Nickelsen J, Leister D (2010) The Arabidopsis thylakoid protein PAM68 is required for efficient D1 biogenesis and photosystem II assembly. Plant Cell 22(10): 3439–3460. doi:10.1105/tpc.110.077453, tpc.110.077453 [pii]
8. Schwenkert S, Legen J, Takami T, Shikanai T, Herrmann RG, Meurer J (2007) Role of the low-molecular-weight subunits PetL, PetG, and PetN in assembly, stability, and dimerization of the cytochrome b6f complex in tobacco. Plant Physiol 144(4):1924–1935. doi:10.1104/pp.107.100131, pp.107.100131 [pii]
9. Blum H, Beier H, Gross HJ (1987) Improved silver staining of plant proteins, RNA and DNA in polyacrylamide gels. Electrophoresis 8:93

Chapter 20

Normalized Quantitative Western Blotting Based on Standardized Fluorescent Labeling

Frederik Faden, Lennart Eschen-Lippold, and Nico Dissmeyer

Abstract

Western blot (WB) analysis is the most widely used method to monitor expression of proteins of interest in protein extracts of high complexity derived from diverse experimental setups. WB allows the rapid and specific detection of a target protein, such as non-tagged endogenous proteins as well as protein–epitope tag fusions depending on the availability of specific antibodies. To generate quantitative data from independent samples within one experiment and to allow accurate inter-experimental quantification, a reliable and reproducible method to standardize and normalize WB data is indispensable. To date, it is a standard procedure to normalize individual bands of immunodetected proteins of interest from a WB lane to other individual bands of so-called housekeeping proteins of the same sample lane. These are usually detected by an independent antibody or colorimetric detection and do not reflect the real total protein of a sample. Housekeeping proteins—assumed to be constitutively expressed mostly independent of developmental and environmental states—can greatly differ in their expression under these various conditions. Therefore, they actually do not represent a reliable reference to normalize the target protein's abundance to the total amount of protein contained in each lane of a blot.

Here, we demonstrate the Smart Protein Layers (SPL) technology, a combination of fluorescent standards and a stain-free fluorescence-based visualization of total protein in gels and after transfer via WB. SPL allows a rapid and highly sensitive protein visualization and quantification with a sensitivity comparable to conventional silver staining with a 1000-fold higher dynamic range. For normalization, standardization and quantification of protein gels and WBs, a sample-dependent bi-fluorescent standard reagent is applied and, for accurate quantification of data derived from different experiments, a second calibration standard is used. Together, the precise quantification of protein expression by lane-to-lane, gel-to-gel, and blot-to-blot comparisons is facilitated especially with respect to experiments in the area of proteostasis dealing with highly variable protein levels and involving protein degradation mutants and treatments modulating protein abundance.

Key words Protein expression, Quantification, Data normalization, Fluorescence labeling, Fluorescent labeling, Fluorescent dye, Stain-free technology, Loading control, Western blot

1 Introduction

Accurate quantification of protein expression is of great interest in research and diagnostics. Techniques like two-dimensional gel electrophoresis and mass spectrometry allow the analyses of

L. Maria Lois and Rune Matthiesen (eds.), *Plant Proteostasis: Methods and Protocols*, Methods in Molecular Biology, vol. 1450,
DOI 10.1007/978-1-4939-3759-2_20, © Springer Science+Business Media New York 2016

complex sample material. Unfortunately, these approaches usually require expensive special equipment not accessible to many labs. Thus, standard one-dimensional western blot (WB) analysis is the method of choice in most cases. It allows the visualization of proteins of interest (POIs) based on detection using specific antibodies. However, the precise standardization, normalization, and quantification of independent protein samples or experiments often prove difficult. To assure that quantitative WB data are not an artifact of technical, i.e. experimental, errors influencing the reference signal, an appropriate method for normalization is indispensable [1]. This is especially important for experiment-to-experiment but also for standard sample-to-sample comparisons. Typically, normalization of the target protein signal is based on a housekeeping protein (HKP) signal (e.g. β-actin, β-tubulin, glyceraldehyde-3-phosphate dehydrogenase, cyclin-dependent kinase or ribulose-1,5-bisphosphate carboxylase/oxygenase). It is important to note that HKP abundance varies strongly under different conditions, e.g. after stress, chemical treatments, growth conditions, etc. [2]. This makes a careful choice of a HKP under the experimental conditions critical. Quantification of total protein amounts allows the normalization completely independent of single HKP signals. However, conventional total protein staining with Coomassie Brilliant Blue R-250 (gels, blots), Ponceau S (blots), amido black (blots), or Fast Green FCF (blots) suffers from low sensitivity, a limited dynamic range, and poor reproducibility depending on staining/destaining intensities and thus only can serve as lane-to-lane loading control within one blot [3, 4]. Recently, new stain-free methods were developed offering a more reliable and accurate total protein quantification [5, 6].

Here, we describe our experiences with a new method for protein expression quantification combining conventional WB with fluorescent labeling and fluorescence detection of total protein. The so-called Smart Protein Layers (SPL) technology is suitable for a standardized, stain-free, quantitative analysis and involves the application of (1) fluorescently labeled standards directly added to each sample allowing both lane-to-lane and inter-experimental comparisons, (2) a fluorescent dye labeling a fraction of the total protein, and (3) a fluorogenic substrate for detection of the actual target protein. Target and total protein abundance in each sample can be simultaneously detected so that quantification artifacts due to washes during the WB procedure are prevented. The SPL *Smartalyzer* (SMA), a fluorescent internal standard protein available in two sizes (12.5 or 80 kDa) that is added to every labeling reaction (*see* **Note 1**), allows correcting for errors in loading of sample volume to each gel pocket. In general, the SMA (*see* **Notes 1** and **2**) is added to each sample, then total protein is labeled with a second fluorescent dye and SDS-PAGE is performed. In the sample protein lysine side chains are covalently labeled by low ratios of the dye, known as Minimal Labeling, e.g. in 2D-DIGE. Approximately one residue per protein and only 3 % of the proteins of one species are labeled. This prevents saturation and allows accurate quantification.

Total protein and sample-dependent standard protein signals in the gel can be detected with a fluorescence scanner or a CCD camera-based fluorescence detection system with high sensitivity within seconds in different detection channels (detection limit is less than 1 ng with a dynamic range of 10^4–10^5). The relation between the quantity of the sample-dependent standard protein corresponding to the total protein quantity allows the precise lane-to-lane protein normalization. Quantification over independent experiments can be performed using an additional fluorescent standard which is applied once to each gel. It consists of three marker proteins (12.5, 25, and 80 kDa) plus one species-dependent 50 kDa marker protein specifically designed to bind the secondary antibody actually used in the experiment. The corresponding signal intensities of both fluorescence label and WB are used for data normalization between experiments. Figure 1 shows a simplified overview over a complete SPL workflow.

2 Materials

Dependent on target detection and available fluorescent imaging device, different kits are available. Here, we focus on the use of one of the SPL kits (SPL Red, NH DyeAGNOSTICS) where total protein is pre-labeled with a red-fluorescent fluorophore. Subsequently, target protein detection has to be done within the blue channel of the detection system. We used the Immuno Blue Western Blotting Substrate (NH DyeAGNOSTICS) which produces a stable blue fluorescent precipitate on the blotting membrane when horseradish peroxidase is used as reporter enzyme linked to the secondary antibody (*see* **Note 2**).

2.1 Protein Extraction

Usually, the SPL technology does not require any changes in protein extraction and quantification protocols. Use your preferred buffers and quantification system. We used the following workflow:

1. Standard 1.5 mL microcentrifuge tubes.
2. Stainless steel beads (Nirosta, 3.175 mm; cat. no. 75306, Mühlmeier Mahltechnik).
3. A modified radioimmunoprecipitation assay (RIPA) extraction buffer consisting of 50 mM Tris–HCl, pH 8; 120 mM NaCl; 1 mM EDTA; 6 mM EGTA; 1% (v/v) Nonidet P-40; the protease inhibitor benzamidine (1 mM); and the two phosphatase inhibitors sodium fluoride (NaF; 20 mM) and sodium pyrophosphate ($Na_4P_2O_7$; 15 mM) or (instead of benzami-

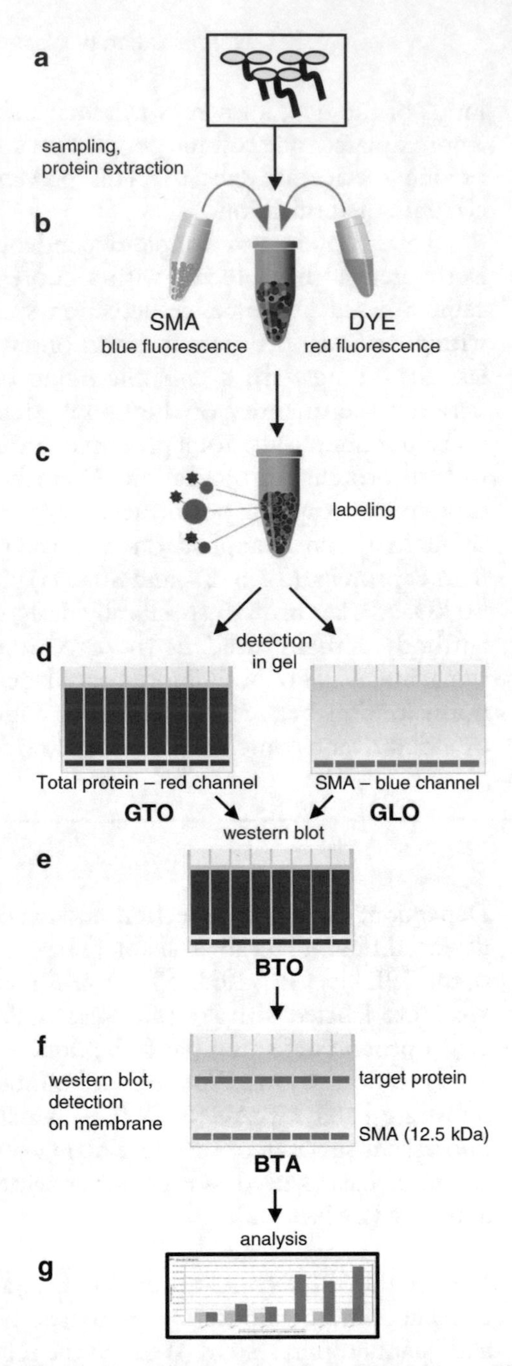

Fig. 1 Complete SPL workflow. (**a**) Protein extraction from sample material and quantification. 20 μg of total protein per sample are transferred to a microcentrifuge tube. (**b**) Addition of red fluorescent dye label and blue fluorescent internal standard (SMA) to each sample. Here, the light 12.5 kDa standard is chosen. An 80 kDa SMA can be chosen if the lighter one interferes with the target size and/or running parameters. Depending on the kit, the dyes and fluorescence signals are different (*see* Table 1). (**c**) Labeling reaction. (**d**–**f**) Fluorescence detection. In our workflow, the labeled total protein is detected in the red channel whereas the blue channel is used to record SMA fluorescence. (**d**) SDS-PAGE. After completion, red and blue channels are recorded with a fluorescence imaging device (GTO = *g*el *to*tal protein; GLO = *g*el *lo*ading). (**e**) Western blotting and immunoprobing. Detection of the red channel (BTO = *b*lot *to*tal protein). (**f**) Development and detection of the target protein with specific antibody and blue fluorescent Immuno Blue secondary reagent (BTA = *b*lot *ta*rget protein), again in the blue channel. Here, the lower band corresponds to the internal 12.5 kDa SMA standard and the higher molecular band corresponds to the target protein. (**g**) Analysis using LabImage software (*see* Subheading 2.8). Adapted from the manufacturer's instruction manuals

Table 1
Available SPL kits depending on the fluorescence detection system

Target detection	SPL kit	Necessary fluorescence channels
Chemiluminescence + HRP-coupled antibody	SPL Red SPL iRed	Red/blue Red/infrared
Immuno Blue + HRP-coupled antibody	SPL Red	Red/blue
Red fluorescently labeled antibody	SPL Blue	Blue/red
Infrared fluorescently labeled antibody	SPL iRed	Red/infrared

dine) a protease inhibitor mix (Protease Inhibitor Cocktail Tablets cOmplete, EDTA-free, Roche Diagnostics, cat. no. 04693132001: each tablet is sufficient for a volume of 50 mL of extraction solution).

4. Bead mill (Retsch).
5. Standard cooling tabletop microcentrifuge (Eppendorf).

2.2 Protein Quantification (*See* Note 5)

1. BCA protein quantification kit (Pierce).
2. Microplate reader (Tecan M200 pro).

2.3 Protein Labeling

1. Appropriate SPL labeling kit (*see* **Note 2**, Table 1).

2.4 SDS-PAGE and Western Blot

1. Any standard SDS-PAGE system, e.g. a Mini-PROTEAN Tetra Cell (Bio-Rad).
2. Any standard blotting system (semi-dry or wet blot; e.g. a Trans-Blot SD Semi-Dry Transfer Cell (Bio-Rad)) or a tank blot apparatus (such as Tankblot SCIE-PLAS EB10).
3. Buffers for electrophoresis and blotting procedure.
4. (Low) fluorescence membrane (PVDF or nitrocellulose), other membranes have to be tested for their background fluorescence. We have successfully used a standard PVDF membrane (Hybond P 0.45 PVDF, GE Healthcare, cat. no. 10600023) that yielded neglectable autofluorescence.

2.5 Antibodies

Appropriate primary and secondary antibody (HRP- or fluorescence-conjugated antibody) depending on target protein abundance and the desired type of detection (*see* **Note 3**). Antibodies tested are listed in Table 2.

2.6 Target Detection by Chemiluminescence

1. Chemiluminescent substrate and detection unit (e.g. G:BOX series, Syngene; OCTOPLUS series, NH DyeAGNOSTICS).
2. X-ray films with a film developing unit (optional for quick check via ECL; *see* **Note 4**).

2.7 Target Detection by Fluorescence

1. Fluorescence imaging devices such as CCD-based cameras or scanners with appropriate LEDs or lasers and appropriate filters for blue and red channel detection (e.g. G:Box series, Syngene; Odyssey, LiCOR; Octoplus series, NH DyeAGNOSTICS; Typhoon FLA 9000, GE Healthcare Life Sciences). In our experiments, we used a Typhoon FLA 9000.
2. Immuno Blue Western Blotting Substrate for fluorescence detection of HRP (NH DyeAGNOSTICS) or fluorescently labeled secondary antibody.

2.8 Quantification

1D analysis software that allows for lane and band detection, background substraction, and determination of lane/band volumes; e.g. LabImage 1D L300 (Kapelan Bio-Imaging; free download at www.kapelan-bioimaging.com). We used the LabImage 1D SPL analysis software (included in the package from NH DyeAGNOSTICS) which included automatic band and lane determination as well as automated normalization and data evaluation. However, also manual analysis of the data is possible using standard image processing and analysis software such as ImageJ (http://imagej.nih.gov/ij/), e.g. by using the workflow described at http://lukemiller.org/index.php/2010/11/analyzing-gels-and-western-blots-with-image-j/.

3 Methods

3.1 Protein Extraction

In general, SPL is compatible with all commonly used extraction protocols. We used the following approach:

1. Harvest sample material in a standard 1.5 mL reaction tube containing three steel beads and immediately freeze in liquid nitrogen.
2. Grind material using a bead mill (1 min, 30 Hz).
3. Add extraction buffer, vortex, and incubate for 20 min shaking at 4 °C.
4. Clear by centrifugation in a pre-cooled standard tabletop microcentrifuge (10 min, 4 °C, >20,000 × *g*). If supernatant is not clear, repeat and prolong centrifugation.
5. Transfer supernatant in a new reaction tube. Store on ice for immediate use or freeze for later analysis.

3.2 Protein Quantification

In general, SPL is compatible with all commonly used quantification protocols. We used the Pierce BCA kit according to the manufacturer's instruction regarding a 96-well plate-based assay (*see* **Note 5**).

3.3 Protein Labeling Reactions

The labeling reaction is carried out according to the manufacturer's instructions (*see* **Note 6**). Make sure to add appropriate controls for inter- and intra-experimental normalization (*see* **Note 1**).

Table 2
Antibodies suitable for SPL

Antigen	Risen in, type	Name	Cat. No.	Supplier	Used for
1° antibodies					
Green fluorescent protein (GFP)	Rabbit, polyclonal	GFP (FL)	sc-8334	Santa Cruz Biotechnology	Western blot 1:1000 dilution in TBST 4% milk
Green fluorescent protein (GFP)	Mouse, monoclonal	GFP (B-2)	sc-9996	Santa Cruz Biotechnology	Western blot 1:1000 dilution in TBST 5% milk
Phospho-p44/42 MAP kinase	Rabbit, polyclonal	Phospho-p44/42 MAPK (Erk1/2) (Thr202/ Tyr204)	9101	Cell Signaling Technology	Western blot 1:1000 dilution in TBST 3% milk
Ubiquitin	Mouse, monoclonal	Ub (P4D1)	sc-8017	Santa Cruz Biotechnology	Western blot 1:1000 dilution in TBST 5% milk
His tag	Rabbit, polyclonal	His-probe (H-15)	sc-803	Santa Cruz Biotechnology	Western blot 1:1000 dilution in TBST 5% milk
HA tag	Mouse, monoclonal	HA.11	MMS-101	Covance or HISS	1:1000 dilution in TBST 4% milk
HA tag	Mouse, monoclonal	HA-probe (F-7)	sc-7392	Santa Cruz Biotechnology	Western blot 1:1000 dilution in TBST 5% milk
HA tag	Rabbit, polyclonal	HA-probe (Y-11)	sc-805	Santa Cruz Biotechnology	Western blot 1:1000 dilution in TBST 5% milk
PSTAIRE peptide epitope from Cyclin-dependent kinases of the Cdk1/2 type	Rabbit, polyclonal	Cdc2 p34 (PSTAIRE)	sc-53	Santa Cruz Biotechnology	Western blot 1:1000 dilution in TBST 5% milk
2° antibodies					
Mouse	Goat, IgG	Anti-mouse IgG-HRP	sc-2005	Santa Cruz Biotechnology	Western blot 1:2500 dilution in TBST 5% milk (1:5000 for anti-HA)
Rabbit	Goat, IgG	Anti-rabbit IgG-HRP	sc-2004	Santa Cruz Biotechnology	Western blot 1:2500 dilution in TBST 5% milk

3.4 SDS-PAGE and Western Blot

1. SDS-PAGE is carried out as suited for your target protein and the chosen SMA standard (*see* **Note 1**). Unbound dye runs within the running front. Therefore, it can be beneficial to let the running front exit the gel or to cut off this part prior to blotting as this will reduce the unspecific but often very strong fluorescent signal on the blot or gel during detection.
2. Scan the gel with both fluorescence channels (*see* Subheading 2.6) before blotting and record GTO and GLO pictures (software) (GTO = *g*el *to*tal protein; GLO = *g*el *lo*ading). Any conventional blotting method can be used and transfer efficiency monitored directly after transfer via fluorescence (Fig. 2).
3. For fluorescence detection, a low fluorescence membrane is recommended. Still we obtained reliable data using a standard PVDF membrane which yielded negligible levels of background fluorescence (*see* Subheading 2.3).

3.5 Antibodies

SPL usually does not require any changes in antibody incubation time. Use your antibodies as usual depending on your experimental design (*see* **Note 7**).

3.6 Target Detection by Chemiluminescence (*See* Note 7)

1. Obtain high-resolution TIFF images using CCD cameras such as in chemiluminescence detection systems (*see* Subheading 2.6). A simply scanned X-ray film is not sufficient due to its limited dynamic range.

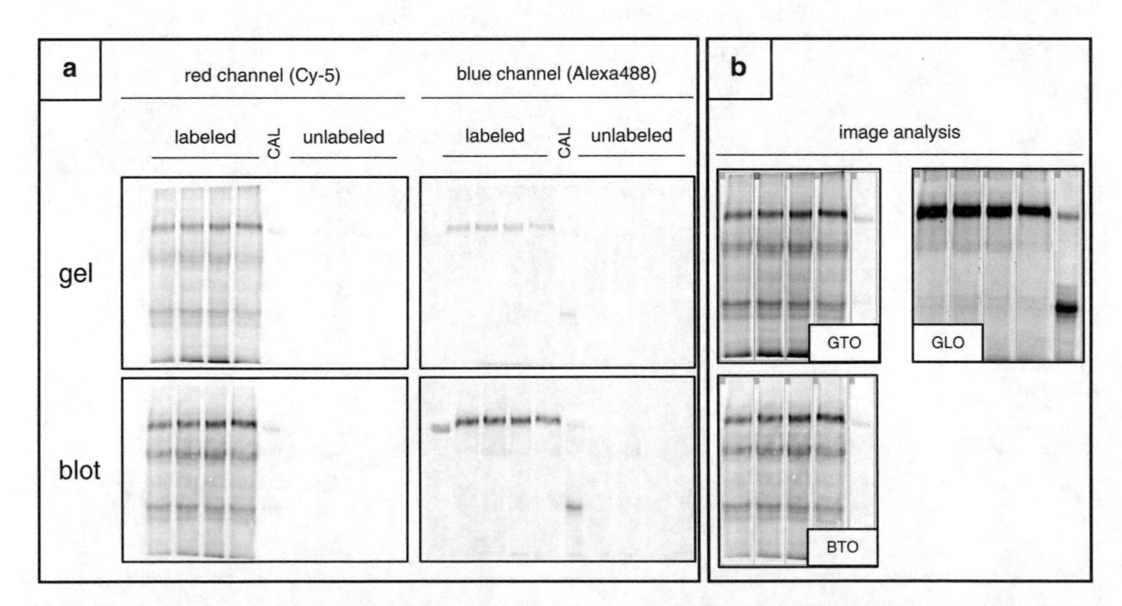

Fig. 2 Monitoring of total protein and SMA standards after SDS-PAGE and western blotting. (**a**) Labeled samples after SDS-PAGE in the gel (*upper panel*) and after western blotting on the PVDF membrane (*lower panel*). Red and blue channels are recorded with a Thyphoon scanner. The two gel pictures together with the red channel picture of the blot are needed for SPL analysis of the samples. (**b**) Screen shots after analysis using the LabImage software (GTO = *g*el *to*tal protein; GLO = *g*el *lo*ading; BTO = *b*lot *to*tal protein)

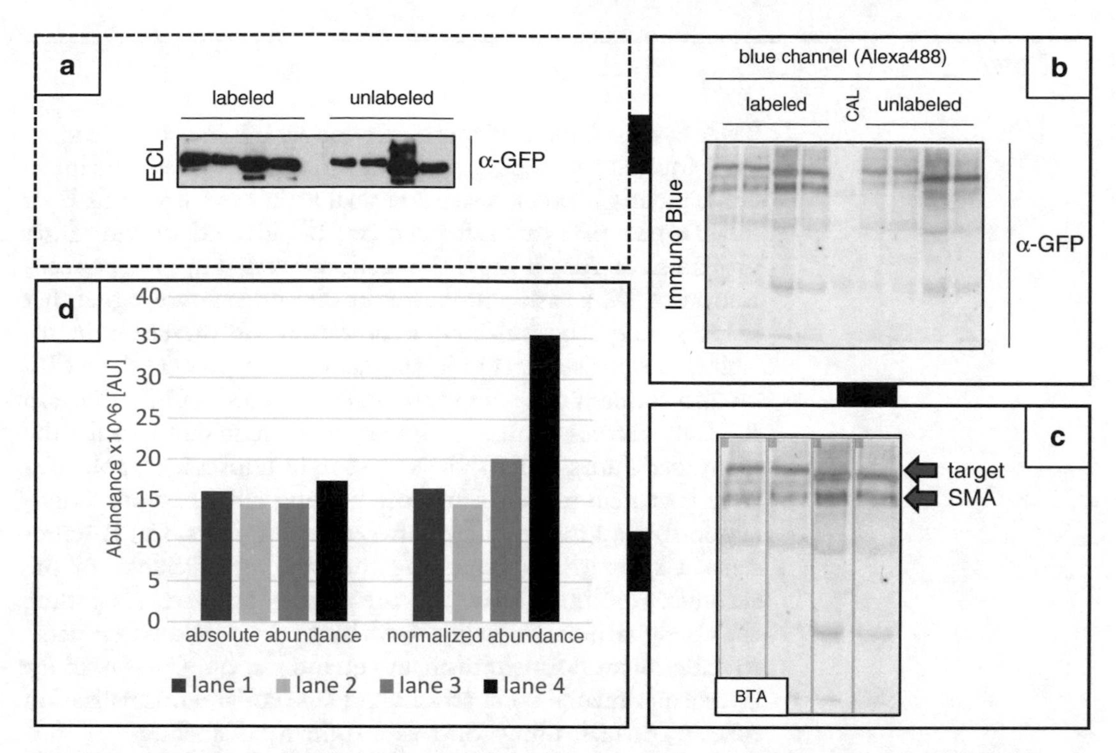

Fig. 3 Target protein detection and data analysis. (**a**) Quick test using a standard ECL substrate after immunoprobing. Both labeled and unlabeled protein is similarly detectable after western blotting indicating that the labeling process has no influence on antibody binding. This part of the SPL workflow serves as a quality control and needs to be done when working with a previously untested antibody. (**b**) Detection with fluorogenic Immuno Blue substrate and a Typhoon scanner. (**c**) Screen shot of the resulting picture (BTA = *b*lot *ta*rget protein) used for SPL analysis of target protein abundance. (**d**) Absolute and relative target abundance after SPL analysis of one single experiment ($n = 1$). Replicate experiments can be compared with this data set

2. ECL can be used to determine signal strength before Immuno Blue staining as a pre-test (Fig. 3a; *see* **Note 3**).

3.7 Target Detection by Fluorescence (*See* Note 7)

1. Apply Immuno Blue according to the manufacturer's instruction and use the blue detection channel of you detection unit (Fig. 3b; *see* **Note 8**).
2. For detection of a fluorescent secondary antibody use the imager with the appropriate filter settings (*see* Subheadings 2.5 and 2.7).

3.8 Quantification

1. Obtain high-resolution TIFF pictures (e.g. for scanners use 100 μm pixel size, for CCD-based cameras use the raw 16 bit data) rather than compressed JP(E)G files.
2. Pictures should not be too low or too saturated in contrast.
3. Excitation strength and exposure time should allow the strongest signal on the blot to be just below its saturation threshold.
4. Use the software of your choice for manual or automated analysis (Fig. 3c, d; *see* Subheading 2.8).

4 Notes

1. SMA basic is added to every reaction as the internal standard and required for loading normalization. The fluorescent internal standard protein SMA is available in two sizes (12.5 or 80 kDa) so that this reference can be adapted to the target protein sizes. Also if the SPL reaction is scaled up, the concentration of SMA basic should not be elevated since a signal that is too strong can result in oversaturation and therefore diminishing of signals detected in the same color channel. The SPL kit also contains a set of experimental standards (Cal A and Cal B). Cal A serves as an experiment-dependent standard for the quantified fluorescence signals and as a bi-fluorescent molecular weight protein marker. Since the binding efficiency of secondary antibodies might differ between experiments, Cal B represents a standard to determine the binding efficiency of the secondary antibody relative to the primary antibody. Together, the Cal signal intensities allow both blot-to-blot and experiment-to-experiment normalization and quantification. Cal is used for inter-experimental comparisons and lane-to-lane normalization relies on SMA loading controls and the total labeling.
2. The choice of the appropriate SPL Kit depends on the accessible fluorescence detection system and as well as the method of target detection. At least the detection of two different fluorescence channels is necessary, i.e. one channel for the detection of the total protein and a second channel for the detection of the basic fluorescence of the sample standard (SMA). To suit different fluorescent imaging devices available on the market and thus the method of target detection, the SPL kit is available in different color combinations (Table 1).
3. In general, the SPL technology is compatible with any kind of antibody depending on your preferred method of detection. We performed the SPL analysis with a series of commercially available primary antibodies (e.g. anti-GFP, Santa Cruz Biotechnology, sc-8334; anti-HA, Covance, MMS-101; further antibodies are listed in Table 1). However, in one experiment, we used a sheep antiserum raised against a peptide epitope that not properly recognized the SPL pre-labeled protein. Still, this seems to be an exception and rather due to the serum as only 3% of the total protein is pre-labeled by the fluorophore. A secondary antibody is chosen according to your needs (e.g. chemiluminescence vs. fluorescence). A standard HRP-coupled antibody offers the highest flexibility but requires an additional detection step not necessary with fluorescently labeled antibodies. Also the choice of the detection mode of the secondary antibody depends mainly on the abundance of the target protein. Fluorescence is less sensitive than

ECL. First, any kind of HRP/AP coupled standard antibody can be used and signals can be quantified with Immuno Blue or chemiluminescence. Despite its short signal stability, chemiluminescence detection should be chosen for very lowly abundant target proteins, whereas Immuno Blue provides signal stability even at room temperature over months. Second, also a secondary fluorescently labeled antibody is the most convenient way in terms of handling and also provides signal stability for months.

4. To obtain quantifiable pictures using standard ECL substrates, a chemiluminescence detection unit is needed (*see* Subheading 2.6). This is advantageous if the target protein migrates in the same range as the sample-dependent standard (SMA basic) since ECL and SMA basic are detected in different channels (ECL signal is a luminescence signal, whereas SMA is fluorescent).

   Since Immuno Blue and ECL substrates are compatible, a WB can first be quickly tested with ECL reagents. This might be desirable because ECL can be more sensitive compared to Immuno Blue dependent on the ECL reagents used. After a quick chemiluminescence detection by X-ray film or camera, the same membrane can be subjected to Immuno Blue staining. An ECL control should be done when using new antibodies to make sure that antibody binding is not influenced by the labeling. Rinse the membrane wetted in ECL reagents briefly in TBST prior to Immuno Blue staining.

5. Any kind of standard protein quantification method is compatible with SPL. Since the system is able to detect and to normalize samples differing in their total protein content, it is only important that the protein content of each sample stays below 20 μg of protein in a volume of up to 10 μL of sample. This is due to the fact that one single labeling reaction only contains enough dye to label 20 μg of total protein. Also, if 10 μL of protein solution in a 20 μL reaction are exceeded this would not leave enough room for a sufficient amount of SPL reaction buffer therefore resulting in incomplete labeling of total protein. Higher protein amounts have to be diluted or require an additional SPL reaction.

6. If 20 μg of total protein in one labeling reaction are not sufficient, e.g. if a very lowly abundant protein is to be analyzed, upscaling can be done almost without any limits. When working with samples with a very high protein content, e.g. 10 μg/μL, it is also possible to upscale reactions in a way that does not follow the initial reaction volume of 10 μg of total protein in a final reaction volume of 20 μL. For example, 30 μg of total protein can be labeled in a total reaction of 40 μL. When upscaling reactions, it is usually not necessary to upscale the sample-dependent standard (SMA basic) since its signal

intensity resulting from the amount used for one reaction is usually sufficient.

7. There are three different detection methods possible depending on your choice of antibody and substrate. In our example, we used an ECL control followed by Immuno Blue fluorescence detection of the target protein. When you detect your POI in the blue channel, you also have to record the red channel to monitor total protein content on the blot at the point of detection (BTO = blot total protein).
8. Immuno Blue is a chromogenic substrate that forms a blue fluorescent precipitate directly on the blot during incubation and allows the detection within the blue channel. Note that Immuno Blue is not compatible with the SPL Blue Kit, where the total protein is labeled in blue.

Acknowledgements

This work was supported by a grant for setting up the junior research group of the *ScienceCampus Halle—Plant-based Bioeconomy* to N.D., by a Ph.D. fellowship of the Landesgraduiertenförderung Sachsen-Anhalt and a grant of the German Academic Exchange Service (DAAD) awarded to F.F. Financial support came from the Leibniz Association, the state of Saxony Anhalt, the Deutsche Forschungsgemeinschaft (DFG) Graduate Training Center GRK1026 "*Conformational Transitions in Macromolecular Interactions*" at Halle, and the Leibniz Institute of Plant Biochemistry (IPB) at Halle, Germany. L. E.-L. is supported by the Protein-Kompetenznetzwerk-Halle "tools, targets & therapeutics"—ProNet-T^3 (03ISO2211B) funded by the Bundesministerium für Bildung und Forschung (BMBF).

References

1. Taylor SC, Posch A (2014) The design of a quantitative western blot experiment. Biomed Res Int 2014, Article ID 361590
2. Ferguson RE, Carrol HP, Harris A, Maher ER, Selby PJ, Banks RE (2005) Housekeeping proteins: a preliminary study illustrating some limitations as useful references in protein expression studies. Proteomics 5:566–572
3. Wilson CM (1979) Studies and critique of amido black 10B, Coomassie blue R, and fast green FCF as stains for proteins after polyacrylamide gel electrophoresis. Anal Biochem 96:263–278
4. Wilson CM (1983) Staining of proteins on gel: comparisons of dyes and procedures. Methods Enzymol 91:236–247
5. Colella AD, Chegenii N, Tea MN, Gibbins IL, Williams KA, Chataway TK (2012) Comparison of stain-free gels with traditional immunoblot loading control methodology. Anal Biochem 430:108–110
6. Gürtler A, Kunz N, Gomolka M, Hornhardt S, Friedl AA, McDonald K, Kohn JE, Posch A (2013) Stain-free technology as a normalization tool in Western blot analysis. Anal Biochem 433:105–111

Part IV

Bioinformatics Analysis

Chapter 21

Sequence Search and Comparative Genomic Analysis of SUMO-Activating Enzymes Using CoGe

Lorenzo Carretero-Paulet and Victor A. Albert

Abstract

The growing number of genome sequences completed during the last few years has made necessary the development of bioinformatics tools for the easy access and retrieval of sequence data, as well as for downstream comparative genomic analyses. Some of these are implemented as online platforms that integrate genomic data produced by different genome sequencing initiatives with data mining tools as well as various comparative genomic and evolutionary analysis possibilities.

Here, we use the online comparative genomics platform CoGe (http://www.genomevolution.org/coge/) (Lyons and Freeling. Plant J 53:661–673, 2008; Tang and Lyons. Front Plant Sci 3:172, 2012) (1) to retrieve the entire complement of orthologous and paralogous genes belonging to the SUMO-Activating Enzymes 1 (SAE1) gene family from a set of species representative of the Brassicaceae plant eudicot family with genomes fully sequenced, and (2) to investigate the history, timing, and molecular mechanisms of the gene duplications driving the evolutionary expansion and functional diversification of the SAE1 family in Brassicaceae.

Key words Comparative genomics, BLAST, Synteny, Brassicaceae, Whole genome duplication, Tandem duplication

1 Introduction

Homologous genes share a common ancestor, from which they have descended usually with divergence. Shared ancestry can be derived from [1] a duplication event, both at the level of entire genomes involving all genes (Whole Genome Duplications or polyplodizations, WGDs) or small regions containing one to a few genes (Small-Scale Genome Duplications, SGDs, including tandem duplications), as in paralogous genes; or [2] speciation, as in orthologous genes. Gene duplication provides new substrate for mutation and selection to act upon. In most cases, a new gene duplicate evolves neutrally, stochastically accumulating mutations and rapidly becoming a pseudogene that will be inactivated or even deleted from the genome. However, a fraction of duplicates might be retained through the acquisition of novel or

L. Maria Lois and Rune Matthiesen (eds.), *Plant Proteostasis: Methods and Protocols*, Methods in Molecular Biology, vol. 1450, DOI 10.1007/978-1-4939-3759-2_21,

specialized functions [1–3]. While orthologs are commonly believed to conserve ancestral functions more frequently than paralogs [4], functional diversification of orthologous genes is also a common phenomenon, and is thought to be driven by similar molecular evolutionary mechanisms [5]. Therefore, accurate classification of genes as orthologs or paralogs is critical for unraveling the evolutionary and functional diversification of specific gene families.

Homologous genes are expected to share some degree of conservation at the sequence level. To identify putative homologs based on sequence identity, different algorithms have been designed, such as Basic Local Alignment Search Tool (BLAST), which is among the most popular [6]. The program uses a nucleotide or amino acid sequence as a query to search against a sequence database for sequences showing a reasonable level of identity within the same or in different species. The algorithm is based on finding regions of local similarity between sequences (high-scoring segment pairs, HSP) and calculates the statistical significance of the resulting matches. To assess whether a given match constitutes evidence for homology, it helps to know how strong an alignment can be expected from chance alone. However, homology cannot be asserted from sequence identity alone, and additional evidences must be provided. One such evidence to help defining two genes as homologous is their occurrence in syntenic genomic blocks. Two genomic blocks (or entire chromosomes) are syntenic, i.e., derive from the same ancestral genomic region, if their genomic features, such as genes, are collinear and conserve significant sequence similarity. The algorithm used by the comparative genomics platform CoGe [7, 8] to identify syntenic regions between genomes is DAGchainer [9]. The DAGchainer software computes chains of syntenic genes found within complete genome sequences. It works by searching some distance between neighboring genes on each genomic block and applying a threshold BLAST *E*-value score between matches. If a number of gene pairs above a threshold are identified, DAGchainer computes and reports maximally scoring chains of ordered gene pairs. These sets of gene pairs are thus interpreted as two syntenic regions, either corresponding to large evolutionary conserved regions between the genomes of two different organisms, or reflecting WGD or SSD within a single genome.

2 Materials

You simply need a computer, a web browser, and a connection to the Internet. Mozilla Firefox is recommended (Google Chrome may also work properly) with the Adobe Flash Player system plug-in installed and Javascript, cookies, and popups enabled. For proper visualization of results, a large and high-resolution computer screen is preferred. Although CoGe can be accessed anonymously, obtaining a user account allows you additional capabilities, for example, exploring your saved analysis history.

3 Methods

3.1 Search for SAE1 Sequences in Selected Plant Genomes. The Case of Arabidopsis Thaliana

1. Go to the **FeatView** tool from **CoGe** and type the name of the *A. thaliana* gene to be used as a query (AtSAE1a: At4g24940, as in ref. 10).
2. In the "Organism Name" search box type *Arabidopsis thaliana*, and scroll down to select Col-0, corresponding to the strain (accession) to be examined. Press the "Search" button.
3. The matches to the query name will appear. There may be more than one version of the genome available. Select the desired version of the genome annotation on the "Genomes" column. Typically, a good selection is an unmasked version with coding-sequence gene models included. *See* Table 1 for a list of the genome versions used.
4. A window summarizing information about the queried gene will appear at the bottom of the page. Selecting **CoGeBLAST** in "CoGe links" will open a new tab to perform **BLAST** analysis on any selected genome, using the selected gene as a query.

Table 1
List of genome annotations used

Genome	CoGe id
Arabidopsis thaliana Col-0 v10.02 unmasked	16911
A. thaliana Bur-0 v1 unmasked (10,001 genomes)	11934
A. thaliana C24 v1 unmasked (10,001 genomes)	11933
A. thaliana Kro-0 v1 unmasked (10,001 genomes)	11935
A. thaliana Ler-1 v1 unmasked (10,001 genomes)	11937
A. thaliana Can-0 v7 unmasked (Wellcome trust)	20359
A. thaliana Ct-1 v7 unmasked (Wellcome trust)	20360
A. thaliana Edi-0 v7 unmasked (Wellcome trust)	20361
A. lyrata v1 unmasked (JGI)	3068
Carica papaya v0.5 unmasked (University of Hawaii)	9198
Brassica rapa v1.5 unmasked (Brassica DB - Chr)	24668
Capsella rubella v0.9 unmasked (JGI)	16754
Schrenkiella parvula (*Thelungiella parvula*, *Eutrema parvulum*) v2 unmasked (UIUC)	12384

Alternatively, you may also try to directly identify syntenic regions against any set of genomes using At4g24940 as a query and the **CoGe** tool **SynFind** (*see* **Note 1**).

5. In "Select Target Genomes" enter the name of the genomes to be searched, i.e., *Arabidopsis thaliana* Col-0. Add the selected version of the genome by clicking on the "+ Add" button (*see* Table 1). Choose the appropriate **BLAST** program to run (e.g., **BLASTN, TBLASTX, TBLASTN**), keeping the default **BLAST** parameters, and click on the "Run CoGe BLAST" button (*see* **Note 2**). The results of the **BLAST** search may be regenerated using this link: https://genomevolution.org/r/ghbw.
6. A sortable table listing an overview of BLAST hits (High-scoring Sequence Pairs; HSPs) in relationship to their genomic locations, coverage, resulting *E*-values and scores, percent ID, quality (*see* **Note 3**), and additional information will be generated. The viewable columns in the table can be set by clicking on the "change viewable columns" button at the bottom right of the table. The locations of the different HSPs in the queried genome(s) can be displayed in the "Genomic HSP Visualization" window, on top of the table. Select from the table the three first HSPs, which have the three *A. thaliana* SAE1 genes as "closest genomic features" (At4g24940, At5g50580, and At5g50680). At5g50580 and At5g50680 are located in close proximity on chromosome 5, likely corresponding to a tandem duplication, whereas At4g24940 is located on chromosome 4, likely arising from a duplicated block descending from a WGD event [10]. Clicking on the "closest genomic features" will open a window summarizing "feature information", including length, location, organism, genome, and "CoGe links".
7. Go to the bottom of the HSP table and *Send selected* genomic features to (1) **FASTAView**, and download the sequences as a FASTA formatted file (*see* **Note 4**), or (2) other tools within **CoGe** (i.e., **GEvo**) by choosing the appropriate item from the select list.
8. Use the **GEvo** (Genome Evolution Analysis) tool for in-depth examination of synteny between the genomic regions containing genes At4g24940, At5g50580, and At5g50680. As genes At5g50580 and At5g50680 are in close vicinity to each other on chromosome 5, we can skip the sequence corresponding to one or the other. To skip one particular sequence, use it as a reference sequence, or other options, click the "Sequence Options" button. **GEvo** lets you change the settings of the alignment algorithm ((B)lastZ for large regions is used by default [11]) and tweak the visualization of the results. For example, you can modify the length of the genomic regions to be

analyzed for each individual sequence by setting different values in the "left sequence" and "right sequence" boxes. You can also select the length of the sequences to be examined to be the same for all genomic regions by using the "Apply distance to all submissions" box at the bottom left of the sequence boxes. Furthermore, **CoGe** includes a TinyURL resource to get permanent links and regenerate results, which can be found at the bottom right of the results (below "Return to this analysis"). *See* results in Fig. 1.

3.2 Analysis of the Brassicaceae-Specific WGD Leading to the Duplicate Gene Pair SAE1a/SAE1b

1. Perform **steps 1–5** from Subheading 3.1.
2. In "Select Target Genomes" enter the name of the genomes to be searched. In this case, we will choose a representative set of five Brassicaceae species with whole genome sequences available (*A. thaliana*, *A. lyrata*, *Brassica rapa*, *Capsella rubella*, and *Thellungiella parvula*), plus *Carica papaya*, belonging to the order Brassicales, as an outgroup (Fig. 2). Add the selected versions of the genomes by clicking on the "+ Add" button (*see* Table 1). Keep the default BLAST parameters and click the

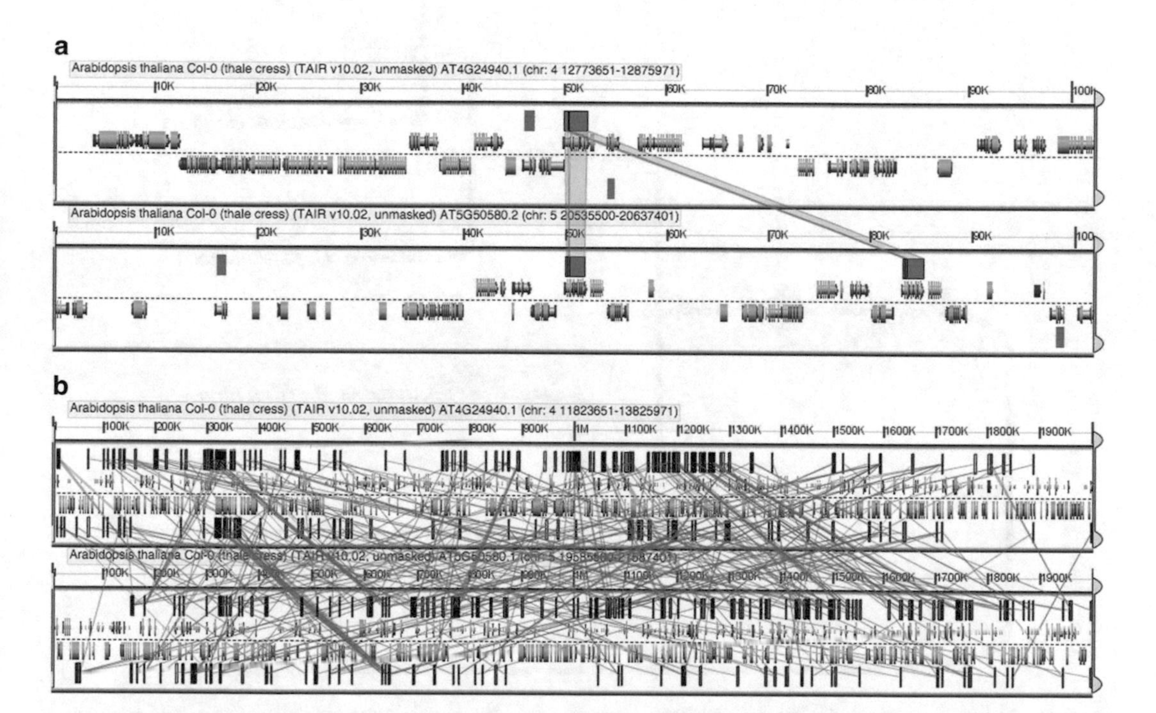

Fig. 1 High-resolution pair-wise comparison of *A. thaliana–A. thaliana* intragenomic syntenic regions of chromosomes 4 and 5 containing SAE1 genes. (**a**) 50 kb view. HSPs between SAE1 genes are shown by *red* connectors. Note the tandem duplication in chromosome 5 leading to At5g50580 and At5g50680 genes. (**b**) 1 Mb view. All HSPs are shown by *red* connectors. Note the series of collinear genes between the two regions suggesting their origin through WGD. These analyses may be regenerated following the links https://genomevolution.org/r/fnjf (50 kb) or https://genomevolution.org/r/fnj8 (1 Mb)

"Run CoGe BLAST" button (follow this link to regenerate results: https://genomevolution.org/r/gf8u).

3. Select the HSPs resulting in significant *E*-values (i.e., >1e−10) (*see* **Note 5**). *A. lyrata*, *C. rubella*, and *S. parvula* have two SAE1 genes each, *B. rapa* has five sequences showing significant similarity with SAE1 genes, while *C. papaya* has only one.
4. "Send selected" genomic features to **FASTAView** and download the sequences as a FASTA formatted file (*see* **Note 4**).
5. Use the **GEvo** tool for detecting synteny between the genomic regions containing SAE1 genes. By opening "Sequence Options", *A. thaliana* Col-0 genes were selected as reference sequences, and noncoding regions were masked from *A. thaliana* Col-0 genomic blocks. For better visualization, all HSPs were drawn on top and masked/unsequenced nucleotides were not colored, using the "Results Visualization Options" menu. Pair-wise comparative genomics analysis between *A. thaliana* Col-0 and the remaining five species are shown in Figs. 3, 4, 5, 6, and 7. Results permit the following

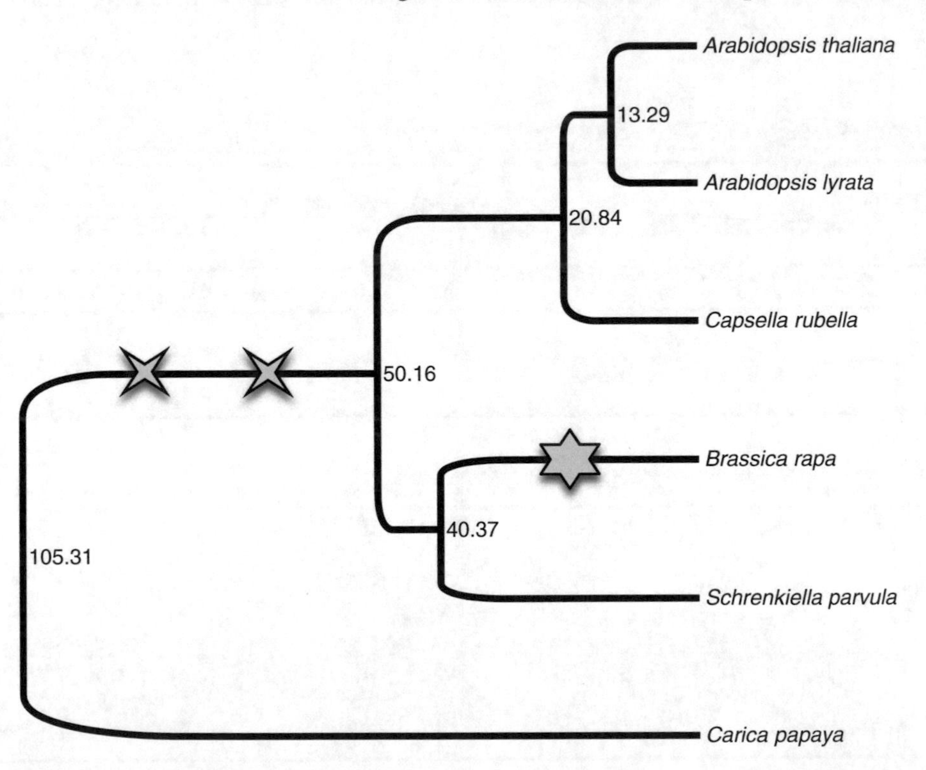

Fig. 2 Taxonomic relationships of five Brassicaceae species. Phylogenetic tree summarizing the taxonomic relationships among five Brassicaceae species with fully sequenced genomes, plus *Carica papaya* from the order Brassicales used as out-group. Branch lengths reflect evolutionary time (in millions of years). Divergence times are shown at internal nodes. For the timing of these events, as well as for the tree topology, we used estimates from [16]. The history of WGDs is mapped onto the tree, with the four-pointed and six-pointed stars representing WGDs and triplications, respectively. The positions of these events are not meant to reflect their precise timings of occurrence

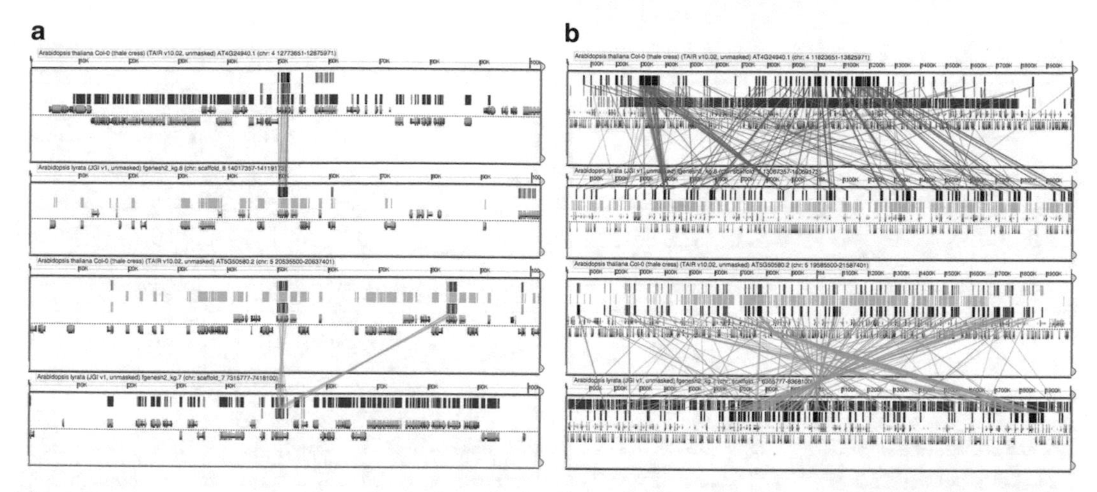

Fig. 3 High-resolution pair-wise comparison of *A. thaliana–A. lyrata* intergenomic syntenic regions containing SAE1 genes. (**a**) 50 kb view. HSPs involving *A. thaliana* SAE1 genes are shown by colored connectors. Note the 1:1 relationship between *A. thaliana* At4g24940 and *A. lyrata* fgenesh2_kg.8__940__AT5G50580, as well as the 2:1 relationship between *A. thaliana* At5g50580 and At5g50680 and *A. lyrata* fgenesh2_kg.7__1777__AT4G24940, suggesting the tandem duplication leading to the duplicate gene pair At5g50580 and At5g50680 occurred after evolutionary divergence between the two species. (**b**) 1 Mb view. All HSPs are shown by colored connectors. Note, in each comparison, the series of collinear genes between the two regions, suggesting their origin through a WGD occurring before the divergence of the two species. These analyses may be regenerated following the links https://genomevolution.org/r/h0zw (50 kb) or https://genomevolution.org/r/h10q (1 Mb)

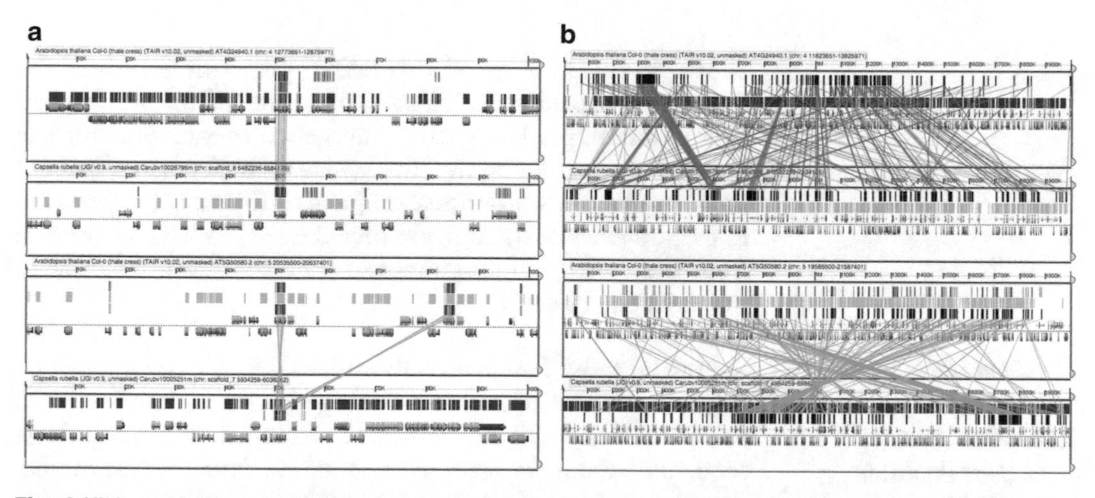

Fig. 4 High-resolution pair-wise comparison of *A. thaliana–Capsella rubella* intergenomic syntenic regions containing SAE1 genes. (**a**) 50 kb view. HSPs involving *A. thaliana* SAE1 genes are shown by colored connectors. Note the 1:1 relationship between *A. thaliana* At4g24940 and *C. rubella* Carubv10026795m.g, as well as the 2:1 relationship between *A. thaliana* At5g50580 and At5g50680 and *C. rubella* Carubv10005251m.g, suggesting the tandem duplication leading to the duplicate gene pair At5g50580 and At5g50680 occurred after evolutionary divergence between the two species. (**b**) 1 Mb view. All HSPs are shown by colored connectors. Note, in each comparison, the series of collinear genes between the two regions, suggesting their origin through a WGD occurring before the divergence of the two species. These analyses may be regenerated following the links https://genomevolution.org/r/h10s (50 kb) or https://genomevolution.org/r/h10v (1 Mb)

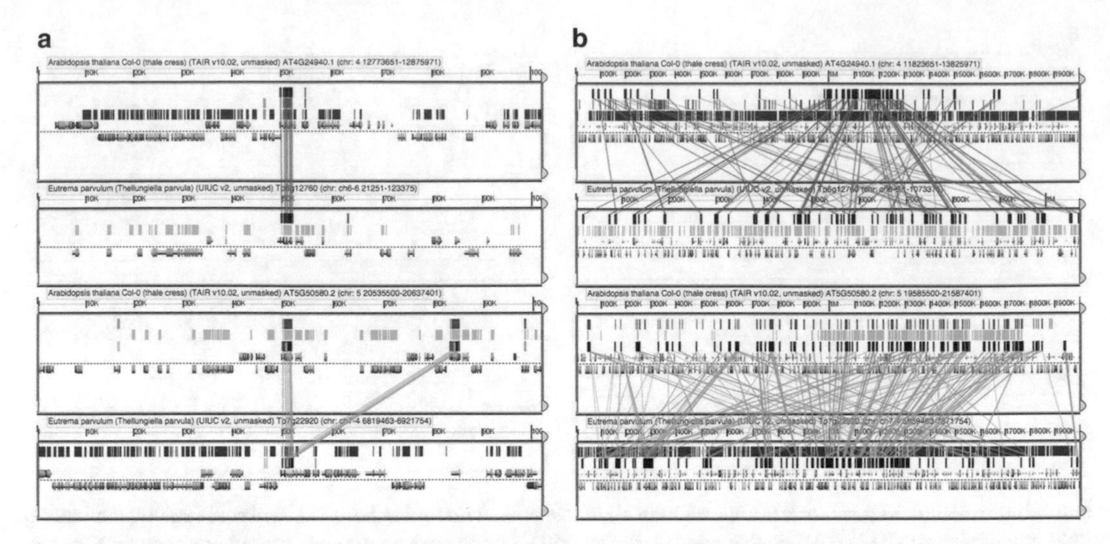

Fig. 5 High-resolution pair-wise comparison of *A. thaliana–Schrenkiella parvula* intergenomic syntenic regions containing SAE1 genes. (**a**) 50 kb view. HSPs involving *A. thaliana* SAE1 genes are shown by colored connectors. Note the 1:1 relationship between *A. thaliana* At4g24940 and *S. parvula* Tp6g12760, as well as the 2:1 relationship between *A. thaliana* At5g50580 and At5g50680 and *S. parvula* Tp7g22920, suggesting the tandem duplication leading to the duplicate gene pair At5g50580 and At5g50680 occurred after evolutionary divergence between the two species. (**b**) 1 Mb view. All HSPs are shown by colored connectors. Note, in each comparison, the series of collinear genes between the two regions, suggesting their origin through a WGD occurring before divergence of the two species. These analyses may be regenerated following the links https://genomevolution.org/r/h10z (50 kb) or https://genomevolution.org/r/h110 (1 Mb)

conclusions: (1) two of the SAE1 genes detected in the Brassicaceae species likely result from the WGD (polyploidization) event predating the emergence of the Brassicaceae lineage, but postdating the divergence with *C. papaya*, and (2) three of *B. rapa*'s additional sequences (likely corresponding to pseudogenes or wrongly predicted gene models) likely arose from the recent whole genome triplication event specific to that lineage [12].

3.3 Analysis of the Tandem Duplication Leading to the Gene Duplicate Pair At5g50580–At5g50680 in A. Thaliana Col-0

1. Following **steps 1–5** from Subheading 3.1, use the nucleotide sequences of *A. thaliana* Col-0 At5g50580 and At5g50680 to perform independent BLASTN searches of the whole sequenced genomes of seven other *A. thaliana* ecotype/accessions (*see* Table 1) [13–15]., In each of the seven genomes, a single region was retrieved as best hit using both queries (follow these links to regenerate results: https://genomevolution.org/r/ghfq and https://genomevolution.org/r/ghfv).
2. Use the **GEvo** tool for detecting synteny between the genomic regions containing SAE1 genes. By using the "Sequence Options" submenu, *A. thaliana* Col-0 genes were selected as reference sequences and noncoding regions were masked from *A. thaliana* Col-0 genomic blocks. Pair-wise comparative

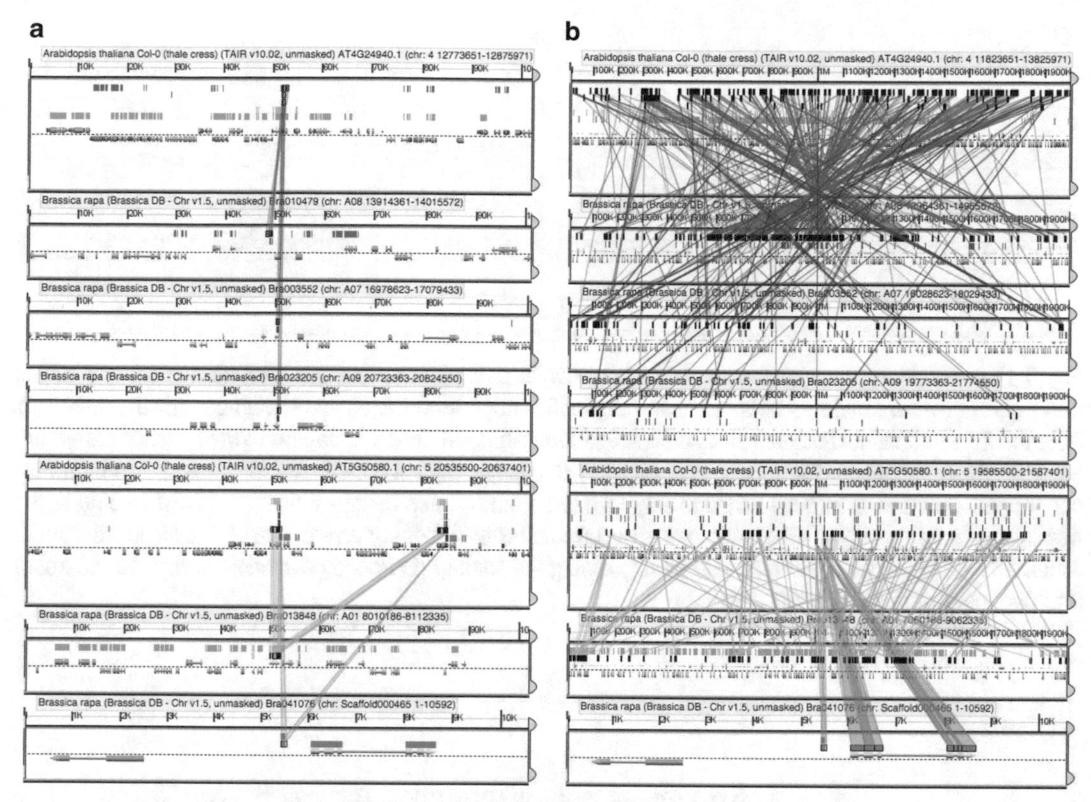

Fig. 6 High-resolution pair-wise comparison of *A. thaliana–Brassica rapa* intergenomic syntenic regions containing SAE1 genes. (**a**) 50 kb view. HSPs involving *A. thaliana* SAE1 genes are shown by colored connectors. Note the 1:3 relationship between *A. thaliana* At4g24940 and *B. rapa* Bra010479, Bra023205, and Bra003552, as well as the 2:1 relationship between *A. thaliana* At5g50580/At5g50680 and *B. rapa* Bra013848 and Bra041076; the latter suggesting the tandem duplication leading to the duplicate gene pair At5g50580/At5g50680 occurred after divergence between the two species. A closer inspection of Bra023205, Bra003552, and Bra041076 reveals that they encode shorter sequences, while Bra010479 encodes a longer sequence, likely corresponding to pseudogenes or wrongly predicted gene models. (**b**) 1 Mb view. All HSPs are shown by colored connectors. Note the series of collinear genes between the genomic region containing At4g24940 and the genomic regions of genes Bra010479, Bra023205, and Bra003552, suggesting the three *B. rapa* genes arose from the recent whole genome triplication event specific to that lineage [12]. Also, note the series of collinear genes between the At5g50580/At5g50680 pair and Bra013848 genomic regions, suggesting their origin through a WGD occurring before evolutionary divergence between the two species. The genomic block containing Bra041076 is only 10,592 bp long, thus synteny with the At5g50580/At5g50680 genomic block cannot be discerned. These analyses may be regenerated following the links https://genomevolution.org/r/h116 (50 kb) or https://genomevolution.org/r/h118 (1 Mb)

genomics analysis between *A. thaliana* and seven different *A. thaliana* ecotype/accessions are shown in Fig. 8. From these results it can be concluded that the tandem duplication giving rise to AT5G50580 and AT5G50680 occurred recently in the Col-0 lineage, after divergence from other closely related accessions/ecotypes. This conclusion is further supported by the fact that both genes have identical coding sequences, revealing their recent origin.

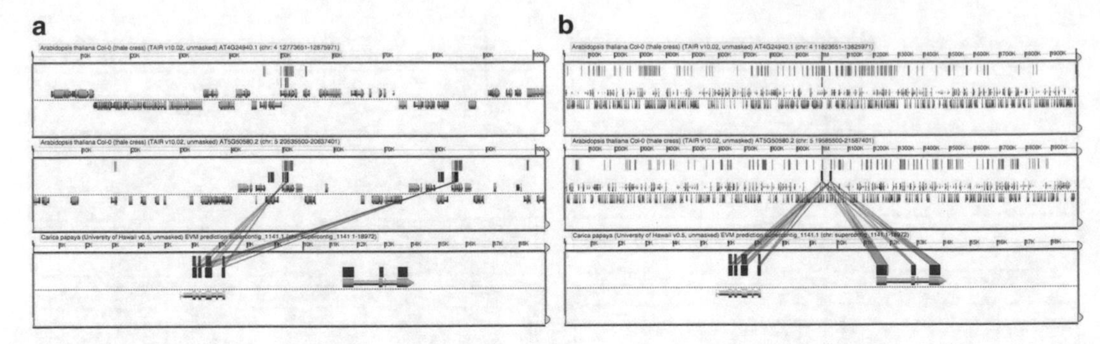

Fig. 7 High-resolution comparison of *A. thaliana–Carica papaya* intergenomic syntenic regions containing SAE1 genes. (**a**) 50 kb view. HSPs involving At5g50580/At5g50680 genes are shown by brown connectors. Note the 2:1 relationship between *A. thaliana* At5g50580/At5g50680 and *C. papaya* EVM prediction supercontig_1141.1. No additional SAE1 gene was found in *C. papaya*. (**b**) 1 Mb view. All HSPs are shown by brown connectors. Note the *C. papaya* contig is only 18,972 bp long. These results suggest the WGD leading to the two genomic regions containing SAE1 genes occurred after evolutionary divergence of Brassicaceae from *C. papaya*. These analyses may be regenerated following the links https://genomevolution.org/r/h11a (50 kb) or https://genomevolution.org/r/h11b (1 Mb)

4 Notes

1. **SynFind** uses an algorithm known as Synteny Score [2] to identify syntenic regions. The results are shown in a table listing all matched regions with their syntenic scores, and whether any syntenic gene was identified in each searched genome. A link to **GEvo** is provided with the results, permitting downstream high-resolution analysis of the detected syntenic regions.
2. Alternatively, you can skip **steps 1–4** by going directly to **CoGeBLAST** and entering the sequence(s) to be used as query in FASTA format in the "Query Sequence(s)" box.
3. For a tutorial on **BLAST** statistics similarity scores, please visit http://www.ncbi.nlm.nih.gov/BLAST/tutorial/Altschul-1.html.
4. The sequence files in FASTA format will be used in the next chapter to perform multiple sequence alignments and phylogenetic analysis.
5. There is no single criterion to define a threshold to infer "homology" from similarity. It depends on the degree of sequence divergence for that particular gene family and the knowledge of the researcher about the gene sequences belonging to the family being examined. For instance, sequences below the selected *E*-value threshold have a length different to what is expected for a SAE1 gene, which may be an indication of that (1) particular sequence is not homologous, or (2) it corresponds to a wrongly annotated gene or pseudogene.

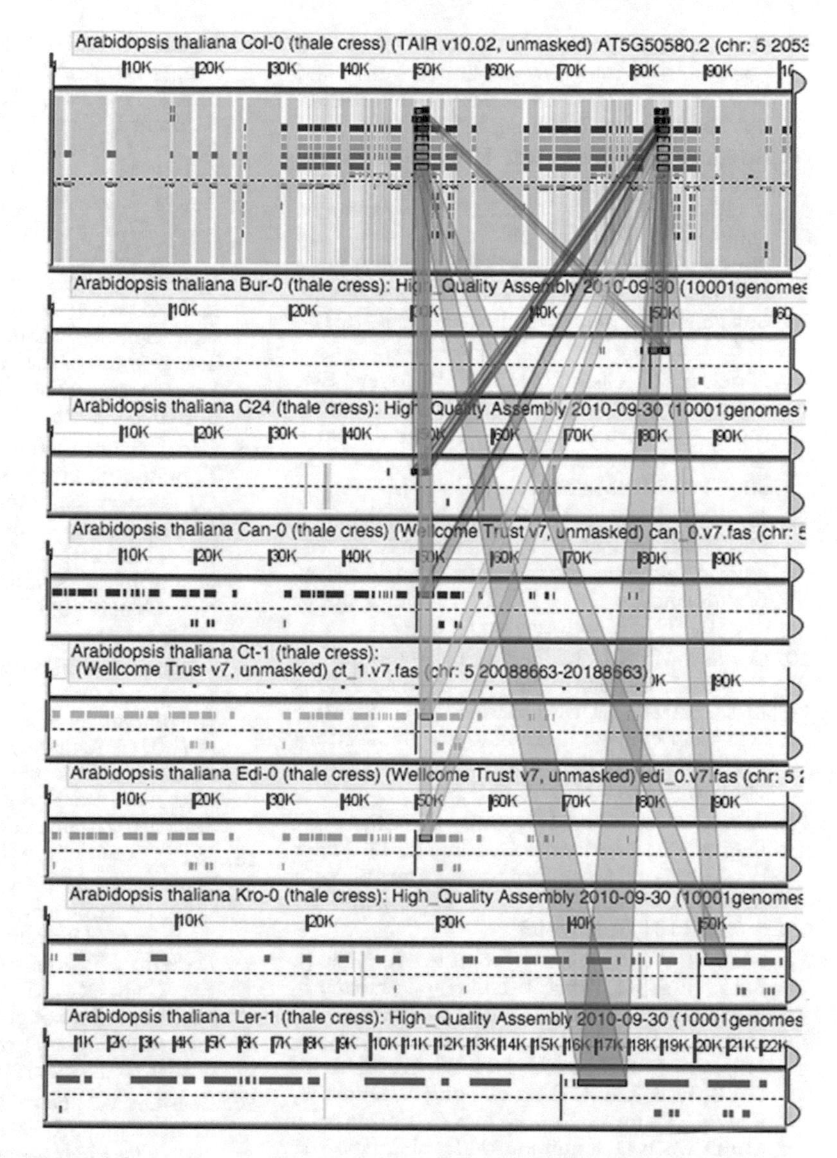

Fig. 8 High-resolution syntenic analysis of *A. thaliana* SAE1b tandem duplication. 50 kb view. HSPs involving *A. thaliana* SAE1 genes are shown by colored connectors. Note the 2:1 relationship between *A. thaliana* Col-0 At5g50580/At5g50680 and single genes in seven other *A. thaliana* accessions (ecotypes), including Bur-0, C24, Kro-0, Ler-1, Can-0, Ct-1, and Edi-0. This analysis may be regenerated following the link https://genomevolution.org/r/gf2d (50 kb)

References

1. Lyons E, Freeling M (2008) How to usefully compare homologous plant genes and chromosomes as DNA sequences. Plant J 53:661–673
2. Tang H, Lyons E (2012) Unleashing the genome of brassica rapa. Front Plant Sci 3:172. doi:10.3389/fpls.2012.00172
3. Ohno S (1970) Evolution by gene duplication. Springer, New York, NY
4. Zhang J (2003) Evolution by gene duplication: an update. Trends Ecol Evol 18(6): 292–298. doi:10.1016/s0169-5347(03)00033-8

5. Conant GC, Wolfe KH (2008) Turning a hobby into a job: how duplicated genes find new functions. Nat Rev Genet 9(12):938–950. doi:10.1038/nrg2482, nrg2482 [pii]
6. Koonin EV (2005) Orthologs, paralogs, and evolutionary genomics. Annu Rev Genet 39:309–338. doi:10.1146/annurev.genet.39.073003.114725
7. Studer RA, Robinson-Rechavi M (2009) How confident can we be that orthologs are similar, but paralogs differ? Trends Genet 25(5):210–216. doi:10.1016/j.tig.2009.03.004
8. Altschul SF, Gish W, Miller W, Myers EW, Lipman DJ (1990) Basic local alignment search tool. J Mol Biol 215(3):403–410. doi:10.1016/S0022-2836(05)80360-2, S0022-2836(05)80360-2 [pii]
9. Haas BJ, Delcher AL, Wortman JR, Salzberg SL (2004) DAGchainer: a tool for mining segmental genome duplications and synteny. Bioinformatics 20(18):3643–3646. doi:10.1093/bioinformatics/bth397
10. Castano-Miquel L, Segui J, Manrique S, Teixeira I, Carretero-Paulet L, Atencio F, Lois LM (2013) Diversification of SUMO-activating enzyme in Arabidopsis: implications in SUMO conjugation. Mol Plant 6(5):1646–1660. doi:10.1093/mp/sst049
11. Schwartz S, Kent WJ, Smit A, Zhang Z, Baertsch R, Hardison RC, Haussler D, Miller W (2003) Human-mouse alignments with BLASTZ. Genome Res 13(1):103–107. doi:10.1101/gr.809403
12. Wang X, Wang H, Wang J, Sun R, Wu J, Liu S, Bai Y, Mun JH, Bancroft I, Cheng F, Huang S, Li X, Hua W, Wang J, Wang X, Freeling M, Pires JC, Paterson AH, Chalhoub B, Wang B, Hayward A, Sharpe AG, Park BS, Weisshaar B, Liu B, Li B, Liu B, Tong C, Song C, Duran C, Peng C, Geng C, Koh C, Lin C, Edwards D, Mu D, Shen D, Soumpourou E, Li F, Fraser F, Conant G, Lassalle G, King GJ, Bonnema G, Tang H, Wang H, Belcram H, Zhou H, Hirakawa H, Abe H, Guo H, Wang H, Jin H, Parkin IA, Batley J, Kim JS, Just J, Li J, Xu J, Deng J, Kim JA, Li J, Yu J, Meng J, Wang J, Min J, Poulain J, Wang J, Hatakeyama K, Wu K, Wang L, Fang L, Trick M, Links MG, Zhao M, Jin M, Ramchiary N, Drou N, Berkman PJ, Cai Q, Huang Q, Li R, Tabata S, Cheng S, Zhang S, Zhang S, Huang S, Sato S, Sun S, Kwon SJ, Choi SR, Lee TH, Fan W, Zhao X, Tan X, Xu X, Wang Y, Qiu Y, Yin Y, Li Y, Du Y, Liao Y, Lim Y, Narusaka Y, Wang Y, Wang Z, Li Z, Wang Z, Xiong Z, Zhang Z, Brassica rapa Genome Sequencing Project C (2011) The genome of the mesopolyploid crop species Brassica rapa. Nat Genet 43(10):1035–1039. doi:10.1038/ng.919
13. Cao J, Schneeberger K, Ossowski S, Gunther T, Bender S, Fitz J, Koenig D, Lanz C, Stegle O, Lippert C, Wang X, Ott F, Muller J, Alonso-Blanco C, Borgwardt K, Schmid KJ, Weigel D (2011) Whole-genome sequencing of multiple Arabidopsis thaliana populations. Nat Genet 43(10):956–963. doi:10.1038/ng.911
14. Schneeberger K, Ossowski S, Ott F, Klein JD, Wang X, Lanz C, Smith LM, Cao J, Fitz J, Warthmann N, Henz SR, Huson DH, Weigel D (2011) Reference-guided assembly of four diverse Arabidopsis thaliana genomes. Proc Natl Acad Sci U S A 108(25):10249–10254. doi:10.1073/pnas.1107739108
15. Gan X, Stegle O, Behr J, Steffen JG, Drewe P, Hildebrand KL, Lyngsoe R, Schultheiss SJ, Osborne EJ, Sreedharan VT, Kahles A, Bohnert R, Jean G, Derwent P, Kersey P, Belfield EJ, Harberd NP, Kemen E, Toomajian C, Kover PX, Clark RM, Ratsch G, Mott R (2011) Multiple reference genomes and transcriptomes for Arabidopsis thaliana. Nature 477:419. doi:10.1038/nature10414, nature10414 [pii]
16. Beilstein MA, Nagalingum NS, Clements MD, Manchester SR, Mathews S (2010) Dated molecular phylogenies indicate a Miocene origin for Arabidopsis thaliana. Proc Natl Acad Sci U S A 107(43):18724–18728. doi:10.1073/pnas.0909766107

Chapter 22

Studying Evolutionary Dynamics of Gene Families Encoding SUMO-Activating Enzymes with SeaView and ProtTest

Lorenzo Carretero-Paulet and Victor A. Albert

Abstract

Molecular evolutionary analysis of gene families commonly involves a sequence of steps including multiple sequence alignment (MSA) and reconstructing phylogenetic trees, using any of the multiple algorithms available. **SeaView** is a multiplatform program that integrates different methods for performing the above tasks, and others, within a friendly and simple-to-use graphical user interface (Gouy et al. Mol Biol Evol 27(2):221–224, 2010). By using **SeaView**, we will investigate the evolutionary relationships among SAE1 genes in Brassicaceae species by means of two alternative methods of phylogenetic reconstruction: Maximum Likelihood (ML) and Neighbor-Joining (NJ). Prior to ML phylogenetic analysis (Guindon and Gascuel. Syst Biol 52(5):696–704, 2003), we will use **ProtTest** to select the best-fit evolutionary model of amino acid substitution for the MSA of SAE1 proteins (Abascal et al. Bioinformatics 21(9):2104–2105, 2005).

Key words Phylogenetic analysis, Multiple sequence alignment, Evolution model, Maximum likelihood, Neighbor joining

1 Introduction

Phylogenetic trees can be used to depict the evolutionary relationship among genes, represented by a set of aligned sequences, i.e., the order in which they are believed to have diverged through evolution. Different methods of phylogenetic reconstruction are available, resulting in potentially different phylogenetic hypothesis that may or may not agree with the true phylogenetic tree. Phylogenetic reconstruction methods can be classified as (1) based on distance-matrices, calculated directly from distances (similarities or identities) counted on pairwise aligned sequences (e.g., the Unweighted Pair Group Method With Arithmetic Means, UPGMA, probably being the simplest and oldest method [1], or NJ [2]); or (2) based on character-states, which consider each position in the alignment as a character, and a particular nucleotide or amino acid site at that position a state, and work by reconstructing the states of ancestral

L. Maria Lois and Rune Matthiesen (eds.), *Plant Proteostasis: Methods and Protocols*, Methods in Molecular Biology, vol. 1450, DOI 10.1007/978-1-4939-3759-2_22, © Springer Science+Business Media New York 2016

nodes (e.g., Maximum Parsimony (MP) [3] or ML [4]). The latter methods generate many possible trees, and compare them by testing them against some optimality criterion. The criterion used by MP to search tree space for the "best" tree is to look for the tree that requires the fewest number of changes during the evolution of the sequences involved. Similarly, the optimality criterion used by ML is to search the most likely tree by comparing the likelihood of any particular tree topology, which is maximized with respect to the branch lengths and the parameters of a probabilistic substitution model. Thus, prior to the ML analysis, the probability model of nucleotide or amino acid substitution (an evolutionary model) that best fits the sequence alignment under study must be selected. The ML algorithm will then compute the probability, under the selected evolutionary model, of observing the data at hand given each possible nucleotide or amino acid reconstruction at every ancestral node, which makes the actual process computationally demanding. A related probabilistic method for constructing phylogenetic trees is Bayesian methodology, which uses Markov chain Monte Carlo (MCMC) approaches to compute model parameters.

Any phylogenetic hypothesis based on nucleotide or amino acid sequence data requires comparing homologous positions (i.e., those that descend from a common ancestral position). Therefore, to obtain meaningful phylogenetic trees, it is essential to generate a reliable MSA in which homologous positions are arranged in columns. Finally, different statistical approaches are used to get some measure of confidence in our inferred tree. Bootstraping by randomly re-sampling columns in the alignment is the most widely used [5]. Different bootstrap replicate trees are generated, and the number of times the original tree topology is retrieved provides an intuitive measure of confidence in that tree. However, because ML methods are computationally intensive, this is sometimes impractical and alternative methods are commonly preferred.

2 Materials

You simply need a computer, a web browser, and a connection to the internet. For proper visualization of results, a large and high-resolution computer screen is preferred. Download the latest version of **SeaView** for your platform (MacOS X, Windows or Linux are available) here: http://doua.prabi.fr/software/seaview. You can download the package with the latest version of the software **ProtTest** here: https://github.com/ddarriba/prottest3 and decompress it in any directory of your computer. It is recommended to read the README file included in the package prior to first execution. The programs are written in JAVA, so they should work in Mac OSX, Windows, and Linux computers with a version of the Java Runtime Environment equal or posterior to 1.6 installed. Please, visit https://github.com/ddarriba/prottest3/blob/master/INSTALL for further instructions.

3 Methods

3.1 Sequence Edition and Multiple Alignment with SeaView

1. Perform **steps 1–4** from Subheading 3.1 of the previous chapter.
2. Rename the file as, e.g., BrassicaSAE1.fna. Open the file with your default text editor and examine the FASTA format. Each sequence begins with a single-line description (defline), followed by lines of sequence data. The defline contains a greater-than (">") symbol in the first column, which distinguishes it from the sequence data (Please *see* **Note 1**). Every new sequence starts with a ">" symbol and its corresponding defline.
3. Open the BrassicaSAE1.fna file with **SeaView** [6]. Using the "Rename Sequence" command from the "Edit" menu, change the sequence names to the new ones in Table 1.

Table 1
List of old and new names of gene sequences used

Organism	Old name	New name
Arabidopsis lyrata	fgenesh2_kg.8__940__AT5G50580.2	AlyrSAE1b
Arabidopsis lyrata	fgenesh2_kg.7__1777__AT4G24940.1	AlyrSAE1a
Arabidopsis thaliana Col-0	AT5G50580.1	AthaSAE1b1
Arabidopsis thaliana Col-0	AT5G50680.1	AthaSAE1b2
Arabidopsis thaliana Col-0	AT4G24940.1	AthaSAE1a
Brassica rapa	Bra013848	BrapSAE1a
Brassica rapa	Bra041076	BrapSAE1b4
Brassica rapa	Bra023205	BrapSAE1b2
Brassica rapa	Bra010479	BrapSAE1b1
Brassica rapa	Bra003552	BrapSAE1b3
Capsella rubella	Carubv10026795m	CrubSAE1b
Capsella rubella	Carubv10005251m	CrubSAE1a
Carica papaya	EVM prediction supercontig_1141.1	CpapSAE1
Eutrema parvulum	Tp6g12760	EparSAE1b
Eutrema parvulum	Tp7g22920	EparSAE1a

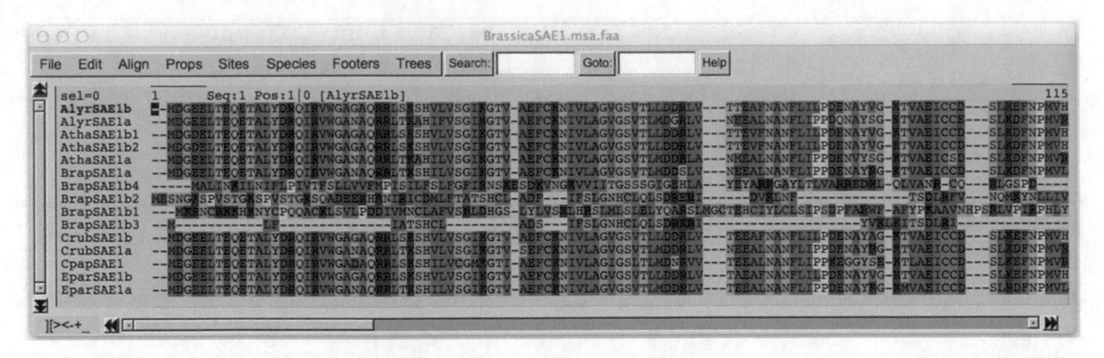

Fig. 1 Screenshot view of the "Alignment panel" in SeaView v4.5.2

4. Click on the "View as proteins" command from the "Props" menu to display nucleotide sequences as translated protein sequences. Save the resulting file as BrassicaSAE1.faa using the "Save prot alignmt" command from the "File" menu.
5. Go to the "Edit" menu and click on "Select all" sequences.
6. In the "Align" menu, go to "Alignment options" and select **muscle** [7] as alignment program. Alternatively, you can choose **clustalo**, which stands for **CLUSTAL-OMEGA** [8], and compare the results obtained using the two different alignment programs. Clicking on "Edit options" will open a command line window where you can enter additional optional arguments, which will be transmitted to the alignment program (please *see* **Note 2**). Click on "Align all" to run the alignment. Save the resulting file as BrassicaSAE1.msa.faa using the "Save prot alignmt" command from the "File" menu.
7. Examine the resulting alignment using the "Alignment panel" (Fig. 1). AthaSAE1b1 (AT5G50580) and AthaSAE1b2 (AT5G50680) show identical sequences, revealing their recent origin through tandem duplication [9]. BrapSAE1b1–4 sequences appear as particularly diverged at the sequence level, likely corresponding to pseudogenes or wrongly predicted gene models. They were selected and discarded by using the "Delete sequence(s)" item from the "Edit" menu. Save the resulting file as BrassicaSAE1bis.msa.faa using the "Save prot alignmt" command from the "File" menu.

3.2 Selection of Best-fit Model of Amino Acid Substitution Using ProtTest

1. Execute the graphical user interface (GUI) version of ProtTest by double-clicking the jar file (**ProtTest.jar**) (*see* **Note 3**) [10]. "Load" the MSA file BrassicaSAE1bis.msa.faa using the "File" menu (*see* **Note 4**). Go to the "Analysis" menu and click on "Compute likelihood scores". A window showing the "Computation Options" will open (Fig. 2).
2. By default, **ProtTest** uses the whole computing resources, i.e., the maximum number of processors (cores) available in your machine. **ProtTest** will select the best-fit model of amino acid

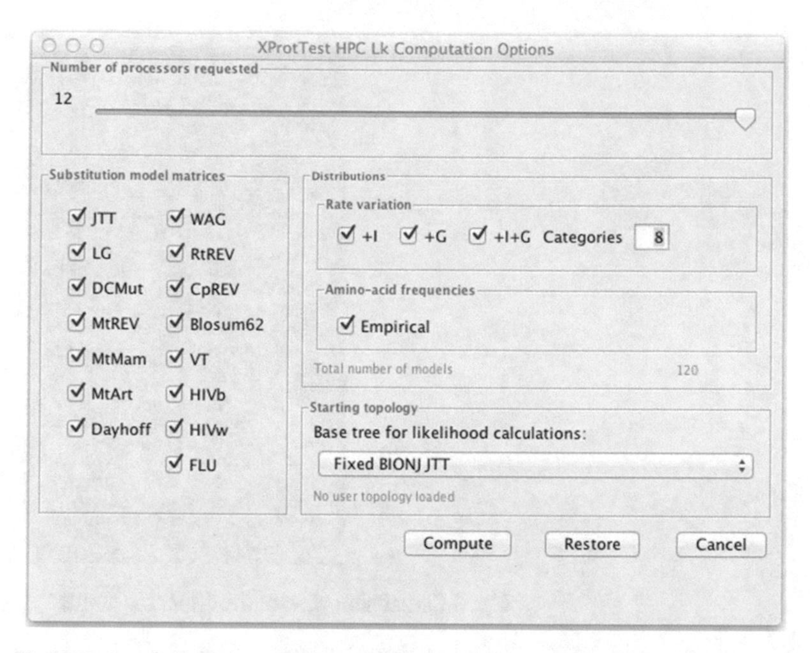

Fig. 2 Screenshot view of the Computation Options panel in ProtTest v3.2

substitution by finding the model in the candidate list with the smallest Akaike Information Criterion (AIC) [11]. Alternative "Model Selection Criteria" available are corrected AIC (AICc) [12], Bayesian Information Criterion (BIC) [13] score, or Decision Theory Criterion (DT) [14]. The version of **ProtTest** used here (3.2) includes by default 15 different amino acid substitution matrices. It is expected that the process of amino acid substitution is heterogeneous across sequences, because of different structural/functional domains of the protein which are subjected to different selective constraints. This evolutionary information can be modeled by **ProtTest** by considering a fraction of amino acids to be invariable (+**I**: invariable sites) [15], assigning each site a probability to belong to given rate categories (+**G**: gamma-distributed rates) [16] and estimating the observed amino acid frequencies (+**F**) [17]. Thus, when we consider +**I**, +**G,** and +**F**, 120 different models can be tested [18]. Optionally, you may select a subset of candidate substitution model matrices by marking/unmarking the desired ones. Additionally, you can also select/unselect to estimate different parameters of the candidate models. We will change the number of categories used to model rate variation to 8 (+**G**), considering models with a proportion of invariant sites (+**I**) and observed amino acid frequencies (+**F**) (*see* **Note 5**).

3. **ProtTest** also allows different options to construct the guide tree. We will keep the default ("Fixed BIONJ JTT") (*see* **Note 6**). Alternatively, you can input a tree topology in Newick format (*see* **Note 7**).

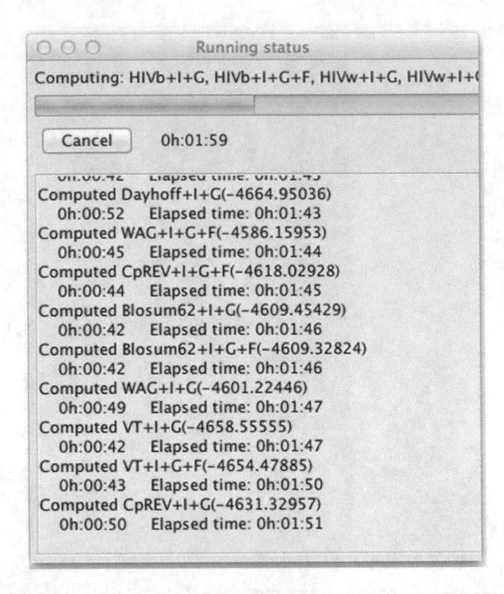

Fig. 3 Screenshot view of the "Running status" window in ProtTest v3.2

4. Click on the "Compute" button. A "Running status" window will pop up while the parameters of the different models are computed under ML (Fig. 3).
5. Once the computations are finished, the results will become accessible under the "Selection Results" menu (Fig. 4). The different selection models can be listed according to their scores under the different "Model Selection Criteria". The **JTT** [19] +**G** (gamma shape -8 rate categories-: 0.872) model was selected under AIC, BIC, AICc, and DT criteria as the best-fit model for the MSA. Every time you select a "Model Selection Criterion" click on "Export to main console" and "Save console" under the "File" menu of the main console as a text file. Take a look at the contents of the file with a text editor and check the parameter estimates of the best-fitting model.

3.3 Phylogenetic Analysis Using PhyML (ML) and NJ

1. Open the MSA file BrassicaSAE1.msa.faa in **SeaView**.
2. Open the "Trees" menu in **SeaView** and click on the "PhyML" item [20]. This will open a "PhyML options" window with different options to reconstruct a ML phylogenetic tree using **PhyML** (Fig. 5a) and the evolutionary model selected by **ProtTest**.
   - *Model*: select the **JTT** evolutionary model, retrieved by **ProtTest** as the best-fitting one.
   - *Branch support*: To estimate the statistical support of the retrieved topology by means of the Shimodaira-Hasegawa-like approximate likelihood ratio test [21] select "aLRT (SH-like)" (*see* **Note 8**).
   - *Nucleotide/Amino acid equilibrium frequencies*: Select "Model-given" to fix amino acid frequencies to the set of

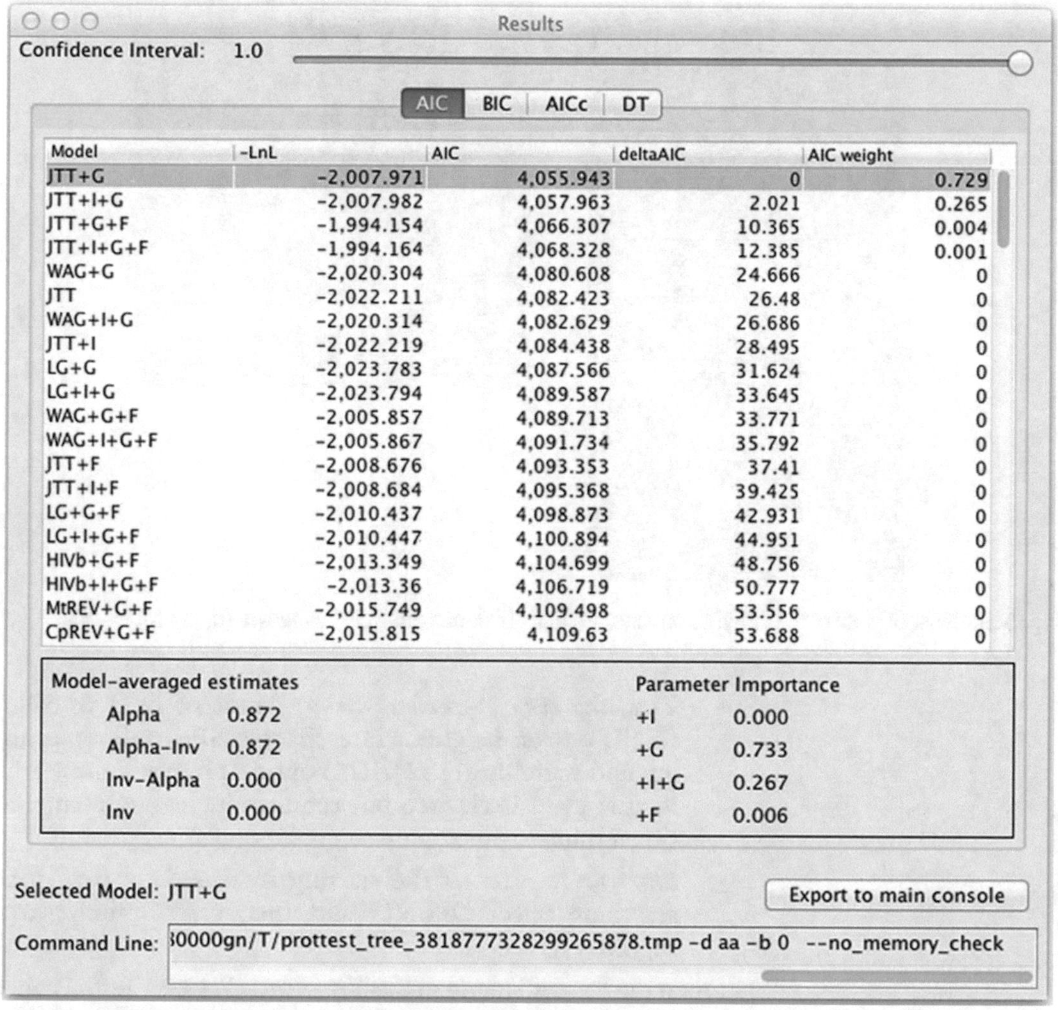

Results

Confidence Interval: 1.0

AIC | BIC | AICc | DT

Model	-LnL	AIC	deltaAIC	AIC weight
JTT+G	-2,007.971	4,055.943	0	0.729
JTT+I+G	-2,007.982	4,057.963	2.021	0.265
JTT+G+F	-1,994.154	4,066.307	10.365	0.004
JTT+I+G+F	-1,994.164	4,068.328	12.385	0.001
WAG+G	-2,020.304	4,080.608	24.666	0
JTT	-2,022.211	4,082.423	26.48	0
WAG+I+G	-2,020.314	4,082.629	26.686	0
JTT+I	-2,022.219	4,084.438	28.495	0
LG+G	-2,023.783	4,087.566	31.624	0
LG+I+G	-2,023.794	4,089.587	33.645	0
WAG+G+F	-2,005.857	4,089.713	33.771	0
WAG+I+G+F	-2,005.867	4,091.734	35.792	0
JTT+F	-2,008.676	4,093.353	37.41	0
JTT+I+F	-2,008.684	4,095.368	39.425	0
LG+G+F	-2,010.437	4,098.873	42.931	0
LG+I+G+F	-2,010.447	4,100.894	44.951	0
HIVb+G+F	-2,013.349	4,104.699	48.756	0
HIVb+I+G+F	-2,013.36	4,106.719	50.777	0
MtREV+G+F	-2,015.749	4,109.498	53.556	0
CpREV+G+F	-2,015.815	4,109.63	53.688	0

Model-averaged estimates		Parameter Importance	
Alpha	0.872	+I	0.000
Alpha-Inv	0.872	+G	0.733
Inv-Alpha	0.000	+I+G	0.267
Inv	0.000	+F	0.006

Selected Model: JTT+G

Export to main console

Command Line: 30000gn/T/prottest_tree_3818777328299265878.tmp -d aa -b 0 --no_memory_check

Fig. 4 Screenshot view of the "Selection Results" menu in ProtTest v3.2

model-given values (**JTT**). Otherwise, if the +**F** parameter has been returned by **ProtTest** in the best-fitting model, the amino acid frequencies can be computed from the MSA by selecting "Empirical".

- *Invariable sites*: Click "Optimized" to optimize the proportion of invariable sites. Otherwise, if the +**I** parameter has been returned by **ProtTest** in the best-fitting model, the proportion estimated by the model can be entered by marking "Fixed".
- *Across site rate variation*: As the +**G** parameter was returned by **ProtTest** in the best-fitting model, we set the alpha parameter of the gamma distribution of rates across sites to 0.872 ("Fixed"), and the "# of rate categories" to 8. Otherwise, select "Optimized".

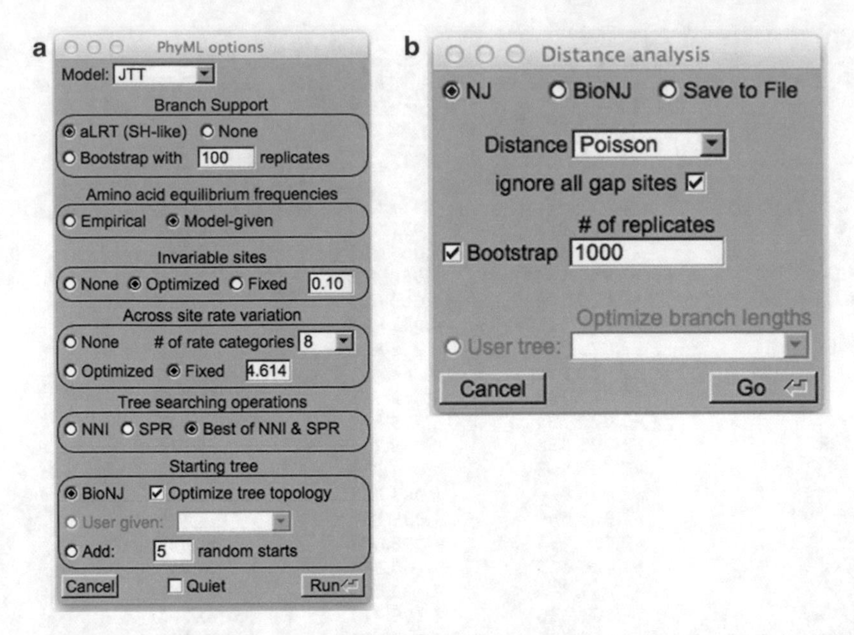

Fig. 5 Screenshot view of the "PhyML options" (**a**) and "Distance analysis" windows (**b**) in ProtTest v3.2

- *Tree searching operations*: Select "best of NNI & SPR" (NNI, nearest-neighbor interchange; SPR, subtree pruning and regrafting) [22]. This option improves the search for the most likely tree but requires increasing computational time.
- *Starting tree*: to set the starting tree used for tree-space searching select "BioNJ" and turn on "Optimize tree topology".

3. Open the "Trees" menu in **SeaView** and click on the "Distance Methods" item. This will open a "Distance analysis" window with different options to reconstruct a NJ phylogenetic tree on a variety of pairwise phylogenetic distances (Fig. 5b).

   NJ/BioNJ: Select "NJ" as tree-building algorithm.

   Distance: select "Poisson" to correct evolutionary distances for multiple substitutions (*see* **Note 9**).

   Ignore all gap sites: if on, all gap-containing sites in the MSA are excluded from analysis; if off, not all sequence pairs use the same set of sites for computation of distances.

   Bootstrap: performs bootstrap evaluation of statistical clade support (can be interrupted).

   User tree: computes least-squares branch lengths for selected user tree topology.

4. The complete tree-building can take a few minutes, mostly depending on the selected "Branch support" option, the size of your dataset, and your computer performance. Once the analysis is finished, click "OK" and a "Trees window" will open.

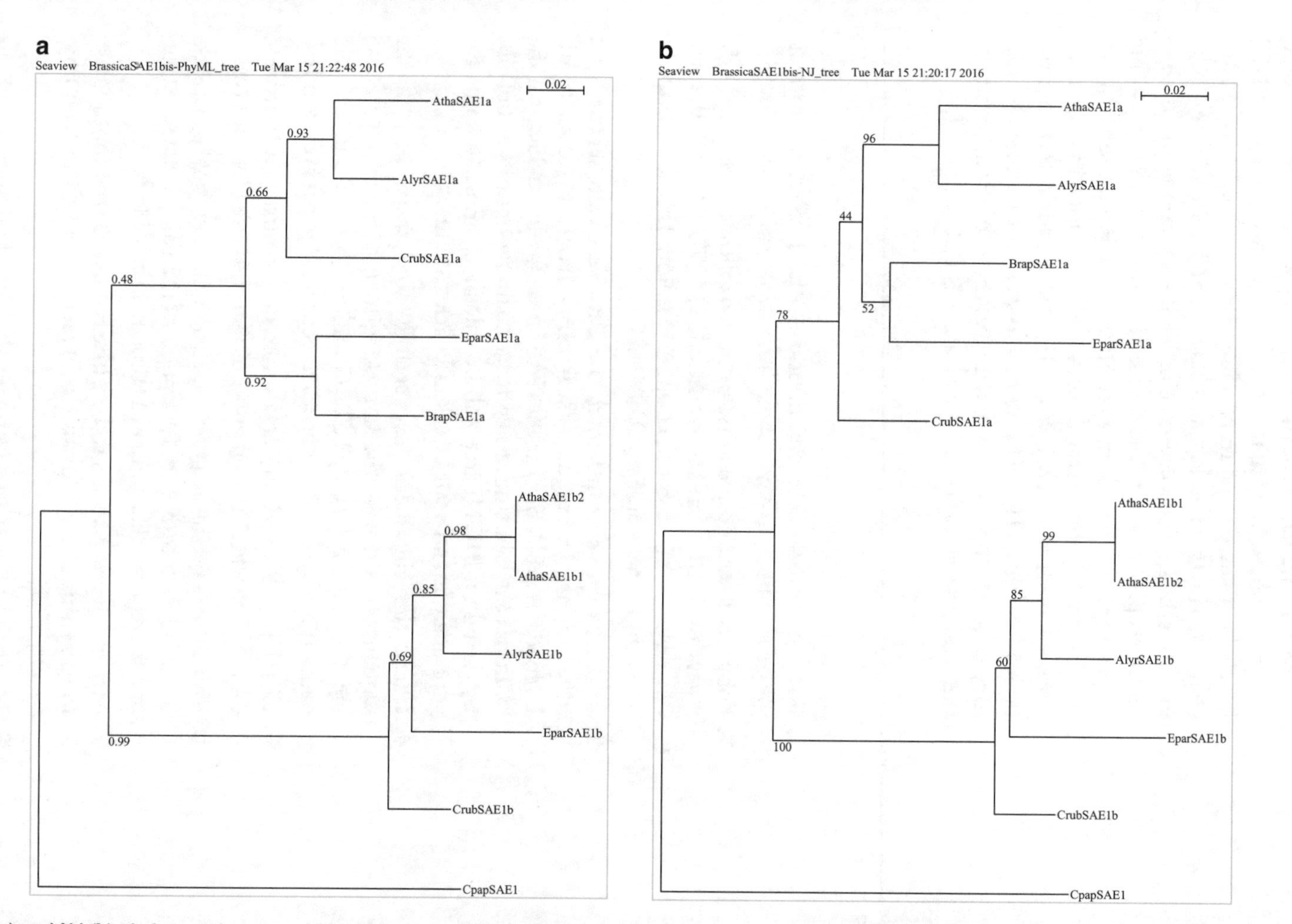

Fig. 6 ML (**a**) and NJ (**b**) phylogenetic trees of Brassicaceae SAE1 sequences in SeaView. Both trees were rooted using the corresponding ortholog from *C. papaya* (CpapSAE1). Values next to nodes indicate statistical support for relevant clades (ML aLRT support values/NJ bootstrap support values). The trees are drawn to scale, with branch lengths (shown above branches) proportional to evolutionary distances between nodes. The scale bar indicates the estimated number of amino acid substitutions per site

5. Root the trees using the *C. papaya* sequence, which is the outgroup. Select "reroot" and click on the node leading to the CpapSAE1 branch. Save the trees in Newick format by using the "save rooted tree" of the "File" menu. Also, "Save as PDF" by clicking on the corresponding item of the "File" menu (Fig. 6). The two paralogous clades of SAE1 sequences resulting from the Brassicaceae-specific WGD (excepting the *B. rapa* SAE1b representatives), as well as the two *A. thaliana*-specific in-paralogs resulting from a recent tandem duplication [9], can be observed in both trees. Topologies are mostly consistent between both trees (ML and NJ), except for a few internal nodes (*see* **Note 10**). The trees allow reconstructing the history of gene duplications underlying the evolutionary expansion and diversification of Brassicaceae SAE1 genes [9].

4 Notes

1. For a more complete description of the FASTA file format, please visit http://www.ncbi.nlm.nih.gov/blast/fasta.shtml.
2. For a full description of muscle options follow this link: http://www.drive5.com/muscle/muscle.html#_Toc81224859. For a full description of clustalo options follow this link: http://www.clustal.org/omega/README.
3. Alternatively, you can launch the **ProtTest web server** here: http://darwin.uvigo.es/software/prottest_server.html. Upload the MSA file BrassicaSAE1.msa.faa by clicking on the "Choose File" button. Keep the remaining options as default. By default, a BIONJ tree will be calculated. Enter a name for your analysis and your email and click "Submit".
4. Check the alignment file carefully for errors, i.e., all the sequences must have the same length and only letters following the standard IUB/IUPAC amino acid and nucleic acid codes (plus hyphens or dashes to represent gaps) are accepted.
5. Check [3, 19], and references therein, for further information on the theoretical background of methods used by ProtTest.
6. Visit http://evolution.genetics.washington.edu/phylip/newicktree.html for a complete description of the Newick Standard format for representing trees in a computer-readable form.
7. Model selection seems to be quite robust to tree topology so long as this is a reasonable representation of the true phylogeny [23].
8. Alternatively, you can estimate the statistical support of the retrieved topology by selecting "Bootstrap" with 1000 replicates, although this will result in a significant increase in computational time.

9. The Poisson correction distance assumes all types of amino-acid substitution are equally likely, i.e., the rate of amino acid substitution at each site follows the Poisson distribution, while correcting for multiple substitutions at the same site [24].
10. For complete graphical viewing and editing of phylogenetic trees, visit FigTree http://tree.bio.ed.ac.uk/software/figtree/ and iTOL (http://itol.embl.de/), which can take trees in Newick format as input.

References

1. Sokal RR, Michener CD (1958) A statistical method for evaluating systematic relationships. Univ Kansas Sci Bull 28:1409–1438
2. Saitou N, Nei M (1987) The neighbor-joining method: a new method for reconstructing phylogenetic trees. Mol Biol Evol 4:406–425
3. Fitch WM (1971) Toward defining the course of evolution: minimum change for a specific tree topology. Syst Biol 20(4):406–416. doi:10.1093/sysbio/20.4.406
4. Felsenstein J (1981) Evolutionary trees from DNA sequences: a maximum likelihood approach. J Mol Evol 17:368–376
5. Felsenstein J (1985) Confidence limits on phylogenies: an approach using the bootstrap. Evolution 39:783
6. Gouy M, Guindon S, Gascuel O (2010) SeaView version 4: a multiplatform graphical user interface for sequence alignment and phylogenetic tree building. Mol Biol Evol 27(2):221–224. doi:10.1093/molbev/msp259
7. Edgar RC (2004) MUSCLE: multiple sequence alignment with high accuracy and high throughput. Nucleic Acids Res 32(5): 1792–1797. doi:10.1093/nar/gkh340, 32/5/1792 [pii]
8. Sievers F, Wilm A, Dineen D, Gibson TJ, Karplus K, Li W, Lopez R, McWilliam H, Remmert M, Soding J, Thompson JD, Higgins DG (2011) Fast, scalable generation of high-quality protein multiple sequence alignments using Clustal Omega. Mol Syst Biol 7:539. doi:10.1038/msb.2011.75
9. Castano-Miquel L, Segui J, Manrique S, Teixeira I, Carretero-Paulet L, Atencio F, Lois LM (2013) Diversification of SUMO-activating enzyme in Arabidopsis: implications in SUMO conjugation. Mol Plant 6(5):1646–1660. doi:10.1093/mp/sst049
10. Abascal F, Zardoya R, Posada D (2005) ProtTest: selection of best-fit models of protein evolution. Bioinformatics 21(9):2104–2105. doi:10.1093/bioinformatics/bti263, bti263 [pii]
11. Akaike H (1974) A new look at the statistical model identification. IEEE Trans Auto Contr 19(6):716–723. doi:10.1109/TAC.1974.1100705
12. Sugiura N (1978) Further analysts of the data by akaike' s information criterion and the finite corrections. Commun Stat Theory Meth 7(1): 13–26. doi:10.1080/03610927808827599
13. Schwarz G (1978) Estimating the dimension of a model. Ann Stat 6:461–464. doi:10.1214/aos/1176344136
14. Minin V, Abdo Z, Joyce P, Sullivan J (2003) Performance-based selection of likelihood models for phylogeny estimation. Syst Biol 52(5):674–683
15. Reeves JH (1992) Heterogeneity in the substitution process of amino acid sites of proteins coded for by mitochondrial DNA. J Mol Evol 35(1):17–31
16. Yang Z (1993) Maximum-likelihood estimation of phylogeny from DNA sequences when substitution rates differ over sites. Mol Biol Evol 10(6):1396–1401
17. Cao Y, Adachi J, Janke A, Paabo S, Hasegawa M (1994) Phylogenetic relationships among eutherian orders estimated from inferred sequences of mitochondrial proteins: instability of a tree based on a single gene. J Mol Evol 39(5):519–527
18. Darriba D, Taboada GL, Doallo R, Posada D (2011) ProtTest 3: fast selection of best-fit models of protein evolution. Bioinformatics 27(8):1164–1165. doi:10.1093/bioinformatics/btr088
19. Jones DT, Taylor WR, Thornton JM (1992) The rapid generation of mutation data matrices from protein sequences. Comput Appl Biosci 8(3):275–282
20. Guindon S, Gascuel O (2003) A simple, fast, and accurate algorithm to estimate large phylogenies by maximum likelihood. Syst Biol 52(5):696–704, 54QHX07WB5K5XCX4 [pii]

21. Anisimova M, Gascuel O (2006) Approximate likelihood-ratio test for branches: a fast, accurate, and powerful alternative. Syst Biol 55(4): 539–552. doi:10.1080/10635150600755453, T808388N86673K61 [pii]
22. Guindon S, Dufayard JF, Lefort V, Anisimova M, Hordijk W, Gascuel O (2010) New algorithms and methods to estimate maximum-likelihood phylogenies: assessing the performance of PhyML 3.0. Syst Biol 59(3):307–321. doi:10.1093/sysbio/syq010
23. Posada D, Crandall KA (2001) Selecting the best-fit model of nucleotide substitution. Syst Biol 50(4):580–601
24. Zuckerkandl E, Pauling L (1965) Molecules as documents of evolutionary history. J Theor Biol 8(2):357–366

Chapter 23

Bioinformatics Tools for Exploring the SUMO Gene Network

Pedro Humberto Castro, Miguel Ângelo Santos, Alexandre Papadopoulos Magalhães, Rui Manuel Tavares, and Herlânder Azevedo

Abstract

Plant sumoylation research has seen significant advances in recent years, particularly since high-throughput proteomics strategies have enabled the discovery of hundreds of potential SUMO targets and interactors of SUMO pathway components. In the present chapter, we introduce the SUMO Gene Network (SGN), a curated assembly of *Arabidopsis thaliana* genes that have been functionally associated with sumoylation, from SUMO pathway components to targets and interactors. The enclosed tutorial helps interpret and manage these datasets, and details bioinformatics tools that can be used for in silico-based hypothesis generation. The latter include tools for sumoylation site prediction, comparative genomics, and gene network analysis.

Key words Arabidopsis, Bioinformatics, Data mining, Functional categorization, Gene expression, Gene network, Post-translational modification, Small ubiquitin-related modifier (SUMO)

1 Introduction

For over 10 years, studies in the model plant *Arabidopsis thaliana* have increased our understanding of the SUMO peptide's role as an important post-translational modification (PTM) mechanism. Studies in both plant and non-plant models have demonstrated an increasing complexity of SUMO pathway components, and functional studies using loss-of-function mutants have specifically implicated sumoylation in many aspects of plant development and the response to external stimuli [1]. The molecular mechanisms underpinning SUMO role in plants started to be unraveled by hypothesis-based identification and subsequent validation of specific SUMO targets. Meanwhile, the introduction of high throughput technologies in protein studies considerably accelerated the discovery of hundreds of SUMO targets and other proteins that interacted with SUMO pathway components [1]. However, the vast majority of these proteins still lack functional validation as targets, nor is the biological context of their interplay with SUMO

L. Maria Lois and Rune Matthiesen (eds.), *Plant Proteostasis: Methods and Protocols*, Methods in Molecular Biology, vol. 1450, DOI 10.1007/978-1-4939-3759-2_23,

pathway components known. To make sense of the increasing amount of genetic evidence on plant sumoylation we compiled the **SUMO Gene Network** (**SGN**), consisting of the collection of genes that have been experimentally linked to the plant sumoylation pathway. Because the overwhelming amount of data on plant sumoylation has been generated in the model plant *Arabidopsis thaliana*, the SGN consists solely of Arabidopsis genes.

The SGN constitutes an excellent resource that can be used to answer a question as simple as *Is my gene-of-interest a known/potential SUMO target*? It is also a powerful tool to drive hypothesis generation, either from the perspective of a seasoned SUMO researcher, or from the point-of-view of an investigator who stumbled upon the field, given the possibility that his study subject might be associated with sumoylation at a molecular level. The sequencing of the *Arabidopsis thaliana* genome and the different large-scale projects that quickly followed (many within the scope of the Arabidopsis 2010 Initiative) generated considerable amounts of data. The even more recent generalization of high-throughput technologies, particularly at the transcript and protein levels, have generated nothing short of a revolution with regards to the way we conduct research in this model species, with beneficial implication to other plant species. This wealth of information has been consistently integrated into freely available, web-based resources and databases. It provides plant researchers with a powerful platform to gather information, provide new context to their biological problems, and formulate new hypothesis, often before going to the wetlab. In silico-based data mining has become a significant resource in current day plant biology, and has been the focus of previous publications [2–4].

In the present chapter, we introduce and detail the SGN. Because the SGN comprises large datasets, we provide an overview of simple strategies to manage this kind of data. Subsequently, we indicate a selection of bioinformatics tools, either in the form of web-based databases and resources, or stand-alone software that can be particularly useful to explore and generate hypothesis within the scope of plant sumoylation. These include resources for plant comparative genomics, functional categorization, and gene network analysis. We use ICE1, a MYC-like bHLH transcriptional activator that has been biochemically established as a bone-fine SUMO target [5], to highlight some of the insights that can be gained by the in silico analysis of the SUMO Gene Network.

2 Materials

The SUMO Gene network was hand curated from the literature, as is available at http://cibio.up.pt/resources-1/details/sgn. An annotation of all bioinformatics tools described in the present chapter is available in Table 1.

Table 1
Summary of bioinformatics resources detailed in the present chapter

Resource	URL	Reference
Dataset management		
Bar Duplicate Remover	http://bar.utoronto.ca/ntools/cgi-bin/ntools_duplicate_remover.cgi	[11]
Bar Venn Generator	http://bar.utoronto.ca/ntools/cgi-bin/ntools_venn_selector.cgi	[11]
Venny	http://bioinfogp.cnb.csic.es/tools/venny/	
TAIR	http://www.arabidopsis.org/	[13]
Comparative plant genomics		
PLAZA	http://plaza.psb.ugent.be/	[14, 15]
Phytozome	http://phytozome.jgi.doe.gov/	[16]
Prediction of sumoylation and SIM sites		
GPS-SUMO	http://sumosp.biocuckoo.org	[17]
SUMOplot	http://www.abgent.com/sumoplot	
SeeSUMO	http://bioinfo.ggc.org/seesumo	[18]
Functional categorization and gene network analysis		
Cytoscape	http://www.cytoscape.org/	[19]
Genemania	http://www.genemania.org/	[12]
BINGO	http://www.psb.ugent.be/cbd/papers/BiNGO/	[20]
ClueGO	http://www.ici.upmc.fr/cluego/	[21]

3 Methods

3.1 The SUMO Gene Network

The SUMO Gene Network has been divided into four different datasets. The first dataset contains the list of current SUMO pathway components present in the *Arabidopsis thaliana* genome (herein **SUMO Path**). Remaining datasets refer to genes that code for proteins that have been functionally linked to SUMO pathway components, by being identified as sumoylation targets (herein **SUMO Target**), or by being capable of protein–protein interactions (PPIs) with either the SUMO peptides themselves (herein **SUMO Interacting Protein** or **SIP**), or SUMO pathway components (**SUMO Path Interact**). Here, we demonstrate how to access and interpret the SUMO Gene Network.

1. Go to http://cibio.up.pt/resources-1/details/sgn and click on **SUMO Gene Network File**.
2. Download the Excel file containing the SUMO Gene Network.
3. Access Spreadsheet 1 (**SUMO Path**). The list details presently known components of the SUMO enzymatic pathway, from SUMO isoforms to the genes involved in the five conserved enzymatic steps that mediate target conjugation/deconjugation to SUMO (SUMO maturation, E1 activation, E2 conjugation, E3 ligation, SUMO deconjugation).

4. Access Spreadsheet 2 (**SUMO Target**). The present list contains genes that have been identified as coding for bona fide SUMO targets by hypothesis-driven research, as well as SUMO targets that have identified via high-throughput approaches (traditionally, isolation of Tag-SUMO conjugates followed by peptide sequencing). Information on the SUMO isoform associated with the target is also provided.
5. Access Spreadsheet 3 (**SIP**). The dataset refers to genes coding for proteins that have been demonstrated to have PPI with SUMO isoforms. Amongst others, these include components of the SUMO pathway, as well as SUMO-Targeted Ubiquitin Ligases (STUbLs), which are important proteins in SUMO/Ubiquitin interplay.
6. Access Spreadsheet 4 (**SUMO Path Interact**). The list contains genes that have been shown to code for interactors of SUMO pathway components other than SUMO peptides. Traditionally, these were identified via yeast two-hybrid (Y2H) assays.

3.2 Managing Datasets

Managing gene datasets is considerably facilitated by the use of a standard annotation code for each gene of a given, fully sequenced, genome. In the *Arabidopsis thaliana* genome, gene identifiers take the form of an Arabidopsis Genome Initiative (AGI) code (At#g#####), where the first number represents one of the five Arabidopsis chromosomes, followed by a five number positional code for each individual gene. Presently, the Arabidopsis genome is in its tenth annotation (TAIR 10; https://www.arabidopsis.org/). The vast majority of Arabidopsis web-based resources rely on the AGI code, which often provides links to additional bioinformatics resources. Here we will overview simple tools that help manage these AGI-based datasets, using SGN datasets as examples (*see* **Note 1**).

3.2.1 Duplicate Removal

Removing duplicates is often required when working with large datasets. In this tutorial we will remove duplicates from the **SUMO Target** dataset, allowing us to have a working list of currently established SUMO targets.

1. Go to the SGN Excel file, Spreadsheet 2 (SUMO Target). Copy the column of AGI identifiers (highlighted in brown).
2. Go to the **Duplicate Remover Tool** (http://bar.utoronto.ca/ntools/cgi-bin/ntools_duplicate_remover.cgi) located at the BAR web-based resource (*see* **Note 2**).
3. Paste the AGI identifiers into the search box and click **Send**.
4. The output will indicate that, out of the 790 entries, 512 unique entries were found (Fig. 1a) (*see* **Note 3**). This constitutes the **No Duplicates SUMO target** dataset.
5. Copy the 512 unique AGI identifiers to a new spreadsheet for further use.

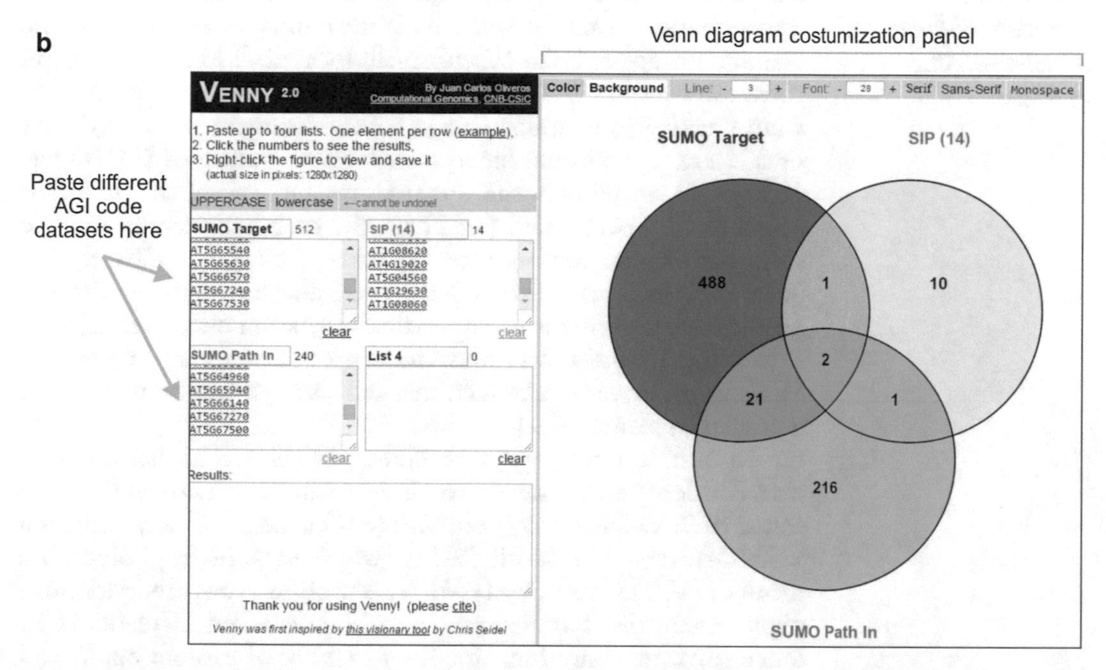

Fig. 1 Tools for dataset management. Output of the Duplicate Remover Tool (**a**). Venn diagrams can be easily generated in Venny (**b**)

6. For convenience, a **No Duplicates** list of each of the four datasets provided in the SGN was created. To access the data, go to the SGN Excel file, **Spreadsheet 5** (**SGN—No Duplicates**).

3.2.2 Generate Venn Diagrams

Cross-referencing datasets is often useful, and Venn diagrams constitute the most informative and visually appealing form of displaying such data. Here, we will estimate how SUMO targets (proteins that are covalently bound to a SUMO peptide) actually match the proteins that have been singled out as interacting with SUMO or sumoylation machinery components.

1. Go to **Venny** (http://bioinfogp.cnb.csic.es/tools/venny/). Venny 2.0 allows for a maximum of four different datasets to be cross-referenced (*see* **Note 4**).

2. Go to the SGN Excel file, Spreadsheet 5 (SGN—No Duplicates), and sequentially copy/paste datasets **SUMO Target**, **SIP**, and **SUMO Path Interact** into Venny lists 1 through 3.
3. An output Venn diagram is automatically generated (Fig. 1b). The Venn diagram can be conveniently edited in several parameters (e.g. font type and size), and its output is suited for publication. Notice how datasets do not overlap significantly, reflecting the different nature of each dataset.

3.3 In Silico Prediction of SUMO Attachment Sites and SIMs

SUMO establishes a covalent interaction with a target protein via an isopeptide bond between its N-terminus G residue and the ε-amino group of a lysine (K), normally located within a sumoylation consensus motif ψKXE (ψ, large hydrophobic residue; X, any amino acid; E, glutamic acid) [6]. Establishment of sumoylation sites is a key step in the functional characterization of SUMO targets. Once established, site-directed mutagenesis of the predicted lysine can be performed (*see* **Note 5**), to biochemically validate sumoylation via, for instance, in vitro studies (*see* Chapter 9). Several bioinformatics tools have been developed to predict if a given protein contains a potential sumoylation site. Presently, all algorithms are based on sumoylation sites of non-plant organisms. However, it is well established that the sumoylation consensus site is similar in plants [7–9].

In addition to isopeptide bonds, SUMO can also establish non-covalent interactions. Proteins that interact with SUMO are called SUMO-Interacting Proteins (SIPs), and normally contain a SUMO-Interacting Motif (SIM), which is a hydrophobic core motif of (V/I)X(V/I)(V/I) [6]. This motif was described for non-plant organisms, but also seems to be conserved in plants [10]. Since SIMs are important for the assembly of protein complexes and for the recognition of STUbLs, prediction of SIM sites is also extremely useful in SUMO research. In this tutorial we will use **GPS-SUMO**, a user-friendly software, to predict both sumoylation and SIM sites in the well characterized SUMO target ICE1 (*see* **Note 6**). Additional resources for SUMO site prediction are available in Table 1.

1. Go to **GPS-SUMO** (http://sumosp.biocuckoo.org/download.php), and download the most updated release of the GPS-SUMO software, choosing also your computer platform of choice.
2. Run the software installer.
3. Go to **UniProt** (http://www.uniprot.org/uniprot/Q9LSE2.fasta) and copy the protein sequence of ICE1 in FASTA format (*see* **Note 7**).
4. Run the GPS-SUMO software. Paste the ICE1 sequence into the **Enter sequence(s) in FASTA format** box. Adjust threshold

values as seen fit (for the present example choose the **Low** setting for both thresholds). Click on **Submit** (*see* **Note 8**).

5. The output will list amino acidic sequences with sumoylation or SIM sites highlighted in red. For ICE1, the software correctly predicts lysine K393, which was previously validated as being a sumoylation site [5], in addition to one SIM at position 220 (Fig. 2). The information can be selected and copied to an Excel spreadsheet.
6. To display all sumoylation sites and SIMs within a protein sequence representation, click over the column of predicted sites with the right mouse button and select the option **Visualize**. The image can be saved by clicking **File** on the header bar. The menu will allow you to select File Type (JPG or PNG), image resolution, and File name/storage destination.
7. Another interesting feature of GPS-SUMO is that it connects to **SUMOdb 1.0**, a database for SUMO-related proteins. However, SUMOdb 1.0 does not contain extensive information on plant proteins. To access the database, select **Tools** on the header bar, and click on **SUMO Database**. You can, for instance, type **bHLH** (the transcription factor class of ICE1) in the **Search** box below. The output will be a **Matched List** of proteins that can be subsequently surveyed for SUMO-associated features.

3.4 Comparative Genomics for Ortholog Identification

SUMO/sumoylation is conserved in eukaryotic organisms, and plants are no exception. SUMO Gene Network components are likely to be conserved among different plant species, and identification of orthologs within a given genome can be an important resource. Here, we will use **PLAZA** as a tool for automated ortholog identification (*see* **Note 9**). PLAZA is an online, user-friendly, platform for plant comparative genomics. It includes over 30 different plant species, contains automatic annotation of gene families, allows the downloading of multiple DNA and protein sequences, and displays a series of tools to further explore this information. Here, we will concentrate on using PLAZA to identify, retrieve, and analyze orthologs of the SUMO target ICE1, also testing if a given sumoylation site is conserved among plant species.

1. Go to **PLAZA** (http://plaza.psb.ugent.be/). Here, you can choose different versions of PLAZA. Versions 3.0 are divided between Dicots and Monocots. Because PLAZA 2.5 displays a more evolutionarily representative selection of plant genomes, we will click on the **Go to PLAZA 2.5** box (*see* **Note 10**).
2. Input the AGI code of your gene of interest in the search bar located at the top of the page. In this case, insert the ICE1 AGI code **AT3G26744**, and click on the **Search** button. The output will be a **General Overview** page, containing

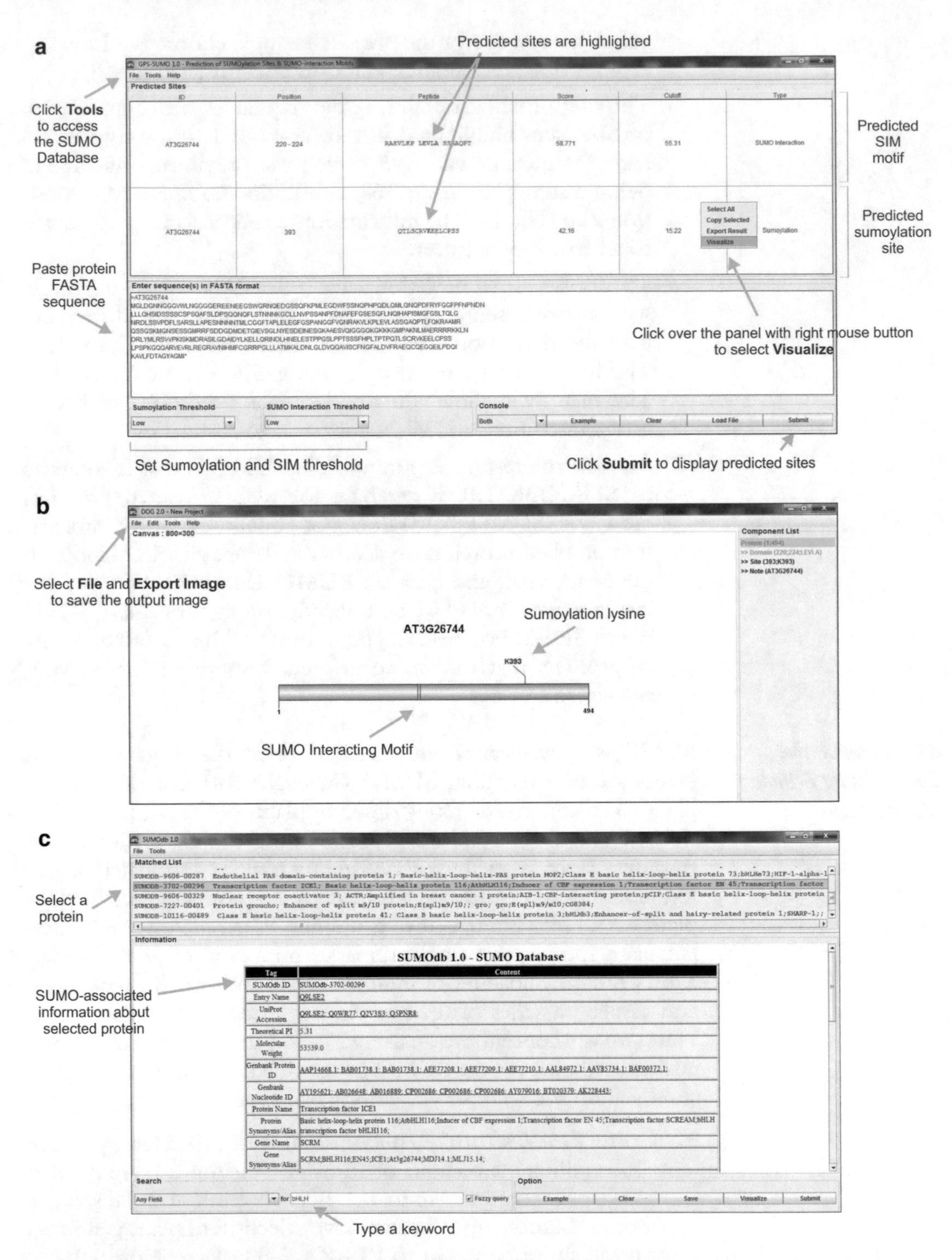

Fig. 2 (**a**) Output of GPS-SUMO for the ICE1 protein sequence. The sumoylation lysine–K–residue is highlighted, as is the SUMO-Interacting Motif. (**b**) The tool Visualize displays a representation of the input sequence, highlighting the predicted sumoylation and SIM sites. (**c**) GPS-SUMO allows access to the SUMO database (SUMOdb 1.0)

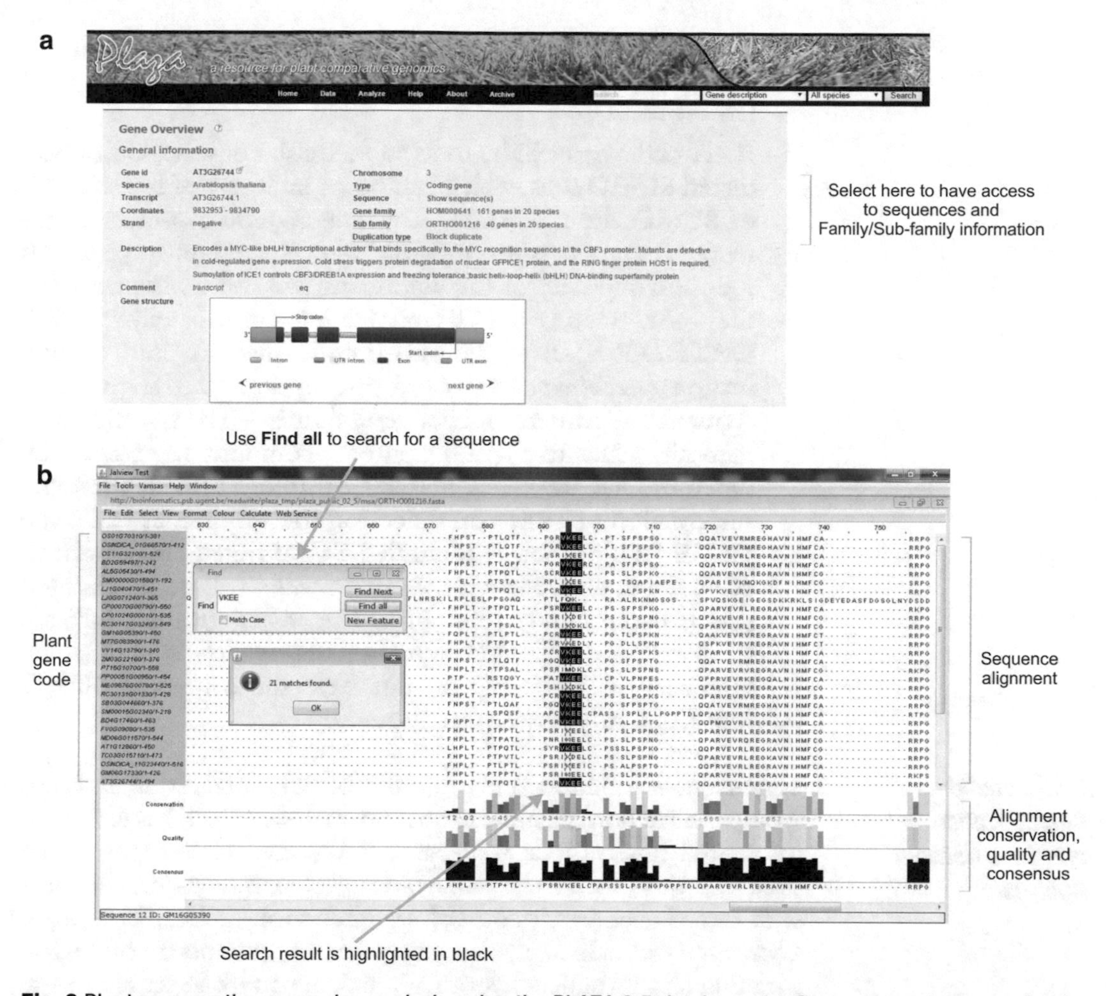

Fig. 3 Plant comparative genomics analysis using the PLAZA 2.5 database. (**a**) General overview page for the gene *ICE1* (AGI code AT3G26744). (**b**) Protein sequence alignment for the ICE protein subfamily (ORTHO001216)

numerous information and Toolbox links to a series of resources associated with the gene-of-interest (Fig. 3a).

3. As a comparative genomics tool, PLAZA assigns a gene family and subfamily for the gene-of-interest. To access information on the ICE1 gene family click on the **Gene family** code **HOM000641**. The output includes a graphic representation of gene family member abundance in the different species. Here, 161 genes were identified in the **HOM000641** gene family, divided between eight sub-families (*see* **Note 11**).

4. An important feature is the retrieval of nucleotide and protein sequences for all the orthologs analyzed. Click on the **Download DNA sequences** or **Download proteins sequences** option. The output will be the FASTA sequence of all protein family members. Copy all the outputted text and paste onto

Notepad or equivalent software, and save as ".txt" file for further use (e.g. phylogenetic analysis, SUMO site prediction, etc.) (*see* **Note 12**).

5. ICE1 orthologs will be used to establish conservation of predicted SUMO sites, which can be an indicator of the strength of SUMO site prediction. This will require a protein alignment. Return to the HOM000641 **Gene family** page. To reduce complexity of the alignment, we will limit analysis to the Arabidopsis ICE1-containing sub-family. Click **ORTHO001216** in the **sub-families** list. In the **Toolbox** section (*see* **Note 13**), select the option "...**the multiple sequence alignment of this gene family**". The site will automatically generate a Java file on your computer (*see* **Note 14**). Open the file. To check if ICE1 orthologs present the same sumoylation site, click on **Select** on the sub-header bar. Type **VKEE** (the previously identified SUMO consensus motif in ICE1), and click **Find all**. Search results will be highlighted (black background) (Fig. 3b). The lower graphics indicate conservation level. Although the conservation is high, some members of this family do not have the conserved SUMO motif, and may not be sumoylated.

3.5 Functional Categorization and Gene Network Analysis

Cytoscape is a stand-alone program for data integration and network visualization, manipulation, and analysis, which has also been integrated into various web-based programs. Cytoscape can be used in various ways to visualize biomolecular interaction networks with ease. Together with other integrated tools or external plugins, Cytoscape can help analyze a given dataset, extrapolate biological meaning and formulate hypothesis, helping to make sense of large datasets like the SGN. As stated, a number of plugins are available that can bring more functionality to the software. One example is the **GeneMANIA** plugin, which identifies the most related genes in a gene set, and groups those terms into networks, taking into account pre-input data already available in Arabidopsis. Data include genetic interactions, physical interactions, predicted interactions, shared protein domains, co-expression, and co-localization. It also integrates into the analysis the enrichment in Gene Ontology (GO) terms (*see* **Note 15**), further enhancing our ability to extract information form a fairly large dataset. Here, we will explore the SGN using Cytoscape as a stand-alone program containing the GeneMANIA plugin (*see* **Note 16**).

1. Go to http://www.cytoscape.org/. Click on the **Download** section of the header bar, and download the software installer (*see* **Note 14**).
2. After installation, open the Cytoscape software. A **Welcome to Cytoscape** window appears. From this window one can create an empty network, or load an existing network from a file or compatible network database.

3. As a stand-alone program, users can upload their data into Cytoscape and generate/manipulate networks to their convenience. However, we will focus on the **GeneMANIA** Cytoscape plugin. To install this plugin go to the header bar and click on **Apps > App Manager**. Search for GeneMANIA amongst available apps (*see* **Note 17**). On the results panel select the GeneMANIA plugin, and click on the **Install** button.
4. GeneMANIA should now be available at the **Apps** menu. To start using this plugin one must first load a dataset. To accomplish this, click on **Apps > Genemania > Choose Another Data Set**... . Choose the latest available dataset, and click on the **core** option, before selecting the **Download** button. The following window allows us to select the species for download; choose the **A. thaliana Arabidopsis** dataset and click **Install**.
5. Once installation is complete, lists of genes can be loaded to begin your specific network analysis within the GeneMANIA plug-in. Click on **Apps > Genemania > Search**... . In the new window, paste your gene list into the **Step 2**: **Choose Genes of Interest** search box. For this tutorial, paste the **SUMO Targets** dataset from the SGN file (Spreadsheet 5).
6. On the option **Step 3**: **Choose Interaction Networks**, choose the relevant types of functional data that you want to integrate into a network. For tutorial purposes, select **Shared protein domains** (which will highlight genes coding proteins from similar functional families), and **Physical interactions** (which will highlight experimentally confirmed interactions between proteins). At the bottom of the window one can choose to include in the network a given number of genes related to our uploaded genes-of-interest. Maintain the default number of 20 genes (*see* **Note 18**), and click on the **Start** button.
7. Results of the analysis are displayed on the main window, which is divided into four sections (Fig. 4a). In the centre section, the main network is represented graphically. On the left section, we have the network selection panel and customization tools. On the bottom section, one can access the attribute tables for node, edge, and network (*see* **Note 19**). The right panel displays a number of tools that are specific to a given plugin. A series of options are available to explore the generated network (*see* **Note 20**). We will highlight some of those options.
8. A number of layouts can be applied to the network. This can result in different clustering patterns that may highlight important gene/protein interactions, otherwise masked by the default layout. Users should experiment different layouts. For tutorial purposes, we will choose an organic layout: go the header bar, and select **Layout > yFiles Layouts > Organic**.
9. To change the functional features displayed in the network, which are presented as different colored edges, go to the right-

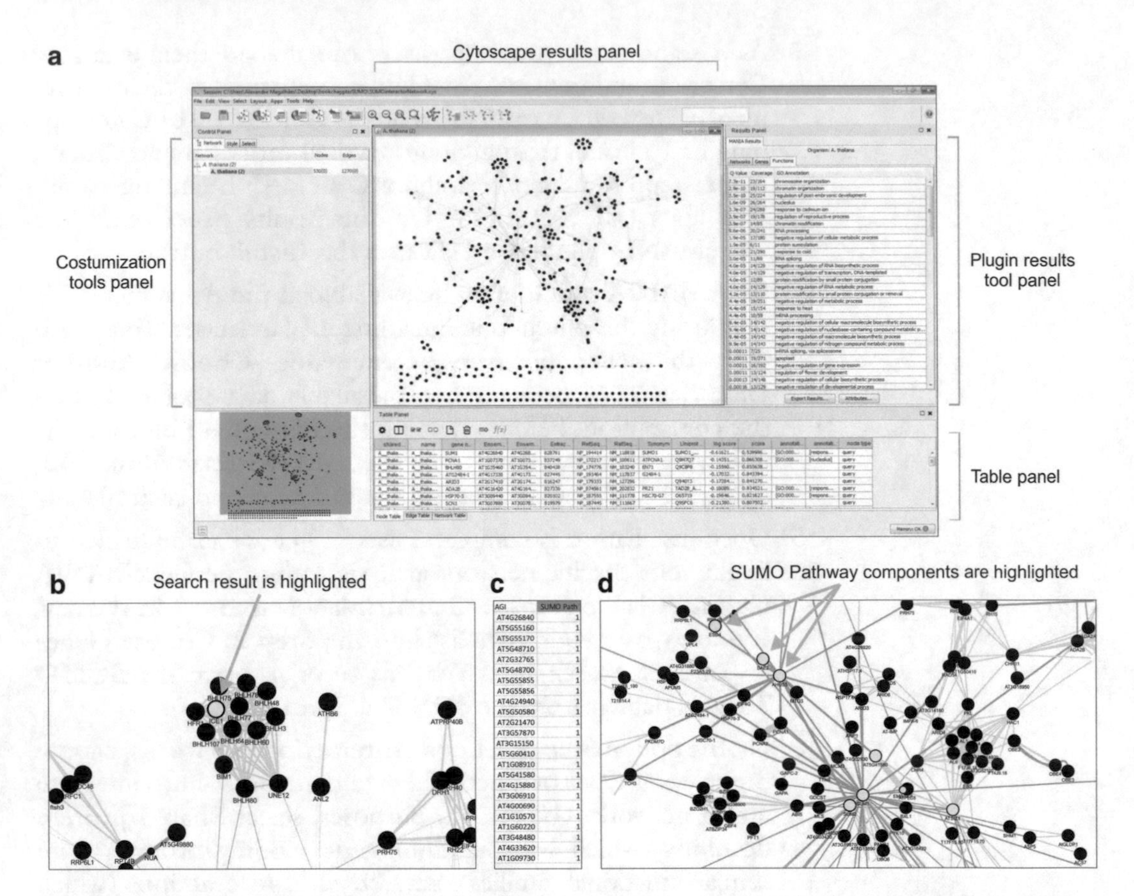

Fig. 4 Gene network analysis using the GeneMANIA plugin at Cytoscape. (**a**) The Cytoscape software environment is divided into four sections, with the network displayed in the centre section. (**b**) The Search option allows for specific genes to be highlighted. (**c**) Example of the data format required to input new information into an existing network. (**d**) SUMO pathway components were inputted into the existing network and now show up as highlighted nodes

hand **MANIA Results** panel, **Networks** section, and click on the different checkboxes.

10. To highlight GO-term categories that are enriched within the dataset, go to the right-hand **MANIA Results** panel, **Functions** section, and select a biological process of interest. Related genes show up in yellow within the network.

11. Search for a specific gene-of-interest using the **Search** box in the top-right section. In this case introduce **ICE1**. The output highlights the gene-of-interest in yellow (Fig. 4b). Notice how the present strategy singled out additional bHLH TFs within the list of SUMO targets.

12. New data can be integrated into an existing network, and used to convey additional information from a graphical point-of-view. Here, we will highlight **SUMO pathway components** within our network. First go to **Spreadsheet 5** of the SGN file,

and paste the **SUMO Path** AGI codes into a new Excel file, adding **1** to the adjoining column of every gene (Fig. 4c); save the file as **Newdata.xls**. Return to Cytoscape. In the header bar, go to **File > Import > Table > File...** . Select the **Newdata.xls** file and click **Open**. In the **Import Columns From Table** window, click on **Key Column for Network**, and select **Ensembl Gene ID**. This will define the attribute in our present network that will be matched with the data we are importing (in this case, Ensembl Gene ID represents the AGI code). Maintain remaining default options and click on **OK**.

13. Go to the left-hand **Control Panel**, click on the **Style** section, select the **Fill Color** property. Click the **Column** dropdown menu, and select the **Value** option. Click the **Mapping type** dropdown menu, and select the **Discrete mapping** option (*see* **Note 21**). Click on the **Edit color** button (represented by the "..." symbol), for the value **1**, and select the color yellow (*see* **Note 22**). The network now contains yellow nodes (SUMO pathway components) and black nodes (remaining genes) (Fig. 4d). Notice how the main Arabidopsis SUMO peptides (SUMO1 and SUMO2) appear as central components of a protein interaction network.

4 Notes

1. Traditionally, bioinformatics resources are not case sensitive with regards to input data. However, one may encounter situations where changes in casing must be enforced in hundreds or thousands of targets. Use the **Replace All** function in Notepad, Excel or equivalent software to perform these changes. For example, replace all "t" with "T" and "g" with "G" to convert all AGI identifiers in a gene list from the "At#g#####" to the "AT#G#####" forms.
2. BAR (http://bar.utoronto.ca/welcome.htm) constitutes an outstanding web-based resource, containing numerous useful tools worth exploring [11]. With regards to managing datasets, a particularly useful tool is the **_at to AGI Conversion Tool** (http://bar.utoronto.ca/ntools/cgi-bin/ntools_agi_converter.cgi). It serves primarily to convert AGI identifiers to probe identifiers of the Affymetrix microarray chip ATH1. However, it can also be used to retrieve the latest annotation, gene name, and UniProt identifier from a given list of AGI codes.
3. In addition to the list of unique entries, the **Duplicate Remover Tool** also provides the list of duplicates (including information on the number of entries for each duplicate), and the list of entries that were unique and not duplicated in the input list. To access these lists, scroll down the output page.

4. There are various web-based resources available for Venn diagram generation (e.g. **Venn Selector** and **VennSuperSelector** tools at BAR; http://bar.utoronto.ca/welcome.htm), most of which are not even biology-driven. Many will not allow for more than three datasets to be compared, but also bear in mind that interpretation of Venn diagrams that cross-reference 6 or more datasets is impractical. **Venny** has the convenience of (1) allowing for up to four datasets to be analyzed, (2) generating an output that is publication-friendly.
5. The lysine residue can be targeted by other PTMs, therefore, in addition to K-directed mutagenesis (usually K to R), also E can be subjected to mutagenesis.
6. Do not assume that a given protein is a SUMO target based solely on the in silico prediction of a sumoylation site. Hundreds of Arabidopsis proteins display potential sumoylation sites, and SUMO modification depends on the cellular context, subcellular localization, tissue expression, etc. Similarly, non-consensus sumoylation sites exist, that rely greatly on the activity of SUMO E3 ligases. SUMO-Interacting Motifs should also be interpreted with caution, as other domains for non-covalent SUMO interaction may exist.
7. FASTA format is a standard, text-based, data format for nucleotide or protein sequences. Any given sequence begins with a first line where the **greater than** symbol (>) is followed by the sequence description. Subsequent lines correspond to the nucleotide or protein sequence.
8. For multiple queries, multiple protein sequences can be pasted in FASTA format.
9. Another powerful plant comparative genomics tools is **Phytozome** (http://phytozome.jgi.doe.gov; Table 1).
10. To select the most appropriate PLAZA version, explore the different plant species available, by entering each database. There you will find an informative phylogenetic tree displaying the evolutionary relationships of the different species of each database.
11. Automated comparative genomics annotation may not match the user's expectation of what constitutes the gene family. For instance, in the SUMO biochemical pathway, gene family HOM001031 corresponds to ULP2-type SUMO proteases, placing ULP1-type proteases in a different family. Conversely, gene families may be too broad, and analysis may require a specific sub-family. Therefore, the user should always manually interpret the significance of the outputted gene family/sub-family assignment.
12. Edit the file to your convenience in Notepad (e.g. remove species from the analysis). An important use of the data is phy-

logenetic reconstruction, whose tools are described in Chapters 21 and 22 in the present book.

13. In the toolbox section, a number of features can be explored, like synteny and phylogenetics. An interesting possibility is to explore gene family expansion from an evolutionary point of view, using the option "...**the expansion/depletion of species in this gene family**". The output is a table that depicts which species and phylogenic clades have above- or below-average number of genes in the family. This tool can be used, for instance, to analyze how SUMO pathway components have evolved in terms of the number of gene copies present in plant genomes.
14. Some analysis runs in the JAVA runtime environment. Be careful to update the JAVA software in your computer, prior to running JAVA files or installing Cytoscape.
15. The **Gene Ontology** (**GO**) project describes gene products with consistent terms across different species. GO terms are organized in a hierarchical structure. A given gene can have assigned several GO terms in the following three categories: biological process, molecular function, and cellular component.
16. **GeneMANIA** has also been converted into a web-based resource (www.genemania.org/), where many of the functionalities that are described in this tutorial are also available. For a tutorial on web-based GeneMANIA within a plant research context, consult [12].
17. **Cytoscape** has other interesting plugins. Examples of commonly used plugins include **ClueGO** and **BINGO**. **ClueGO** generates a network from a gene dataset, using GO and pathway enrichment analysis based on BioGRID, Gene Ontology, BioCyc, and KEGG databases. **BINGO** is used to do a GO enrichment analysis and visualize the GO tree structure in a network fashion.
18. If you do not wish to include any extra genes into the network, type **0** on the text box next to that particular option. You may also select the number and type of attributes to be considered in the analysis. Begin by running default settings. You may latter choose to run the analysis placing more weight into specific attributes (e.g. GO cellular component), depending on your biological question.
19. **Node** represents a gene and **edge** represents the line (common attribute) connecting nodes.
20. **Cytoscape** documentation is available at http://wiki.cytoscape.org/. There, one has access to more in-depth tutorials, manuals and extensive information on other capabilities of the software, as well as technical descriptions of the available options within **Cytoscape**.

21. There are other mapping types available. For instance, to visually input information on gene expression, use **Continuous Mapping**, generating a color gradient that will match gene expression values.

22. You can add different colors to different sets of genes, by assigning different numbers to genes in the **Newdata.xls** file.

Acknowledgments

This work was supported by FEDER through the Operational Competitiveness Program—COMPETE—and by national funds through the Foundation for Science and Technology—FCT—within the scope of project "SUMOdulator" [Refs. FCOMP-01-0124-FEDER-028459 and PTDC/BIA-PLA/3850/2012]. PHC was supported by the Fundação para a Ciência e a Tecnologia (FCT) [grant ref. SFRH/BD/44484/2008 and PTDC/BIA-PLA/3850/2012]. HA was supported by the "Genomics and Evolutionary Biology" project, co-financed by North Portugal Regional Operational Programme 2007/2013 (ON.2 – O Novo Norte), under the National Strategic Reference Framework (NSRF), through the European Regional Development Fund (ERDF).

References

1. Castro PH, Tavares RM, Bejarano ER et al (2012) SUMO, a heavyweight player in plant abiotic stress responses. Cell Mol Life Sci 69:3269–3283. doi:10.1007/s00018-012-1094-2
2. de Lucas M, Provart NJ, Brady SM (2014) Bioinformatic tools in Arabidopsis research. Methods Mol Biol 1062:97–136. doi:10.1007/978-1-62703-580-4_5
3. Brady SM, Provart NJ (2009) Web-queryable large-scale data sets for hypothesis generation in plant biology. Plant Cell 21:1034–1051. doi:10.1105/tpc.109.066050
4. Usadel B, Obayashi T, Mutwil M et al (2009) Co-expression tools for plant biology: opportunities for hypothesis generation and caveats. Plant Cell Environ 32:1633–1651. doi:10.1111/j.1365-3040.2009.02040.x
5. Miura K, Jin JB, Lee J et al (2007) SIZ1-mediated sumoylation of ICE1 controls *CBF3/DREB1A* expression and freezing tolerance in Arabidopsis. Plant Cell 19:1403–1414. doi:10.1105/tpc.106.048397
6. Gareau JR, Lima CD (2010) The SUMO pathway: emerging mechanisms that shape specificity, conjugation and recognition. Nat Rev Mol Cell Biol 11:861–871. doi:10.1038/nrm3011
7. Miller MJ, Barrett-Wilt GA, Hua Z et al (2010) Proteomic analyses identify a diverse array of nuclear processes affected by small ubiquitin-like modifier conjugation in Arabidopsis. Proc Natl Acad Sci U S A 107:16512–16517. doi:10.1073/pnas.1004181107
8. Miller MJ, Scalf M, Rytz TC et al (2013) Quantitative proteomics reveals factors regulating RNA biology as dynamic targets of stress-induced SUMOylation in Arabidopsis. Mol Cell Proteomics 12:449–463. doi:10.1074/mcp.M112.025056M112.025056
9. Elrouby N, Coupland G (2010) Proteome-wide screens for small ubiquitin-like modifier (SUMO) substrates identify Arabidopsis proteins implicated in diverse biological processes. Proc Natl Acad Sci U S A 107:17415–17420. doi:10.1073/pnas.1005452107
10. Elrouby N, Bonequi MV, Porri A et al (2013) Identification of Arabidopsis SUMO-interacting proteins that regulate chromatin activity and developmental transitions. Proc Natl Acad Sci U S A 110:19956–19961. doi:10.1073/pnas.1319985110
11. Toufighi K, Brady SM, Austin R et al (2005) The botany array resource: e-northerns, expression angling, and promoter analyses.

Plant J 43:153–163. doi:10.1111/j.1365-313X.2005.02437.x

12. Warde-Farley D, Donaldson SL, Comes O et al (2010) The GeneMANIA prediction server: biological network integration for gene prioritization and predicting gene function. Nucleic Acids Res 38:W214–W220. doi:10.1093/nar/gkq537
13. Rhee SY, Beavis W, Berardini TZ et al (2003) The Arabidopsis Information Resource (TAIR): a model organism database providing a centralized, curated gateway to Arabidopsis biology, research materials and community. Nucleic Acids Res 31:224–228. doi:10.1093/nar/gkg076
14. Proost S, Van Bel M, Vaneechoutte D et al (2015) PLAZA 3.0: an access point for plant comparative genomics. Nucleic Acids Res 43:D974–D981. doi:10.1093/nar/gku986
15. Proost S, Van Bel M, Sterck L et al (2009) PLAZA: a comparative genomics resource to study gene and genome evolution in plants. Plant Cell 21:3718–3731. doi:10.1105/tpc.109.071506
16. Goodstein DM, Shu S, Howson R et al (2012) Phytozome: a comparative platform for green plant genomics. Nucleic Acids Res 40:D1178–D1186. doi:10.1093/nar/gkr944
17. Zhao Q, Xie Y, Zheng Y et al (2014) GPS-SUMO: a tool for the prediction of sumoylation sites and SUMO-interaction motifs. Nucleic Acids Res 42:W325–W330. doi:10.1093/nar/gku383gku383
18. Teng S, Luo H, Wang L (2012) Predicting protein sumoylation sites from sequence features. Amino Acids 43:447–455. doi:10.1007/s00726-011-1100-2
19. Smoot ME, Ono K, Ruscheinski J et al (2011) Cytoscape 2.8: new features for data integration and network visualization. Bioinformatics 27:431–432. doi:10.1093/bioinformatics/btq675
20. Maere S, Heymans K, Kuiper M (2005) BiNGO: a Cytoscape plugin to assess overrepresentation of gene ontology categories in biological networks. Bioinformatics 21:3448–3449. doi:10.1093/bioinformatics/bti551
21. Bindea G, Mlecnik B, Hackl H et al (2009) ClueGO: a cytoscape plug-in to decipher functionally grouped gene ontology and pathway annotation networks. Bioinformatics 25:1091–1093. doi:10.1093/bioinformatics/btp101btp101

INDEX

A

B

C

D

L. Maria Lois and Rune Matthiesen (eds.), *Plant Proteostasis: Methods and Protocols*, Methods in Molecular Biology, vol. 1450, DOI 10.1007/978-1-4939-3759-2, © Springer Science+Business Media New York 2016

Printed in the United States
By Bookmasters